American Mathematical Society
Colloquium Publications
Volume 12

Topology

Solomon Lefschetz

American Mathematical Society
Providence, Rhode Island

2000 *Mathematics Subject Classification.* Primary 54-02, 55-02.

Library of Congress Cataloging-in-Publication Data

Lefschetz, Solomon.
Topology.
New York, The American mathematical society, 1930.
1. Topology. I. Title.

QA611.L45 31010604

ISBN 978-0-8218-4603-2

Reprinted by the American Mathematical Society, 2008
Printed in the United States of America.

∞ The paper used in this book is acid-free and falls within the guidelines
established to ensure permanence and durability.
Visit the AMS home page at `http://www.ams.org/`

10 9 8 7 6 5 4 3 2 13 12 11 10 09

PREFACE TO THE SECOND EDITION

This second edition only differs from the first in the insertion of a collection of errata from a large list provided by Mr. R. S. Pieters. A couple of the corrections which it was too difficult to insert in the text have been relegated to a list of addenda at the end of the volume. These are marked in the text by the superscript [a]. The author wishes to express his deep appreciation to Mr. Pieters for his list, and likewise to the Chelsea Publishing Company for having undertaken the arduous task of preparing this second edition.

UNIVERSIDAD NACIONAL DE MÉXICO
September, 1956

PREFACE

Perhaps on no branch of mathematics did Poincaré lay his stamp more indelibly than on topology. To him we owe the basic notion of complex, the boundary relations and related numerical invariants, the first duality theorem. But Poincaré quickly centered his efforts upon the classification of manifolds and other questions, leaving the foundations in rather unstable equilibrium. It is largely to Veblen and Alexander that we owe the remedy for this state of affairs, and the present improved situation. A date marks the transition: 1922, when there appeared Veblen's excellent *Cambridge Colloquium Lectures: Analysis Situs,* which has deservedly become the standard work on the subject. The ground being thus well prepared, new developments came rapidly, and it is with the new phases of the subject that we shall be chiefly concerned here.

Our central topic is the theory of manifolds for which the first two chapters are the preparation. The manifolds are of Alexander's combinatorial type. While more general than the usual type their treatment is scarcely more difficult, and in addition it has been possible to prove that they are topologically invariant (van Kampen). They may be open or closed complexes, and in the former case no boundary conditions are imposed. The principal questions treated will include the general theory of the homology characters, their duality theorems, the intersection theory, the study of continuous transformations, their coincidences and fixed points, the applications to analytical and algebraic varieties. In all these directions we are taking up again questions with which we have dealt in recent papers (Transactions of the American Mathematical Society, 1926, 1927, Annals of Mathematics, 1927–1930), recasting and greatly extending our results. In the duality theory it will be recalled that two unrelated basic formulas, due to Poincaré and Alexander, were reduced to a single formula for relative invariants. This work of unification is carried still further, with new results, giving considerably greater reach to our formulas for coincidences and fixed points. We have succeeded, in fact, in extending them to arbitrary compact metric spaces, a fundamental result for real analysis.

Regarding the intersection theory, our first treatment was based on intersections of convex cells. This is an excellent procedure from the geometric standpoint and for the applications. Howewer, for the basic invariance proofs a procedure followed by Veblen-Weyl in the simplest case (isolated intersections) is to be preferred and we have adopted it here. Unfortunately, in their scheme the intersecting elements are in a very special mutual relationship, which makes it awkward in practise. We have therefore introduced a third and more general type of intersections, based on Brouwer's looping coefficients, which includes the other two types as special cases, and thus acts as the unifying element.

Side by side with general topological invariance, one may consider more strictly combinatorial invariance, or invariance under subdivision of the cells of a complex. In its treatment continuity should play no part, and its study constitutes combinatorial topology proper. While we have made a mild effort to separate the two theories, and to establish combinatorial invariance wherever readily possible, we have taken as our program the study of complexes and manifolds from any point of view whatever. In short, in the combinatorial theory, we have not adopted the abstract and postulational point of view of Dehn-Heegaard in their Encyklopädie article. We have taken configurations immersed in some Euclidean number space, and made free use of the elementary properties of the space as regards intersections of its linear subspaces. These properties, and the elementary properties of the rational number system, are all that we assume when we deal with the combinatorial theory.

Much of the topological part proper has been put in the seventh chapter, devoted to the infinite complex. It seems to be just what is wanted for the best treatment of that type of question. Infinite manifolds have already been studied by Kerékjártó, largely in the spirit of the classical theory of finite two dimensional manifolds. We are giving here however for the first time a general theory of infinite complexes and manifolds with their applications to metric spaces. Closely related recent contributions are the approximations to compact metric spaces by finite complexes introduced by Alexandroff, and the homology theory of these spaces due to Vietoris.

Perhaps the most novel feature of this book is the rôle played by the so-called "relative" concepts. It appears that almost all boundary relations and related homologies may be replaced by others in which we omit everything that is not on a certain configuration A, thus obtaining analogous relations relative A. For example it is advantageous to consider an open segment as a relative one-circuit, a plane polygonal region as a relative manifold. This is entirely in line with the "relative" concepts of point set theory as defined, say, by Hausdorff.

When the relative concepts are systematically introduced, it is found that a good part of the combinatorial theory of Poincaré, Veblen, Alexander, also our own results on continuous transformations, etc. have a greatly increased range of application. As a minor, but highly interesting feature, Brouwer's classical results on the invariance of dimensionality and regionality are found to be intimately tied up with the invariance of the homology characters.

References are indicated by the author's name, followed by a number in square brackets which corresponds to the bibliography at the end.

Valuable assistance was given me by several past or present young members of the Princeton group. Dr. D. A. Flanders was my collaborator in connection with the first two chapters and Dr. L. W. Cohen with Ch. VII § 4. The fairly complete bibliography is largely due to the efforts of Dr. A. B. Brown who also carefully read the proofs, and in this connection made many valuable suggestions. Messrs. W. W. Flexner, J. Singer, and A. W. Tucker helped me also in various ways, especially with the first three chapters, and in addition Mr. Singer made all the drawings. To one and all of them I wish to address here my sincere thanks.

PRINCETON, N. J.
April 1930.

TABLE OF CONTENTS

Section Page

CHAPTER VII

INFINITE COMPLEXES AND THEIR APPLICATIONS

CHAPTER VIII

APPLICATIONS TO ANALYTICAL AND ALGEBRAIC VARIETIES

INTRODUCTION

1. A rapid glance over the foundations of our subject, besides its intrinsic value, will also provide an excellent occasion for introducing various important concepts needed later. These foundations rest upon the theory of abstract spaces founded by Fréchet* [1]. On these questions the reader will consult with profit the interesting papers by Chittenden [1, 2].

2. *Topology* or *Analysis Situs* is usually defined as the study of properties of spaces or their configurations invariant under continuous transformations. But what are *spaces* and their *continuous* transformations?

Whatever a *space* may be it is difficult to reconcile it with anything conforming with our "spatial" intuition unless it is endowed with the following property, the most salient and primitive possessed by the familiar types: With each point there goes a portion of the space in which it is imbedded. This leads to the conception of an abstract space $\mathfrak{R}$ as a set of elements $\{x\}$, its *points*, together with an aggregate of subsets $\{N\}$, the *neighborhoods*, so chosen as to satisfy:

---

* The general symbolism of the theory of sets will be used throughout. A set of elements, any one of which is a, is designated by $\{a\}$. Let A, B be two sets. Then the set of all elements

in A or in B $= A + B =$ the *sum* of A and B;

in A but not in $B = A - B =$ the *complement* of B with respect to A;

in A and also in $B = A \cdot B =$ the *intersection* of A and B.

$A \supset B$, or $B \subset A$ mean that B is a subset of A. When B coincides with A we write $B = A$. Negation of a relation is designated by drawing a line through its symbol, as $A \not\subset B$, $A \neq B$, etc.

Axiom I. Between the sets $\{x\}$, $\{N\}$, there exists a correspondence such that to each x there corresponds one or more N's, the neighborhoods N^x of x. Every $N^x \supset x$.

This is the first of four axioms due to Hausdorff. The other three are given below and are of a more restrictive nature.

3. We keep $\mathfrak{R}$ fixed for the present and consider certain attributes of its subsets. It is surprising how far it is possible to go even on such a slender foundation as Axiom I.

Let A be a subset of $\mathfrak{R}$. The point x is an *inner* point of A whenever there is an $N^x \subset A$, which by Axiom I implies $x \subset A$.

A point x is a *boundary* point of A if every N^x includes points of both A and of its complement $\mathfrak{R} - A$.

The *boundary* $F(A)$ of A is the sum of the boundary points of A. The set A is *open* when $F(A) \subset \mathfrak{R} - A$, *closed* when $F(A) \subset A$. An open set consists solely of inner points. A closed set is the complement of an open set.

If $A \subset N \subset \mathfrak{R}$, A closed, N open, then N is called a *neighborhood* of A, and often designated by N^A, $\mathfrak{R}^A$, etc.

The sum $A + F(A)$ is called the *closure* of A and is designated by $\bar{A}$. It is a set function defined for *all* subsets of $\mathfrak{R}$ and will play a particularly important part in the sequel.

A set A is *dense* on $\mathfrak{R}$ whenever there are points of A on every N. An equivalent definition is that $\bar{A} = \mathfrak{R}$.

The point x is a *limit point* of A whenever A has points on every $N^x - x$. The closure $\bar{A} = A +$ all its limits points.

The infinite sequence $\{x_n\}$ is said to *converge to the limit* x whenever each N^x contains all but a finite number of points of the sequence.

Any given subset of $\mathfrak{R}$, say A, may be turned into a space by agreeing to choose as neighborhoods for A its intersections $N \cdot A$ with those of $\mathfrak{R}$. This convention will be adhered to in the sequel. It is then possible to introduce the sets open or closed *on* A, etc.; these are the well known *relative* concepts of point-set theory.

4. A *transformation* T of a space $\mathfrak{R}$ into another $\mathfrak{R}'$ is the assignment of a correspondence between their points such that to every point x of $\mathfrak{R}$ there corresponds at least one point $x' = T \cdot x$ of $\mathfrak{R}'$. The transform $T \cdot A$ of a subset A of $\mathfrak{R}$ is the set of all points $T \cdot x$ for $x \subset A$. The transformation is:

Single-valued whenever $x' = T \cdot x$ is unique for every x.

One-to-one when it assigns also a unique x to every x', thus pairing off in a unique way the points of $\mathfrak{R}$ and $\mathfrak{R}'$.

Continuous if whenever the transform $T \cdot A$ is relatively open for $T \cdot \mathfrak{R}$, then A is open for $\mathfrak{R}$.

Homeomorphic when it is one-to-one and continuous both ways.

We have now all the elements needed for a formal

DEFINITION OF TOPOLOGY: *It is the study of the properties of spaces that are preserved under homeomorphism.*

Any property of a space $\mathfrak{R}$ which belongs to every homeomorph of $\mathfrak{R}$ is said to be *topologically invariant.* A numerical function attached to $\mathfrak{R}$ which has the same value for all its homeomorphs is a *numerical topological invariant.* As an example, when $\mathfrak{R}$ is a sphere the fact that any simple closed curve on it decomposes it into two regions is topologically invariant, the number *two* is a numerical topological invariant.

The concepts introduced in No. 3 are topologically invariant: Thus when $\mathfrak{R}$ is homeomorphically transformed into $\mathfrak{R}'$, its closed (open) sets, etc. go over into similar sets for $\mathfrak{R}'$.

5. Given an abstract set $\mathfrak{R}$, it may be possible to convert it into a space by two distinct choices of subsets as neighborhoods, say $\{N\}$ and $\{N'\}$. We have thus two different spaces $\mathfrak{R}, \mathfrak{R}'$, built up out of the same abstract set of elements. The two spaces are considered as *identical* whenever the identical point-to-point transformation between them is homeomorphic. This means that any subset A of $\mathfrak{R}$ which is open when the N's are the fundamental neighborhoods remains open when they are replaced by the N''s and conversely. The two sets of neighborhoods are then called *equivalent.* The space $\mathfrak{R}$ is not so much characterized by a single ag-

gregate of neighborhoods as by a whole class of equivalent sets of them.

Since identical spaces are homeomorphic the concepts introduced in No. 3 are unchanged when a set of neighborhoods is replaced by an equivalent set. From the very general type of space considered so far we proceed to what might be termed "working spaces" by imposing restrictive axioms upon the neighborhoods. The set in widest use today goes back to Hausdorff [1] and characterizes what we shall call a *Hausdorff space.* It consists of Axiom I plus the following three:

Axiom II. If N, N' are any two neighborhoods of x, there exists a third neighborhood of x, $N'' \subset N \cdot N'$.

Axiom III. If the point $y \subset N^x$, there is an $N^y \subset N^x$.

Axiom IV. If $x \neq y$ there exist N^x, N^y without common points.

Among the Hausdorff spaces are included the usual simple types: projective, Euclidean, function, n-spaces, with simplexes (generalized triangles) as the neighborhoods.

A necessary and sufficient condition for the equivalence of two coincident Hausdorff spaces $\mathfrak{R}, \mathfrak{R}'$, defined by two distinct sets of neighborhoods $\{N\}$, $\{N'\}$ is that every N^x carry an N'^x and conversely.

6. A Hausdorff space $\mathfrak{R}$ is:

Regular whenever Axiom III may be replaced by the stronger condition: When $y \subset N^x$, there is an $\bar{N}^y \subset N^x$.

Separable (Fréchet) whenever it is identical with a space whose set of neighborhoods is enumerable.

Compact (Fréchet) whenever every infinite subset of $\mathfrak{R}$ has a limit point on $\mathfrak{R}$.

An important consequence of separability is the presence on $\mathfrak{R}$ of an enumerable dense subset: For if $\{N^i\}$ is the enumerable set of neighborhoods, we can take a point P^i on N^i and the sequence $\{P^i\}$ is dense on $\mathfrak{R}$.

The space whose points are all the real numbers with $\{N\}$ as the intervals whose end points are rational, is separable but not compact, since the set of all integers has no limit point.

7. The definition of a space by means of neighborhoods is, so to speak, purely qualitative. However, one of the most important properties of the more common spaces is that they can serve as a basis for making measurements, that a *metric* can be affixed to them. If the notion of neighborhood deserves logically the first place, nevertheless in all applications of mathematics the existence of a metric is the essential element.

If we take the metric as the basic factor the natural definition of a space $\mathfrak{R}$ assumes the following form: It is an abstract set of elements, its points, such that to their pairs x, y there may be attached a *distance function,* $d(x, y)$, or real function having the following properties of Euclidean distances:

I. $d(x, y) = 0$ when and only when $x = y$.

II. For any three points x, y, z of $\mathfrak{R}$ the *triangle* inequality holds, i. e.

$$d(y, z) \leqq d(x, y) + d(x, z).$$

From these two properties one may deduce, as observed by Lindenbaum [1], the other two usually ascribed to the distance:

III. $d(x, y) \geqq 0$.

It is obtained by making $z = y$ in II and taking into account I.

IV. $d(x, y) = d(y, x)$.

Making $z = x$ in II we have $d(y, x) \leqq d(x, y)$. Similarly $d(x, y) \leqq d(y, x)$, from which IV follows.

The *distance* $d(A, B)$ between two sets $A, B \subset \mathfrak{R}$, is the greatest lower bound of $d(x, y)$ for $x \subset A, y \subset B$. The *diameter* of A, $d(A)$, is the least upper bound of $d(x, y)$ for $x, y \subset A$. By *spheroid of center A and radius* ϱ, $\mathfrak{S}(A; \varrho)$, we understand the set of all points x such that $d(A; x) < \varrho$.

The spheroids $\mathfrak{S}(x, \varrho)$ *whose centers are the points* x *of* $\mathfrak{R}$ *constitute a set of neighborhoods which turn* $\mathfrak{R}$ *into a regular Hausdorff space.* The verification is instantaneous. The Hausdorff space thus defined is considered as identical with $\mathfrak{R}$, and continues to be designated by $\mathfrak{R}$. A practical result of importance is that *on a metric space the limit concept can be* *put in the same form as in the familiar ε, δ procedure of elementary analysis.*

For a metric space separability is equivalent to the existence of an enumerable dense subset: We have seen (No. 6), that such a set is present on every separable space. On the other hand, $\mathfrak{R}$ being metric, if $A = \{x_n\}$ is dense on it, the spheroids $\mathfrak{S}(x_n; \varrho_m)$, ϱ_m rational, are enumerable and their set is equivalent to $\{\mathfrak{S}(x; \varrho)\}$, x any point of $\mathfrak{R}$, ϱ arbitrary. Hence $\mathfrak{R}$ is separable.

The fundamental question that must be answered in order to bridge the two points of view: metric and qualitative, is then: Under what conditions is a Hausdorff space *metrisable* (susceptible of having a distance function affixed to it)? In the light of the preceding remarks it is largely answered for all spaces that offer any interest by the following fundamental theorem due to Urysohn [1, 2], with an important complement regarding the condition of regularity by Tychonoff [1]:

In order that a separable Hausdorff space be metrisable, it is necessary and sufficient that it be regular.

As a rapid Corollary (Urysohn):

In order that a compact Hausdorff space be metrisable it is necessary and sufficient that it be separable.

We are then in possession today of a strictly qualitative characterization of metric spaces with an enumerable dense subset (separable metric spaces). These spaces include the most general type to be considered in any connection whatever in the present work. The preceding sketch is a sufficient description of their foundations to cover all our needs.

8. An important result obtained in recent years is the proof by Alexandroff [10] that compact metric spaces can be indefinitely approximated by certain polyhedral configurations, the *complexes.* The topological theory of complexes acquires thus a fundamental importance; it is in truth the necessary first step in a general study of metric spaces.

CHAPTER I

ELEMENTARY COMBINATORIAL THEORY OF COMPLEXES

For all purposes of the present chapter the configurations that are to be studied may be considered as immersed in Euclidean number spaces. It might therefore logically and properly constitute a chapter in an extension of such a work as the *Projective Geometry* of Veblen and Young. Moreover while properties involving limiting processes are introduced and mentioned, *they are actually never used in the proofs,* in which all steps are strictly combinatorial. Thus everything that is taken up would hold if we only allowed in our spaces points with rational coördinates.

The justification for the introduction of terms which apparently involve limiting processes in their definitions is first that it is not advisable to repeat and amplify definitions several times, next that it is practical and worth while to become accustomed to them as early as possible.

§ 1. The Simplex, the Cell, the Sphere and the Complex

1. Let S_n* denote an n-dimensional Euclidean space referred to Cartesian coördinates $x_1, \cdots, x_n$ with the usual distance formula, $d(x, x') = \sqrt{\sum (x_i - x'_i)^2}$. We assume known the most elementary properties of the S_k's on S_n, in particular

* The dimension of our configurations is indicated throughout by the subscripts.

regarding their intersections and the like. For the present purpose they may be considered as clothing in geometrical language the elementary properties of systems of linear equations in n variables.

Let A^i $(i = 0, 1, \cdots, k)$ be $k+1$ *independent* points of S_n, or points determining a unique S_k through them, and let a_{ij} be the coördinates of A^i. If $(x_1, \cdots, x_n)$ is any point of that S_k, among the equations in the y's

$$(1) \qquad \begin{cases} \sum a_{ij}\, y_i = x_j, \\ \sum y_i = 1, \end{cases}$$

precisely $k+1$ are linearly independent and yield finite values of these variables. The y's are the *barycentric* coördinates of the point (x) of S_k. They are independent of the choice of coördinate system for S_n. Furthermore, if S_n is considered as a subspace of S_m, $m>n$, all that is necessary is to increase the number of variables x and of equations in (1), but the values of the y's for a given point will remain the same.

2. Let us then assume $k=n$, so that we now have $n+1$ independent points of S_n and the y's become barycentric coördinates for that space itself.

We understand by an *n-simplex*, σ_n, with vertices $A^0, A^1, \cdots, A^n$, the set of all points of S_n whose barycentric coördinates relatively to the A's satisfy the inequalities $0<y_i<1$. This set is independent of the choice of coördinates for S_n, or for whatever space in which it may happen to be immersed. σ_n is a bounded region. Its boundary points are obtained by making one or more of the y's equal to 0 or 1. It is therefore the sum of the various simplexes of less than n dimensions which also have A's for vertices, together with the A's themselves. Since each n-simplex $(n>1)$ has $n+1$ distinct $(n-1)$-simplexes on its boundary, it is convenient to define a *zero-simplex* as a single point, so that a 1-simplex has two 0-simplexes on its boundary. Thus the boundary of an n-simplex $(n>0)$ is composed of the $\binom{n+1}{p+1}$ p-simplexes $(p=0, 1, \cdots, n-1)$

determined by all possible combinations of the $n+1$ vertices taken $p+1$ at a time. The p-simplexes thus determined are the *p-faces* of the n-simplex. For $n = 0, 1, 2, 3$, σ_n is respectively the point, the segment, the triangle, the tetrahedron.

3. The *n-cell*, E_n, is the basic configuration of analysis situs. It is by definition the homeomorph of the n-simplex; the homeomorph of the boundary of σ_n is called the *(n—1)-sphere*, H_{n-1}. The justification of the latter name is found in the fact that a spherical region of S_n, $\sum x_i^2 < 1$, is homeomorphic to σ_n, while the sphere, its boundary, is the homeomorph of the boundary of σ_n. This is a special case of an example considered presently. It is to be noted that E_n is not closed. As a matter of convenience we shall often say "the closed cell" for "the closure of the cell".

A simple type of E_n which includes the spherical region is a bounded region $\mathfrak{R}$ of S_n possessing an interior point P such that every ray through P meets the boundary in just one point. For let σ_n be an arbitrary simplex $\supset P$, and let a ray PT issuing from P meet the boundaries of $\mathfrak{R}$ and σ_n at the respective points T and T'. Then, Q being any point of PT, let Q' be the point of the ray such that $PQ' = \frac{PT'}{PT} PQ$. The correspondence between Q and Q' defines a homeomorphism between σ_n and $\mathfrak{R}$. A simple example of a region $\mathfrak{R}$ is a *convex region*, i.e. a region such that when P, Q belong to it so does the segment PQ.

4. A *convex polyhedral region* of an S_n, or a *convex polyhedral n-region*, is defined by recurrence thus: for $n = 0$ it is a single point; for $n > 0$ it is a finite convex region $\mathfrak{R}$ on an S_n, whose boundary consists of a finite number of convex polyhedral regions on spaces of lower dimensions. It is evidently an n-cell. A convex polyhedral part of the boundary which is a region of an S_k, is a k-cell, called *k-face* of the n-cell.

Let us take the intersection with a plane π not meeting the faces of less than $n-2$ dimensions. It cuts $\mathfrak{R}$ in a convex polygonal region bounded by the intersections of the $(n-1)$- and $(n-2)$-faces of $\mathfrak{R}$ with π, which give rise to the sides and vertices of the polygon. The latter are in the same mutual relations of incidence as the faces in question. Since each vertex of the polygon is an end-point of two and only two sides,* each E_{n-2} is on the boundary of two and only two E_{n-1}'s of the boundary of E_n. Furthermore there are at least three E_{n-1}'s, and in fact at least $n+1$, the minimum being reached only for a σ_n, a fact which we do not stop to prove.

A simple example is obtained by taking the interior E_3 of a simple regular polyhedron of ordinary space and its faces. Thus for the cube we would have six 2-cells, twelve 1-cells and eight 0-cells (all convex and polyhedral) on its boundary.

5. In order not to confront too many difficulties at the outset we shall consider for the present a relatively simple type of complex and later extend our results to more general types.

We first define the *simplicial polyhedron* Π_n: it is merely a set of a finite number of simplexes σ_p^i, of S_m, $m \geqq n$, of dimensions up to and including n, such that (a) no two simplexes of the set intersect; (b) if σ_p^i belongs to the set so do all of its faces. The σ's are the *faces* of Π_n. In view of (a) they do not cross one another, while (b) implies that when $\sigma \subset \Pi_n$, likewise $\bar{\sigma} \subset \Pi_n$, so that Π_n is closed.

The simplicial n-complex, K_n, is an aggregate of point sets in one-to-one correspondence with the faces of Π_n (one set for each face and *vice-versa*), whose sum is a point set homeomorphic to Π_n. The sets in question are thus cells of the same dimension as the corresponding simplexes of Π_n. They are the 0, 1, 2, $\cdots$, n-cells of K_n. They constitute a subdivision of K_n with the same structure as the subdivision of Π_n into its simplexes. Thus K_n has a certain number of zero-cells for end points, etc.: on the boundary of each q-cell of K_n there are $q+1$ cells of dimension $q-1$. We shall omit to mention "simplicial" for the present since until further notice no other type of cell or complex will be considered.

* For the system has the same structure as the set of arcs cut out on a circumference by a finite number of points plus those end-points. This is shown by a construction such as that of No. 3.

Two cells of K_n such that one is on the boundary of the other are said to be *incident* with one another. The *incidence* relations of the complex are for the present the list of pairs of incident cells. Later they will also be understood to include certain orientation information.

Simple examples of a K_n are a σ_n plus its boundary, and the sum of two simplexes σ_n, σ_n' having a common k-face, plus all their boundary faces. A projective plane, and a sphere H_2, suitably triangulated, are 2-complexes.

For the sake of brevity we shall extend to configurations on K_n the terminology which, strictly speaking, is applicable only to Π_n. Thus we shall speak of segments, simplexes, polyhedral configurations on K_n, meaning thereby the images of such elements on K_n under the homeomorphism existing between them. In particular the barycentric coördinates of any point on the cell E_p^i of K_n are by definition those of its image point on the corresponding face σ_p^i of Π_n.

6. It may be asked at once, why not stop at Π_n and leave out K_n altogether? The answer is that K_n serves to put the emphasis on the primordial structural elements, the cells, and their incidence relations, as against the properties of linear spaces peculiar to Π and its geometry. We shall make very large use of linear elements, or polyhedral configurations, but only as means to an end, not as an end in itself. This is somewhat analogous to the relationship between S_n and its coördinate systems. Whenever one studies the geometry of S_n one may find coördinates exceedingly convenient, but the theorems to be sought must be independent of them.

To take an example consider the boundary of a hemispherical region of S_3. Join a point A^0 not on the circumference bounding the circular part to three points A^1, A^2, A^3 of that circumference by appropriate arcs $A^0 A^i$. There results a K_2 homeomorphic to the boundary of a σ_3 and with the same structure, i. e. the K_2 and the boundary of a σ_3 correspond to one another cell for cell and the incidence relations are the same. The two figures are very different indeed, yet they have many properties in common; and it is these properties that will attract our attention to the exclusion of all others. K^n may thus be said to embody the structural attributes of Π_n to the exclusion of those properties which pertain only to the domain of projective or Euclidean geometry.

7. The bearing of conditions (a), (b) must be kept in mind. They imply, for example, that if we remove an end point from a closed segment, a vertex or a side from a triangle, what is left is not a complex. Similarly, let PQ be a segment meeting the complex K_2 of the preceding Number at a point R

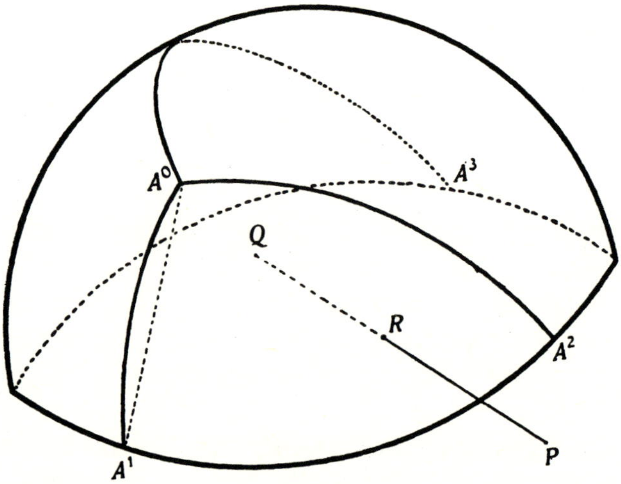

Fig. 1.

on a 2-cell of K_2, while P and Q are not on K_2. Then K_2 plus the segment do not constitute a complex. On the other hand, assuming A^0 on the spherical part of K_2, let us draw the segment A^0A^1. That segment plus K_2 constitute a new K_2.*

* One restriction which our definition imposes on the simplicial complex is of a more arbitrary nature; viz., that the complex be actually representable as a polyhedron in Euclidean space of sufficiently high dimension. Less restrictive conditions are usually imposed which, for example, allow a circle to be represented as a 1-complex by two 1-cells incident with two 0-cells. We have imposed this restriction for the sake of the added convenience of applying to our complexes the notions of straightness and distance implicit in their Euclidean prototypes. As a matter of fact it is removed together with many others in the generalisation of complexes at the end of Ch. II.

8. A zero-complex consists merely of a finite number of points. The first case of interest is therefore the 1-complex, K_1, also called *linear graph* or *graph.* There is an extensive literature dealing with the topology of graphs and its application to certain mathematical problems and pastimes.

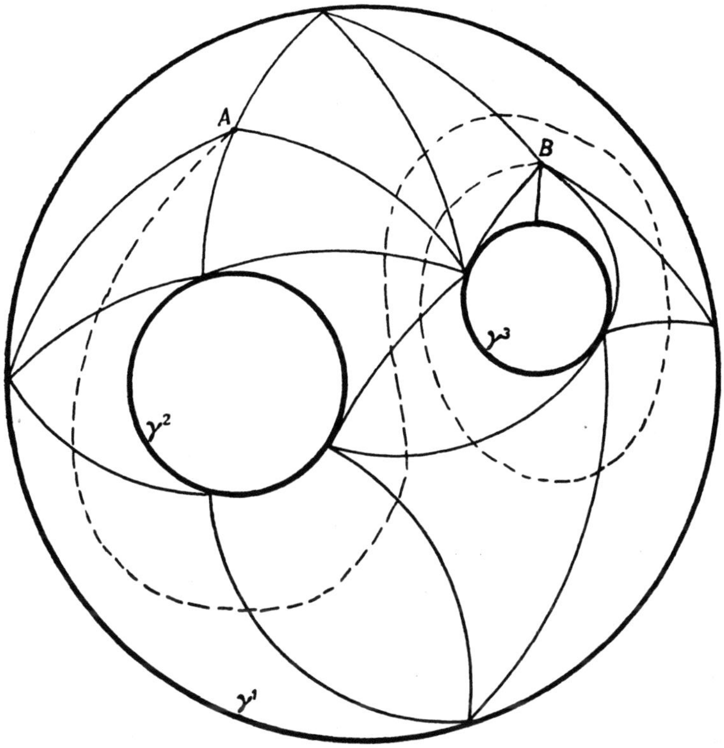

Fig. 2.

Strictly speaking, the graph as usually understood consists of a finite number of non-intersecting arcs plus their end-points, whereas our K_1 may include in addition a finite number of isolated points. Their presence, however, does not have any serious influence on the topology.

Next to $\bar{E}_1$ the simplest graph is the homeomorph of the one-sphere *(Jordan curve, simple closed curve)*, exhaustively studied in the last quarter century.

Much less is known concerning the most general K_2, but a special type, the *two-dimensional manifold* which includes Riemann surfaces, has been well covered in the literature.

9. *A simplicial complex K_n is completely determined when the incidences between its cells are specified.* To be more precise, let K_n' consist of a set $\{E_p'^i\}$ of cells, corresponding cell for cell to those of K_n and with the property that when E_p^i and E_{p-1}^j are incident so are $E_p'^i$ and $E_{p-1}'^j$; then K_n' is homeomorphic to K_n. For every point on K_n or K_n' is uniquely determined when the vertices of the carrying cell and the barycentric coördinates of the point relatively to that cell are given. Therefore the correspondence between K_n and K_n' associating points of every pair E_p^i, $E_p'^i$ with the same barycentric coördinates is a homeomorphism.

10. In many applications the problem arises of decomposing the points of a known configuration into the cells of a complex. The complex is then said to *cover* the configuration. Take for example a plane region $\mathfrak{R}$ bounded by three circles $\gamma^1, \gamma^2, \gamma^3$ where γ^2 and γ^3 are exterior to one another but interior to γ^1. Figure 2 illustrates a decomposition of $\mathfrak{R}$ into curvilinear triangles giving rise to a covering K_2 for $\mathfrak{R}$. It is quite evident that these covering complexes can be varied *ad infinitum.* But they are all homeomorphic to one another and for the structure of $\mathfrak{R}$ the essential properties will be those common to them all. In the preceding example it is a simple matter to construct a suitable Π_2. Indeed we may actually construct a Π_2 lying wholly in the plane.

§ 2. Orientation, Chains, Cycles

11. Just as there may be constructed a theory of linear sets without the notion of order, so there may be developed a theory of complexes without the theory of orientation. This has been done in fact by Veblen and Alexander [22], very extensively in Veblen's Colloquium Lectures [5]. Such a theory presents indeed a good deal of merit on the score of simplicity. But on the one hand its theorems are as

a rule special cases of theorems of the theory of oriented complexes, or *chains*, on the other it is the latter that one needs in the major applications, notably in analysis. Therefore it is with them that we shall deal almost exclusively in this book.

We begin with the orientation of a simplex σ_n, $n > 0$. Let its vertices be $A^0, A^1, \cdots, A^n$, and put into one class the modes of ordering them derived from the normal order by an even permutation, into another class the other modes. We designate by $+\sigma_n$ the association of σ_n with one of the two classes, by $-\sigma_n$ its association with the other. We also write $+\sigma_n = A^0A^1\cdots A^n = A^2A^0A^1A^3\cdots A_n = -A^1A^0A^2\cdots A^n\cdots$, and $-\sigma_n = A^1A^0A^2\cdots A^n = -A^0A^1\cdots A^n\cdots$. This choice of class is described as *orienting* the simplex.

Let us orient the $(n-1)$-faces of σ_n as we please and let σ^i_{n-1} $(i = 0, 1, \cdots, n)$ be the face of which A^i is not a vertex, together with its orientation. We can thus write

$$\sigma^i_{n-1} = \varepsilon A^0 \cdots A^{i-1} A^{i+1} \cdots A^n, \quad \varepsilon = \pm 1.$$

We also have

$$\sigma_n = \varepsilon' A^i A^0 \cdots A^{i-1} A^{i+1} \cdots A^n.$$

We shall say that σ^i_{n-1} is *positively* or *negatively* related to σ_n according as $\varepsilon\varepsilon' = +1$ or -1. Denote its value by η_i. The incidence relations between σ_n and its boundary $(n-1)$-simplexes are summed up in a symbolical boundary relation:

$$\sigma_n \to \sum \eta_i \, \sigma^i_{n-1}. \tag{2}$$

The *oriented boundary* $F(\sigma_n)$ of the σ_n is by definition the set of faces on the boundary of the simplex associated with the sum in (2).

Relations such as (2) and their generalizations are due to Poincaré [2]. However, he used $\equiv$ in place of the very convenient $\to$ introduced by Alexander [19].

Strictly speaking, what we have said has no application to zero-simplexes. It is convenient, however, to extend the conventions to this special case. The reasonable starting point is to write down the equivalent of (2) when $n = 1$. Let $\sigma_1 = A^0 A^1$. We call A^0 the *initial* point, A^1 the *terminal* point of the one-simplex or segment, and express the relation to σ_1 by $\sigma_1 \to A^1 - A^0$. It leads us to introduce into our boundary relations symbols such as $+A^i$, $-A^i$. We call these symbols associated with the point or zero-cell A^i, *positively* or *negatively oriented* points. Finally, $\sigma_0 \to 0$, always.

12. In comparing oriented simplexes it will often be convenient to use the following notation: Let $\sigma_n = A^0 A^1 \cdots A^n$, $\sigma_q = A^p A^{p+1} \cdots A^{p+q}$, it being understood that both simplexes are oriented. Then we shall also write $\sigma_n = A^0 A^1 \cdots A^{p-1} \sigma_q A^{p+q+1} \cdots A^n$. More generally, if for example $\sigma_{p-1} = A^0 A^1 \cdots A^{p-1}$ we shall also write $\sigma_n = \sigma_{p-1}\sigma_q A^{p+q+1} \cdots A^n$, and so on: the general scheme is obvious.

13. Let us return to the simplicial K_n of § 1. As it is to be the fundamental space throughout most of this book, it will be simpler to designate it by K, omitting the dimensionality index n. Everything just stated for any σ is directly applicable to any cell homeomorphic to it. We orient then each cell of K in a definite way and the set so oriented is called an *oriented complex*, K. For each E_p^i of K we shall then have a boundary relation such as (2),

$$E_p^i \to \sum \eta_{ij}^{p-1} E_{p-1}^j, \tag{3}$$

where the only cells at the right are those on the boundary of E_p^i. Therefore $\eta_{ij}^{p-1} = 0$ when E_p^i and E_{p-1}^j are not incident; when they are incident its value is $+1$ if they are positively related, -1 otherwise. The matrices $\|\eta_{ij}^p\|$ give then a complete description of the incidence relations between the oriented cells.

The n matrices just introduced, called *incidence matrices*, completely determine K, in the sense that if they are the same for K_n, K_n', the two complexes are necessarily homemorphic. This follows at once from No. 9.

14. We now introduce a most important entity aptly called by Alexander the *chain*. The p-chain, C_p, is the

association of a linear expression having integral coefficients, $\sum t_i E_p^i$, with the closed cells $\bar{E}_p^i$ that actually figure in it. The cells E_p^i and those on their boundaries are the cells of C_p, which thus coincides with a complex made up of cells of K. We write symbolically

$$C_p = \sum t_i E_p^i.$$

The *boundary*, $F(C_p)$, of C_p, is by definition the $(p-1)$-chain

$$F(C_p) = \sum \eta_j E_{p-1}^j; \quad \eta_j = \sum t_i \eta_{ij}^{p-1},$$

and we write the associated boundary relations

$$C_p \to \sum \eta_j E_{p-1}^j, \quad C_p \to F(C_p).$$

The simplest p-chain is a closed p-simplex.

It will be convenient to call *subcomplex, subchain*, of K or C, a complex or chain whose cells belong to K or C. Thus, the above C_p is a p-subchain of K.

When the boundary of C_p is identically zero, that is, when every $\eta_j = 0$, the chain is a *p-cycle*, Γ_p. Symbolically the cycle is characterized by the boundary relation:

$$\Gamma_p \to 0.$$

REMARK. The term *boundary* is used in two different senses: (a) Boundary of a point set as in the *Introduction.* (b) Boundary of a chain, as above. Since a chain also includes a point set it may conceivably have two different boundaries in certain cases. To avoid ambiguity we shall agree to call boundary of a p-chain only the $(p-1)$-chain that we have just defined.

The notions of chain, cycle and the boundary relations can be extended in three different ways.

(a) We neglect everywhere the cells of a definite subcomplex L of K. Then we have *chains, cycles, boundary relations*, mod L (= modulo L), or *relative* chains, cycles, boundary relations for $K-L$, and we write

$$C_p \to F(C_p), \quad \Gamma_p \to 0, \quad C_{p-1} = F(C_p), \text{ mod } L.$$

The true situation is as follows: We take any finite set of cells which do not meet one another, say $\mathfrak{K}$, whose sum is open in K, then designate as the boundary of any chain the sum of the boundary cells which belong to $\mathfrak{K}$ and proceed as before. Thus a cycle becomes merely a chain which has no boundary cells among those of $\mathfrak{K}$, etc.

It is quite clear above that $\mathfrak{K} = K - L$ if we define K as the sum of $\mathfrak{K}$ and those of its boundary cells that do not belong to $\mathfrak{K}$.

As regards the set $\mathfrak{K}$, a suggestive term is "open complex", K itself being then called "closed complex". There are many advantages in holding to a definite fundamental complex throughout and to have it closed.

We have here a double terminology: Relative relations for $K - L$ or $\mathfrak{K}$; relations mod L to denote the neglect of whatever is on L. The first expression is more adequate when we describe intrinsic properties of $K - L$ or $\mathfrak{K}$ alone, and is in line with the terminology of point set theory. The second expression is more natural and usual when associated with the modular systems and linear relations to be met presently.

(b) Another extension is obtained whenever in the expression of any chain we omit every cell whose coefficient is divisible by a fixed integer m. Then we have chains, cycles, boundary relations mod m and write relations similar to the above with "mod m" in place of "mod L"; for example

$$\Gamma_p \to 0 \mod m.$$

(c) The two preceding schemes can be combined and we then have chains, etc. mod (L, m).

Sometimes we shall designate the ordinary chains, cycles, boundary relations, by *absolute* chains, etc., to distinguish them from the three new types which will be called *relative* or *modular*.

Historically, absolute chains and their theory go back to Poincaré [2, 3, 4]. Boundary relations and chains mod 2 were first considered by Tietze [1], then by Veblen-Alexander [22] and their theory systematically developed by Veblen [5]. They present many features of exceptional simplicity due mainly to the fact that in dealing with them orientation may be disregarded. The case of the general m was introduced and studied at length by Alexander [19]. Chains and boundary relations mod L or mod (L, m) were first considered in our papers [12, 14]. It will be found that practically the whole of known Combinatorial Analysis Situs can be extended to the most general modular system.

15. Chains and cycles (modular or absolute) have the following basic properties:

I. *p-chains are additive.* That is if

$$C_p = \sum t_i E_p^i, \qquad C_p' = \sum t_i' E_p^i$$

then

$$C_p'' = \sum (st_i + s' t_i') E_p^i$$

is a chain whatever the integers s, s'. We write symbolically $C_p'' = sC_p + s' C_p'$.

II. *Chain-boundaries are linear functions of the chains.* That is

$$F(sC_p + s' C_p') = sF(C_p) + s' F(C_p').$$

III. *p-cycles have the additive property.* This follows from I and II.

IV. *The boundary of a p-simplex is an absolute* $(p-1)$*-cycle.* To prove this we merely have to show that

$$F(F(\sigma_p)) \equiv 0, \text{ i. e. } F(\sigma_p) \rightarrow 0. \tag{4}$$

We may write $\sigma_p = A^0 A^1 \sigma_{p-2}$. Therefore

$$\sigma_p \rightarrow -A^0 \sigma_{p-2} + A^1 \sigma_{p-2} - \cdots = F(\sigma_p),$$

the missing terms representing faces not having σ_{p-2} on their boundaries. In $F(F(\sigma_p))$ the terms in σ_{p-2} can only come from $F(A^1 \sigma_{p-2} - A^0 \sigma_{p-2})$. But

$$A^0 \sigma_{p-2} \rightarrow \sigma_{p-2} + \cdots,$$

and

$$A^1\sigma_{p-2} \to \sigma_{p-2} + \cdots,$$

the missing terms being simplexes other than σ_{p-2}. Therefore σ_{p-2} does not appear in $F(A^1\sigma_{p-2} - A^0\sigma_{p-2})$, and (4) follows.

By combining II and IV we have:

V. *The boundary of an absolute or modular p-chain is a* $(p-1)$*-cycle of the same type.*

The last two properties provide simple examples of cycles. But one must not conclude that the converse of V holds: *A cycle may well not be the boundary of a chain.* Thus in Fig. 2 by suitably orienting the 1-cells of γ^1 we can express this circle as a Γ_1, but there is no C_2 whose sole boundary is Γ_1.

16. A simple application will throw some light on the preceding notions and make their purpose clearer. Let $\mathfrak{R}$ be the plane region of No. 10 and let

$$u = \int P dx + Q dy, \qquad \frac{\partial P}{\partial y} = \frac{\partial Q}{\partial x},$$

be an integral of total differentials taken on $\mathfrak{R}$. We assume that P, Q and the two partial derivatives are single-valued and continuous on $\mathfrak{R}$. What can be said concerning the totality of all integrals taken along a path of $\mathfrak{R}$ joining two given points A, B? Construct the covering K_2 so as to include A and B among its vertices and let l denote the segment AB, oriented as written. If we deform any path from A to B continuously on $\overline{\mathfrak{R}}$ keeping its end points fixed, the value of u does not change. We see intuitively that the path can thus be deformed into another which will consist of one-cells of K_2 each taken a certain number of times. It will then be a subchain C_1 of K_2 with the sole boundary points A, B, each taken once. It is then a one-cycle relative to the points A, B which play here the role of L above. On closer examination we find as a matter of fact that $C_1 - l$ will be an absolute cycle Γ_1. Therefore

$$u = \int_l + \int_{\Gamma_1}.$$

Designate as before by E_2^i the cells of K_2 oriented, and suppose that there exists a chain

$$C_2 = \sum t_i E_2^i \rightarrow \Gamma_1.$$

We have then

$$\Gamma_1 = \sum t_i \Gamma_1^i, \qquad \Gamma_1^i = F(E_2^i).$$

Hence

$$\int_{\Gamma_1} = \sum t_i \int_{\Gamma_1^i}.$$

But Γ_1^i can be continuously deformed into a point on $\overline{\mathfrak{R}}$, hence

$$\int_{\Gamma_1^i} = 0$$

and therefore likewise

$$\int_{\Gamma_1} = 0.$$

Thus in the expression of any Γ_1 as a chain we may neglect any chain which is the boundary of a C_2. Let us agree to write $\Gamma_1 \sim 0$ whenever Γ_1 is an $F(C_2)$, and more generally $\Gamma_1 \sim \Gamma_1'$ whenever $\Gamma_1 - \Gamma_1' \sim 0$. The two cycles Γ_1, Γ_1' are said to be *homologous.* We define a set of cycles Γ_1^i as *independent* whenever

$$\sum t_i \Gamma_1^i \sim 0$$

implies that every $t_i = 0$, such a relation being called a *homology.* Then it is readily shown that the circles γ^1, γ^2 carry cycles, which we denote by $\overset{1}{\varDelta}_1, \overset{2}{\varDelta}_1$, such that every

$$\Gamma_1 \sim p_1 \overset{1}{\varDelta}_1 + p_2 \overset{2}{\varDelta}_1,$$

while furthermore the two $\varDelta$'s are independent. Each $\varDelta$ is in fact the sum of the one-cells on its circle all sensed alike. The relation just written may be intuitively verified by observing that every sensed closed curve on $\mathfrak{R}$ is continuously deformable into a sum of multiples of the two cycles $\varDelta$ plus some arc joining γ^1 to γ^2 taken as many times in one direction as in the other.

Since the integral has the same value for two homologous cycles

$$\int_{\Gamma_1} = p_1 \int_{\Delta_1^1} + p_2 \int_{\Delta_1^2},$$

$$u = \int_A^B = \omega + p_1 \omega_1 + p_2 \omega_2; \quad \omega = \int_l, \quad \omega_i = \int_{\Delta_1^i}.$$

In other words: The complete set of determinations of u is obtained by adding to any fixed one, ω, some linear combination with integer coefficients of two numbers ω_1, ω_2, sometimes called the *periods* of the integral, independent of the end points A, B.

17. The preceding treatment can be greatly generalized. It applies for example to the determinations of the integral of a function $f(z)$ of the complex variable taken over a path on its region of holomorphism. We may also consider integrals of total differentials in S_3, double integrals of Stokes type, etc., and there are many applications to algebraic geometry.

Let us bring out the important features of the discussion:

(a) The analytical problem refers to some configuration which may be replaced by a subchain C of a suitable K.

(b) The chain is a cycle or relative cycle arbitrarily taken in a linear family $\mathfrak{M}$.

(c) Among the chains of $\mathfrak{M}$ our analytical problem leads us to neglecting the chains of a linear subfamily $\mathfrak{M}^0$ of $\mathfrak{M}$, determined by some sort of condition of bounding.

(d) We have been able to single out a basic set of chains of $\mathfrak{M}$, namely here Δ_1^1, Δ_1^2, or chains such that every other is a linear combination of them plus a chain of $\mathfrak{M}^0$.

(e) The existence of the base gave us the immediate solution of the analytical problem.

(f) In the course of the discussion we encounter a basic integer, the number of independent cycles, equal to *two* in the case considered above. We must expect one for each dimension $p \leqq n$, and we shall in fact exhibit a whole series of similar integers.

From the geometric standpoint the chief elements to be retained are (b), (c), (d), (f). Their general treatment is the chief object of the rest of this chapter, all questions of topological invariance being left for the next.

§ 3. ARITHMETICAL DIGRESSION ON MATRICES AND MODULI

18. We collect in this section some arithmetical propositions which are of constant use in the sequel. The first at least is classical, but owing to its cardinal importance in our subject a rapid proof of it may not be amiss.

NOTATIONS FOR MATRICES. They shall be substantially the same as in Bôcher's *Introduction to Higher Algebra*. In particular we write $\mathbf{a}$ for $\|a_{ij}\|$, $\mathbf{a}'$ for its transverse; when $\mathbf{a}$ is square its determinant is designated by $|\mathbf{a}|$, and its inverse, when there is one, by $\mathbf{a}^{-1}$. A unit matrix of any order is written $\mathbf{1}$; the order is always readily inferred from the context and as a matter of fact, is usually immaterial. The terms of a unit matrix are designated by δ_{ij}. They are the Kronecker symbols: $\delta_{ij} = 0$, $i \neq j$; $\delta_{ii} = 1$. A square matrix whose terms are integers and determinant ± 1 is called *unimodular*. A matrix whose terms are all zero is designated by $\mathbf{0}$.

We shall also have occasion to consider at times matrices whose terms are themselves matrices: $\|\mathbf{a}_{ij}\|$. They obey the ordinary laws of combinations of matrices except that in taking the transverses each term must be replaced by its transverse, or:

$$\|\mathbf{a}_{ij}\|' = \|\mathbf{a}'_{ji}\|$$

19. Let $\mathbf{a}$ be a matrix with p rows, q columns, and integer terms. If $\mathbf{b}$, $\mathbf{c}$ are unimodular square matrices of order p, q, respectively, $\mathbf{b\,a\,c}$ is said to be *equivalent* to $\mathbf{a}$. The problem is to find a matrix equivalent to $\mathbf{a}$ and in a certain canonical form to be given below.

It is well known that the following operations on a matrix transform it into an equivalent one: (a) Change of sign of all elements of a row (column). (b) Adding to a given row (column) the elements of another row (column) multiplied by the same integer. Another operation having the same effect is the interchange of two rows or columns, but it can be obtained by a repetition of (a) and (b). It follows that we can transform a given matrix $\boldsymbol{a}$ into an equivalent one in which an element a_{kh} of $\boldsymbol{a}$ has been replaced by $a_{kh}+t\,a_{kj}$ without affecting the remaining elements of the k-th row. Under all such operations a factor common to all elements of $\boldsymbol{a}$ is common to all elements derived from $\boldsymbol{a}$. For the first two do not affect the factors of the elements, and as for the third, any factor common to a_{kh} and a_{kj} is a factor of $a_{kh}+t\,a_{kj}$. Also we can choose t so that $a_{kh}+t\,a_{kj}$ is the remainder from dividing a_{kh} by a_{kj}.

Let r be the absolute value of the non-zero terms of $\boldsymbol{a}$ of least absolute value. If the value of r can be reduced by a finite number of operations of equivalence, it can only be reduced a finite number of times, since it is always a positive integer. Hence we can obtain a matrix $\boldsymbol{b}$ equivalent to $\boldsymbol{a}$ which has in its upper left hand corner a positive term $b_{11}(=d_1)$ whose absolute value is as small as that of any term of any matrix which can be derived from $\boldsymbol{a}$ by operations of equivalence.

In this matrix every b_{i1} and b_{1i} must be divisible by d_1, since otherwise one of them could be replaced by a term whose absolute value was less than d_1. Thus $\boldsymbol{b}$ may be written

$$\left\|\begin{array}{c|ccc} d_1 & 0 & \cdots & 0 \\ \hline 0 & & & \\ \vdots & & \boldsymbol{c} & \\ 0 & & & \end{array}\right\|,$$

where $\boldsymbol{c}$ is a matrix of $p-1$ rows and $q-1$ columns. The terms of $\boldsymbol{c}$ must all be divisible by d_1, else by adding one of the other rows of $\boldsymbol{b}$ to the first we would have again the situation already ruled out. We can repeat the same reasoning for $\boldsymbol{c}$. The operations used at this stage do not affect the first row and column. Hence we shall have a new reduced matrix

$$\left\Vert \begin{array}{ccc|c} d_1 & 0 & \cdots & 0 \\ 0 & d_2 & \cdots & 0 \\ \vdots & \vdots & & \\ 0 & 0 & & \boldsymbol{e} \end{array} \right\Vert .$$

Since every term of $\boldsymbol{c}$ is divisible by d_1, by a remark above d_2 is divisible by d_1. This can be continued and we have finally a matrix which is in the canonical form

$$(5) \qquad \left\Vert \begin{array}{cccccccccc} d_1 & 0 & \cdot & \cdot & \cdot & \cdot & \cdot & \cdot & \cdot \\ 0 & d_2 & \cdot & \cdot & \cdot & \cdot & \cdot & \cdot & \cdot \\ \cdot & \cdot & \cdot & \cdot & \cdot & \cdot & \cdot & \cdot & \cdot \\ \cdot & \cdot & \cdot & \cdot & \cdot & \cdot & d_r & \cdot & \cdot \\ \cdot & \cdot & \cdot & \cdot & \cdot & \cdot & \cdot & \cdot & \cdot \end{array} \right\Vert ,$$

in which all terms are zero except the first r along the main diagonal and $0 < d_1 \leqq d_2 \leqq \cdots \leqq d_r$, $\dfrac{d_{i+1}}{d_i}$ an integer.

Now if $\boldsymbol{a}$, $\boldsymbol{b}$ are equivalent, a simple calculation shows that the minors of given order s of the one are linear integral multiples of those of the same order of the other. Therefore the H. C. F. δ_s of these minors is the same for $\boldsymbol{a}$ and for $\boldsymbol{b}$, and the matrices have the same *rank* ϱ (least s for which not all minors are zero). The rank of (5) is r, hence $\varrho = r$. Furthermore the δ_s of (5) is $d_1 d_2 \cdots d_s$, $s \leqq \varrho$. Hence $d_1 = \delta_1$, the H. C. F. of all terms of $\boldsymbol{a}$, and every other $d_s = \dfrac{\delta_s}{\delta_{s-1}}$. The δ's are called the *elementary divisors* of $\boldsymbol{a}$, the d's its *invariant factors*. We have then:

THEOREM OF FROBENIUS: *A matrix* $\boldsymbol{a}$ *with integer terms is equivalent to a unique canonical matrix (5) in which the d's are the invariant factors of* $\boldsymbol{a}$ *and their number is its rank.*

20. From what precedes we can derive a noteworthy result concerning the simultaneous canonical reduction of a pair of matrices $\boldsymbol{a}, \boldsymbol{b}$ such that $\boldsymbol{a}\boldsymbol{b} = 0$. It is implicit in a topological proposition given by Veblen [5]. Suppose indeed that

$$\boldsymbol{xay} = \alpha = \left\|\begin{array}{ccc|c} d_1 & & & \\ & \cdot & & \\ & & \cdot & \mathbf{0} \\ & & \cdot & \\ & & d_r & \\ \hline & \mathbf{0} & & \mathbf{0} \end{array}\right\|$$

is the canonical matrix derived from $\boldsymbol{a}$, so that r is the rank and the d's the invariant factors of $\boldsymbol{a}$.

Since $\boldsymbol{y}$ is unimodular so is $\boldsymbol{y}^{-1}$ and if $\beta = \boldsymbol{y}^{-1}\boldsymbol{b}$, manifestly $\alpha\beta = 0$. Owing to the form of α we find at once that the first r rows of β consist entirely of zeros, that is

$$\beta = \left\|\begin{matrix} \mathbf{0} \\ \gamma \end{matrix}\right\|,$$

where γ is the matrix of the remaining rows of β.

Let $\boldsymbol{z}\gamma\boldsymbol{t} = \delta = \left\|\begin{matrix} \boldsymbol{\theta} & \mathbf{0} \\ \mathbf{0} & \mathbf{0} \end{matrix}\right\|$ in the canonical form, so that $\boldsymbol{\theta}$ is the diagonal matrix in the canonical form of $\boldsymbol{b}$. Therefore its rank r' is that of $\boldsymbol{b}$ and its terms are the invariant factors of the latter. We find by direct multiplication

$$\alpha = \alpha \left\|\begin{matrix} \mathbf{1} & \mathbf{0} \\ \mathbf{0} & \boldsymbol{z}^{-1} \end{matrix}\right\|,$$

where $\mathbf{1}$ stands for the unit matrix of order r.

Also

$$\varepsilon = \left\|\begin{matrix} \mathbf{1} & \mathbf{0} \\ \mathbf{0} & \boldsymbol{z} \end{matrix}\right\| \cdot \left\|\begin{matrix} \mathbf{0} \\ \gamma \end{matrix}\right\| \cdot \boldsymbol{t} = \left\|\begin{matrix} \mathbf{0} \\ \delta \end{matrix}\right\| = \left\|\begin{matrix} \mathbf{0} & \mathbf{0} \\ \boldsymbol{\theta} & \mathbf{0} \\ \mathbf{0} & \mathbf{0} \end{matrix}\right\|$$

and we still have $\alpha\varepsilon = 0$.

The matrix ε equivalent to $\boldsymbol{b}$, is not in the canonical form but may be reduced to it by a permutation of rows. This amounts to a multiplication at the left by a unimodular matrix

$$\left\|\begin{array}{cc|c} \mathbf{0} & \mathbf{1} & \\ \mathbf{1} & \mathbf{0} & \mathbf{0} \\ \hline & \mathbf{0} & \mathbf{1} \end{array}\right\|.$$

Let us call *quasi-canonical* a matrix derived from a canonical matrix by a mere permutation of rows or columns. They are obtained by means of right or left handed multiplication by unimodular matrices derived from the identity by mere permutation of rows or columns.

The reasoning just applied would prove in exactly the same way that if $\boldsymbol{a}\boldsymbol{b} = \boldsymbol{0}$ and $\boldsymbol{a}$ is reduced to any quasi-canonical matrix $\alpha = \boldsymbol{x}\boldsymbol{a}\boldsymbol{y}$, arbitrary among those equivalent to $\boldsymbol{a}$, the reduction can be so carried out that there exists a quasi-canonical matrix $\beta = \boldsymbol{y}^{-1}\boldsymbol{b}\boldsymbol{z}$ equivalent to $\boldsymbol{b}$. It is also applicable with the parts of $\boldsymbol{a}$ and $\boldsymbol{b}$ interchanged. From this follows the

THEOREM. *Let $\boldsymbol{a}_1, \cdots, \boldsymbol{a}_n$ be n matrices such that $\boldsymbol{a}_i \boldsymbol{a}_{i+1} = 0$ for every i. It is possible to find unimodular matrices $\boldsymbol{x}_i$ such that every $\alpha_i = \boldsymbol{x}_i^{-1} \boldsymbol{a}_i \boldsymbol{x}_{i+1}$ is quasi-canonical, one of the α's being of preassigned type. The new matrices likewise have the property $\alpha_i \alpha_{i+1} = 0$.*

It may be added that we can always so reduce the system that in one assigned pair α_i, α_{i+1} one term is canonical and the other of the same type as ε or its transverse according as α_i or α_{i+1} is the canonical matrix.

21. We now pass to the consideration of moduli. As we shall understand the term for the present, a *modulus* $\mathfrak{M}$ is a set $\{x\}$ of elements such that if $x_1, \cdots, x_n$ belong to $\mathfrak{M}$ so does $\sum t_i x_i$, whatever the integers t_i. The subchains of fixed dimension of a complex K evidently form a modulus, as do its subcycles of fixed dimension.

Given $\mathfrak{M}$, let $\mathfrak{M}^0$ be any modulus such that it is possible to determine whether each element of $\mathfrak{M}$ belongs also to $\mathfrak{M}^0$. Since we shall be concerned only with those elements of $\mathfrak{M}^0$ which are also in $\mathfrak{M}$, and since these elements themselves evidently form a modulus, we may as well disregard the remaining elements of $\mathfrak{M}^0$ and let $\mathfrak{M}^0$ represent the subset contained in $\mathfrak{M}$.

We express the fact that the difference of two elements of $\mathfrak{M}$ is an element of $\mathfrak{M}^0$ by the symbolic *homologies*

$$x_1 - x_2 \sim 0 \quad \text{mod}\, \mathfrak{M}^0,$$
$$x_1 \sim x_2 \quad \text{mod}\, \mathfrak{M}^0,$$

which carries with it the notation

$$x \sim 0 \quad \text{mod}\, \mathfrak{M}^0$$

to indicate that the element x of $\mathfrak{M}$ belongs to $\mathfrak{M}^0$.

Homologies may be added and subtracted like algebraic equations and may be multiplied (but not necessarily divided, even though the coefficients possess a common integral factor) throughout by the same integer.

We shall say that a set of elements $\{y\}$ of $\mathfrak{M}$ constitutes a *base* of $\mathfrak{M}$ (mod $\mathfrak{M}^0$) if every

$$x \sim \sum t_i\, y_i,$$

where only a finite number of elements y need appear in the expression of any particular x. Every modulus with which we shall deal in the first six chapters will possess a finite base.

A set of homologies $\{\sum a_j\, x_j \sim 0\}$ (where the x's of one homology are not necessarily those of another and not all a's are zero) will be said to constitute a *fundamental set* if each homology can be expressed as a linear combination with integer coefficients of a finite number of them.

A base or a fundamental set is *minimal* if no subset forms a base or a fundamental set, respectively.

When the base $y_1, \cdots, y_q$ *is finite there is also a finite fundamental set for the homologies between its elements.* For let a non-redundent set of homologies

$$A_h = \sum a_{hi}\, y_i \sim 0$$

contain an infinite number of homologies with $a_{h1} \neq 0$. Since the a's are integers there is at least one homology whose $|a_{h1}|$ assumes its least positive value. Let it be

$$B = b_1\, y_1 + \cdots \sim 0.$$

Then b_1 is a divisor of every a_{h1}. For otherwise $a_{h1} = b_1\, \alpha + c$, $|c| < b_1$ and hence

$$\sum (a_{hi} - \alpha\, b_i)\, y_i \sim 0$$

is a homology with c as the coefficient of y_1. As b_1 is assumed to be the least in absolute value non-zero coefficient of y_1 in any homology, $c = 0$ and a_{h1} is divisible by b_1. Hence $A_h - \alpha B = D$, a linear expression in $y_2, \cdots y_q$. Therefore: (a) If $q = 1$, $B \sim 0$ is a fundamental set for all homologies. (b) If $q > 1$, a fundamental set is obtained by adjoining to $B \sim 0$ one for all homologies in $y_2, \cdots, y_q$ alone. Assuming as we now may, that our statement is proved for bases with less than q terms, there will be a finite fundamental set for the homologies without y_1 and therefore also for the total set.

The connection with the theory of abstract groups is clear. The elements of $\mathfrak{M}$ constitute the elements of an abstract Abelian group, the product of two elements being denoted by the sign $+$ instead of the customary product sign. $\mathfrak{M}^0$ is an Abelian subgroup of $\mathfrak{M}$. The elements reduced mod $\mathfrak{M}^0$ are then merely the elements of the quotient group $\mathfrak{M} : \mathfrak{M}^0$. Indeed everything that follows in this section can be, and frequently is, translated in terms of the theory of groups. It is of course a mere question of a different terminology.

22. We are now in position to prove two fundamental theorems concerning homologies.

THEOREM I. *Let*

(6) $$x_1, \cdots, x_p$$

be a base of $\mathfrak{M}$ *mod* $\mathfrak{M}^0$ *and*

(7) $$u_i = \sum a_{ij} x_j \sim 0 \quad (i = 1, 2, \cdots, s)$$

be a fundamental set of homologies connecting the x*'s. Then we can find a minimal base of* $\mathfrak{M}$ *mod* $\mathfrak{M}^0$,

$$y_1, \cdots, y_q,$$

and a set of positive integers

$$\theta_1, \cdots, \theta_r \ (0 \leqq r \leqq q)$$

such that $1 < \theta_1 \leqq \theta_2 \leqq \cdots \leqq \theta_r$, *every* $\dfrac{\theta_{i+1}}{\theta_i}$ *is an integer, and the homologies*

$$\theta_i y_i \sim 0 \qquad (i = 1, 2, \cdots, r)$$

form a minimal fundamental set.

Without loss of generality we may suppose that the rank of $\boldsymbol{a}$ is $s \leqq p$. Any set of elements

$$z_k \sim \sum b_{kj} x_j, \ \boldsymbol{b} \text{ unimodular},$$

is a base, since we can express the x's in terms of the z's by

$$x_j \sim \sum B_{jk} z_k, \quad \boldsymbol{B} = \boldsymbol{b}^{-1}.$$

Similarly, any set of homologies

$$v_g = \sum c_{gi} u_i \sim 0, \quad \boldsymbol{c} \text{ unimodular},$$

forms a fundamental set. By No. 19 we can determine $\boldsymbol{b}$ and $\boldsymbol{c}$ so that $\boldsymbol{cab}^{-1}$ is in the canonical form $\| \delta_{gk} \zeta_k \|$, where no ζ is zero since the rank of $\boldsymbol{a}$ is s. Hence the homologies v expressed in terms of the elements z are

$$v_g = \zeta_g z_g \sim 0.$$

The first w, say, of the ζ's may be unity. The corresponding homologies,

(8) $$z_i \sim 0 \qquad (i = 1, \cdots, w),$$

state that these elements belong to $\mathfrak{M}^0$ and hence are redundant in the base set. If we let

$$d_i = \zeta_{w+i} \quad (i = 1, 2, \cdots, r;\ w + r = s),$$

and

$$y_i = z_{w+i} \quad (i = 1, 2, \cdots, q;\ w + q = p),$$

we shall have a new relative base,

(9) $$y_1, \cdots, y_q,$$

and a new fundamental set of homologies

(10) $$d_i y_i \sim 0 \qquad (i = 1, 2, \cdots, r),$$

both of which we assert are minimal.

For if any homology involved the elements $y_{r+1}, \cdots, y_q$, it would, as a consequence, introduce a new homology (8), which would in turn lead back to the addition of a new homology to the original set $u_i \sim 0$, the new homology being linearly independent of the others. Again, no element y_j of the first r y's is $\sim \sum a_i y_i \, (i \neq j)$, since such a homology, having 1 as the coefficient of y_j, would not be a linear combination of the homologies (10). Hence the base is minimal. Hence also the fundamental set (10) is also minimal. The other conditions of the theorem have already been satisfied.

23. THEOREM II. *For given moduli* $\mathfrak{M}$ *and* $\mathfrak{M}^0$ *the coefficients* $d_1, \cdots, d_r$ *of Theorem I are independent of the choice of base and fundamental set of homologies.*

(a) Consider a base (6) with the fundamental homologies (7) and suppose in the first place that we add to (7) a new homology

$$\sum b_i x_i \sim 0.$$

From the assumptions on (7) it must be a sum of multiples of its homologies. Therefore the new matrix

$$\begin{Vmatrix} \boldsymbol{a} \\ \boldsymbol{b} \end{Vmatrix}$$

has the same invariant factors as $\boldsymbol{a}$.

(b) Next let us increase the base by a new element z. We have

$$z \sim \sum c_i x_i \tag{11}$$

and the relations analogous to (7) for the new base are (7) plus (11), and in place of $\boldsymbol{a}$ we have now

$$\left\Vert \begin{array}{c|c} -1 & c_1 \cdots c_p \\ \hline \mathbf{0} & \boldsymbol{a} \end{array} \right\Vert,$$

which is easily reduced to

$$\begin{Vmatrix} 1, & \mathbf{0} \\ \mathbf{0}, & \boldsymbol{a} \end{Vmatrix}.$$

Hence the δ_s of the new matrix is equal to the δ_{s-1} of $\boldsymbol{a}$, while its $\delta_1 = 1$, since one of its terms is 1. Therefore again the invariant factors > 1 are unchanged.

Suppose now that we have a new base $z_1, \cdots, z_q$ with fundamental homologies

$$\sum b_{ij} z_j \sim 0.$$

Since the z's can be expressed in terms of the x's and *vice versa* we have

$$z_i \sim \sum c_{ij} x_j,$$
$$x_i \sim \sum f_{ij} z_j.$$

Let us adjoin in succession the elements $z_1, z_2, \cdots$ to the base $x_1, \cdots, x_p$. Ultimately we shall arrive at the new base $x_1, \cdots, x_p, z_1, \cdots, z_q$ and by repeated application of (b) it will follow that the matrix $\boldsymbol{a}$ has the same invariant factors > 1 as a certain matrix

$$\left\| \begin{matrix} \boldsymbol{a} & \boldsymbol{0} \\ \boldsymbol{c} & -\boldsymbol{1} \end{matrix} \right\|,$$

where in the array each term stands for an obvious matrix. Then by applying (a) we find in turn that this matrix and hence $\boldsymbol{a}$ itself has the same invariant factors as

$$\left\| \begin{matrix} \boldsymbol{a} & \boldsymbol{0} \\ \boldsymbol{c} & -\boldsymbol{1} \\ \boldsymbol{0} & \boldsymbol{b} \\ -\boldsymbol{1} & \boldsymbol{f} \end{matrix} \right\|.$$

But by reversing the parts played by the two bases we find that the last matrix has the same invariant factors > 1 as $\boldsymbol{b}$. Hence $\boldsymbol{a}$ and $\boldsymbol{b}$ have the same invariant factors > 1 which proves the theorem.

24. The elements $y_1, \cdots, y_r$ corresponding to the invariant factors d_i may be termed *zero-divisors,* since they are not themselves ~ 0, but multiples of them are. For any x we have a homology

$$x \sim \sum t_i y_i.$$

If it is a zero-divisor, $t_i = 0$ for $i > r$, since the last $(q - r)$ y's are independent. Furthermore t_i, $i \leqq r$, may be replaced by its least positive residue mod d_i, since we can suppress by (10) any multiple of $d_i y_i$. It follows that every zero-divisor

$$(12) \qquad x \sim \sum s_i y_i, \ 0 \leq s_i < d_i.$$

Every such x is a zero-divisor or else ~ 0.

But as every homology between the y's must be a sum of multiples of the homologies (10), x is a true zero-divisor unless every $s_i = 0$. Furthermore every zero-divisor is represented once and only once by (12).

The zero-divisors may be considered as the elements of a subgroup G of the group considered in No. 21. From the representation (12) for the elements of G, it follows that the group is finite and of order $n = d_1 \cdots d_r$. The y's constitute a *base* for G, and the d's are the orders of the elements of that base. Theorem II is then a statement of a well known theorem on finite Abelian groups, but most proofs given for it are rather involved and indirect. The very simple one just given has recently been published by Alexander [20].

25. It is frequently convenient to enlarge $\mathfrak{M}^0$ by adding to it all zero-divisors and to write $x \approx 0 \bmod \mathfrak{M}^0$ to indicate that $x \sim$ a zero-divisor or $x \sim 0$. In the field of action of $\approx$, division of all coefficients in a homology by a common factor is always permissible, while it is not in the field of $\sim$. For $tx \sim 0$ merely implies that x is a zero-divisor, not that it is ~ 0. In many questions of analysis situs this introduces great simplifications, as exemplified in our paper [14].

Another notion of great importance is that of the independence of the elements of $\mathfrak{M} \bmod \mathfrak{M}^0$. Elements $x_1, \cdots, x_n$ of $\mathfrak{M}$ are said to be *independent* mod $\mathfrak{M}^0$ provided

$$\sum t_i x_i \sim 0$$

if and only if every t_i is zero. The maximum number of independent elements of $\mathfrak{M} \bmod \mathfrak{M}^0$ is called its dimension

$R(\mathfrak{M})$ mod $\mathfrak{M}^0$, or merely $R(\mathfrak{M})$. Its value is evidently $q - r$, i. e. the number of elements of a minimal base minus the number of fundamental homologies connecting them.

It is to be noted that the definition of independence mod $\mathfrak{M}^0$ ceases to have significance when the coefficients of all homologies are reduced by an arithmetic modulus m. In that case we shall replace the condition that the t's be zero by the condition that no linear combination be a sum of zero-divisors, that is of terms having a multiple $< m$, ~ 0 mod m.

REMARK. The notion of modulus may be extended to linear combinations with rational coefficients. This can be done freely when dealing with $\approx$ and shall be done exclusively in Ch. VII.

§ 4. THE BETTI AND TORSION NUMBERS OF A COMPLEX AND THEIR CALCULATION

26. We have already outlined in an example (No. 16) the direction in which the preceding arithmetical discussion is to be applied. The treatment follows closely that of Veblen [5]. It rests essentially upon a proper choice of moduli $\mathfrak{M}$, $\mathfrak{M}^0$. There are four different choices according to the particular type of modular relation chosen.

I. *Absolute chains.* We take for $\mathfrak{M}$ the set of all absolute p-cycles of K, and for $\mathfrak{M}^0$ the *bounding* cycles or the cycles of the form $F(C_{p+1})$. The corresponding homologies are written

$$\Gamma_p \sim 0 \text{ on } K.$$

The notation "on K" is needed whenever several complexes enter simultaneously in the discussion. The reduced homologies corresponding to a *canonical* base, or base for which the matrix $\boldsymbol{a}$ of the homologies (7) is canonical, will be written

$$\zeta_i^p \Gamma_p^i \sim 0 \text{ on } K.$$

Those invariant factors ζ_i^p which are > 1 are called the *p-th torsion coefficients* and will be denoted by θ_i^p. They were first introduced by Poincaré [3]. The dimension of $\mathfrak{M}$ mod $\mathfrak{M}^0$ is designated by $R_p(K)$ and will be called the *p-th Betti number of K.*

II. *Relative chains, modular chains, and their combination.* Everything that we have said applies to them except that $\mathfrak{M}$ consists respectively of the p-cycles mod L, or mod m, or mod (L, m), and $\mathfrak{M}^0$ of those that have the form $F(C_{p+1})$ mod L, etc. (and in the case of mod m or (L, m), those that are $\equiv 0$ mod m). The homologies are

$$\Gamma_p \sim 0 \bmod L, \quad \text{or} \bmod m, \quad \text{or} \bmod (L, m), \quad \text{on } K,$$
$$\theta_i^p \Gamma_p^i \sim 0 \bmod L, \quad \text{or} \bmod m, \quad \text{or} \bmod (L, m), \quad \text{on } K.$$

The Betti numbers are designated by $R_p(K; L)$, $R_p(K; m)$, $R_p(K; L, m)$. When several different sets of torsion coefficients appear simultaneously we may likewise distinguish them by some such notation as $\theta_i^p(K; L)$, etc.

We shall apply the generic term *combinatorial homology characters* to all these numbers.

REMARK. There has been considerable confusion in terminology and definition of the quantities which we have called Betti numbers. Poincaré, who dealt only with absolute chains, defined no such quantity for the dimension 0, and defined as Betti numbers certain quantities P_i $(i > 0)$ related to ours by $P_i - 1 = R_i$. Veblen uses Poincaré's definition and terminology in the absolute case but introduces a number P_0 identical with our R_0, owing to a different definition of 0-cycle. For the corresponding quantities mod 2 Veblen uses the term "connectivity number" and the symbols R_i, so that his $R_0(K)$ equals our $R_0(K; 2)$ while otherwise $R_i - 1 = R_i(K; 2)$. Alexander follows our definitions but calls the numbers, both absolute and mod m, "connectivity numbers", so that his P_i is identical with our R_i, whether absolute or mod m.

We have chosen the term "Betti number" to cover all cases as being a sufficiently brief term which is historically connected with the same order of ideas (though in a more restricted sense). The particular definitions which we have chosen, involving the slight differences in value noted above, arise naturally from the abstract theory of matrices and simplify some formulas without complicating the others.

While we are on this matter of terminology, let us observe that Poincaré and practically all authors after him have written "mod K" where we have "on K". Veblen and Alexander use "mod m" as we do here. The departure from Poincaré's terminology seems imperative when L comes into play.

27. In each of the four cases with which we have to deal we can recognize the following common elements: (a) The elements of $\mathfrak{M}$ are derived from a modulus $\mathfrak{M}'$ having as base elements a set of p-cells, the base varying with the different cases. (b) The elements of $\mathfrak{M}$ have as base certain elements of $\mathfrak{M}'$ defined by means of the notion of bounding, which also varies from case to case. (c) $\mathfrak{M}^0$ is composed of all those elements of $\mathfrak{M}$ which bound according to the definition of bounding in each case. The treatment of homologies on $K-L$ evidently includes the case of homologies on K by letting L be identically zero, so that we shall have to deal explicitly only with the cases of homologies mod L and mod (L, m).

Let E_p^i, $i=1, 2, \cdots, \alpha_p$, designate the p-cells of $K-L$. We have the basic incidence relations

$$\left.\begin{aligned} &(13) \qquad E_{p+1}^i \to \sum \eta_{ij}^p \; E_p^j \\ &(14) \qquad E_p^j \;\;\to \sum \eta_{jk}^{p-1} \; E_{p-1}^k \end{aligned}\right\} \bmod L.$$

Since L is a closed complex the boundaries of its cells are cells of L and we have

$$F(E_{p+1}^i) \to \sum \eta_{ij}^p \, \eta_{jk}^{p-1} \, E_{p-1}^k \quad \bmod L.$$

The part of the boundary of $F(E_{p+1}^i)$ which is on L does not include any E_{p-1}^k. Since the total boundary of a boundary is $\equiv 0$, the coefficients of the cells in the sum must be zero, that is

$$\sum_j \eta_{ij}^p \, \eta_{jk}^{p-1} = 0,$$

which is equivalent to the unique matrix equation

$$\eta^p \, \eta^{p-1} = 0. \tag{15}$$

Suppose now that we introduce new chains

$$C_p^i = \sum a_{ig}^p \, E_p^g, \qquad \boldsymbol{a}^p \text{ unimodular.} \tag{16}$$

The C_p's constitute new bases for the p-chains. The new boundary relations are

$$C_{p+1}^i \to \sum \zeta_{ij}^p C_p^j \mod L; \qquad \zeta^p = \boldsymbol{a}^{p+1} \eta^p (\boldsymbol{a}^p)^{-1}. \tag{17}$$

The situation is then as before except that the C's have taken the place of the cells. Since (13) is a fundamental set of incidence relations the same property holds for (17). It is also clear that the matrices $\boldsymbol{a}^p$ are entirely arbitrary unimodular matrices.

According to No. 20 we can so select the matrices $\boldsymbol{a}^p$ that the ζ's will be quasi-canonical; in particular, any given consecutive pair can be written, ϱ_p being the rank of η^p:

$$\left.\begin{array}{lll}
\text{(18a)} & C_{p+1}^{1h} \to \zeta_h^p C_p^{1h}, & h = 1, 2, \cdots, \varrho_p, \\
\text{(18b)} & C_{p+1}^{2i} \to 0, & i = \varrho_p + 1, \cdots, \alpha_{p+1}, \\
\hline
\text{(19a)} & C_p^{1h} \to 0, & \\
\text{(19b)} & C_p^{2j} \to \zeta_j^{p-1} C_{p-1}^j, & j = \varrho_p + 1, \cdots, \varrho_{p-1} + \varrho_p, \\
\text{(19c)} & C_p^{3k} \to 0, & k = \varrho_{p-1} + \varrho_p + 1, \cdots, \alpha_p,
\end{array}\right\} \mod L.$$

Up to this point we have the canonical reduction of the modulus of all p-chains with the vacuous set as the zero-modulus. We are now going to examine the consequences for $\mathfrak{M}$, the modulus of all p-cycles, $\mathfrak{M}^0$ being then the sub modulus of all bounding cycles.

(a) Every p-cycle mod L, that is every element of $\mathfrak{M}$, is a linear combination of the chains C_p^{1h}, C_p^{3k}, since any other

p-chain will have an actual boundary mod L. These two sets constitute a base for the modulus of p-cycles.

(b) The fundamental homologies for the base are

$$\zeta_h^p C_p^{1h} \sim 0 \quad \text{mod}\, L. \tag{20}$$

Every boundary mod L will have to be a combination of the left hand sides of these homologies. The cycles C_p^{1h}, constitute a base for the zero-modulus. Those corresponding to a $\zeta_h^p > 1$ are the *zero-divisors* for the dimension p, and the numbers $\zeta_h^p > 1$ are the p-th *torsion coefficients* mod L.

The Betti-number for the dimension p is then the number of chains C_p^{3k}. Hence

$$R_p(K;\, L) = \alpha_p - \varrho_{p-1} - \varrho_p. \tag{21}$$

In calculating R_n and R_0 we must remember to make $\varrho_{-1} = \varrho_n = 0$, since on the one hand every $C_0 \to 0$, on the other there is no η^n.

We have also proved incidentally:

THEOREM I. *Every p-cycle mod L is homologous to a linear combination of the cycles C_p^{3k} plus a zero-divisor or bounding cycle mod L.*

An equivalent statement is: *The cycles C_p^{3k} constitute a base for the homologies $\approx$* (with division allowed). The base is in fact *minimal* since these cycles are independent. We have then for every p-cycle mod L:

$$\Gamma_p \approx \sum t_k C_p^{3k} \quad \text{mod}\, L. \tag{22}$$

THEOREM II. *Let $\zeta_h^p > 1$ for $h > r_p$. Then the cycles C_p^{1h}, $h > r_p$, and C_p^{3k} together constitute a minimal base for ordinary homologies.*

For every

$$\Gamma_p \sim \sum_{h=r_p+1}^{\varrho_p} s_h C_p^{1h} + \sum_k t_k C_p^{3k}. \tag{23}$$

The preceding reduction is essentially Poincaré's [3, 4], with the important improvement, due to Veblen [5], that all incidence matrices are simultaneously reduced to the quasi-canonical form. The application to relative boundary relations is found here for the first time.

28. There is a simple and noteworthy geometric interpretation for the numbers R_0. We will say for the present that two vertices A, B of K can be connected on a subset of K, if they are coincident points of the subset or else end points of a closed arc l which is a sum of closed one-cells on the subset. We can prove in succession the following simple properties:

(a) If A and B can be connected on K they are homologous zero-cycles on K. For if l is the joining arc suitably oriented, $l \rightarrow B - A$, hence $A \sim B$.

(b) If A can be connected to L (that is, to a point of L) on K, it is $\sim 0 \bmod L$.

(c) If none of the vertices $A^1, \cdots, A^r$ can be connected to one another or to L they are independent mod L, as we proceed to show.

Suppose that

$$C_1 \rightarrow \sum t_i A^i \bmod L. \tag{24}$$

Let it be possible to describe on C_1 a simple broken line whose end points are on L or else coincide. Orient the one-cells of that line in such manner that whenever any two have a common vertex they are oppositely related to it and let C_1' be the cycle mod L thus obtained, E_1 one of its cells not on L. We shall have $C_1 = sE_1 + \cdots$, $C_1' = E_1 + \cdots \rightarrow 0$, mod L, where the cells omitted do not include E_1. Therefore $C_1 - s\,C_1' \rightarrow \sum t_i A^i \bmod L$. The effect on C_1 is to replace it by a chain with one cell less than before. Proceeding thus we shall ultimately have a C_1 with no associated C_1'. Suppose then that $t_1 \neq 0$. Starting on that C_1 from A^1 along one of the one-cells issued from it and so on, we can never return to a vertex crossed unless we meet L. But meeting L is ruled out owing to our assumption on the A's. Therefore the path described must stop at some point A^0. This point will be the vertex of a unique cell of C_1 and not on L, hence it must actually appear in $F(C_1)$ mod L. It must then be one of the points $A^2 \cdots$ so that A^1 can be connected to one of them. But this is again contrary to assumption. Therefore the A's are independent.

CONCLUSIONS. I. *$R_0\ (K;\ L)$ is the number of connected subsets of K which do not meet L.*

II. *$R_0\ (K)$ is the number of connected subsets of K.*

III. *There are no torsion coefficients for the dimension zero.*

For again let

$$t \sum t_i A^i \sim 0 \text{ mod } L, \qquad t > 1. \tag{25}$$

If A^i is connected to L, it is ~ 0, and hence can be suppressed. If it is connected to A^1 it is $\sim A^1$, hence we can suppress it without changing the zero-divisor provided that we replace t_1 by $t_1 + t_i$. Proceeding thus we have ultimately a homology (25) such that the A's are not connected to one another nor to L on K, which can only be if every $t_i = 0$. In other words we do not have a true zero-divisor.

REMARK. From (a) and (c) follows this noteworthy result: If the A's are on a connected subset of K, $\sum t_i A^i \sim 0$ when and only when $\sum t_i = 0$. For by (a)

$$\sum t_i A^i = (\sum t_i) A^1 \sim 0,$$

by (c) when and only when the coefficient of A^1 is zero.

29. **Homology characters mod (L, m).** In calculating these characters it is important to recall that when a chain $= 0$ mod $(L,\ m)$, i. e. has all coefficients of its cells not on L divisible by m, it is also to be considered as ~ 0 mod $(L,\ m)$.

The chains that served as a base for all chains mod L constitute evidently also a base for those mod (L, m). Hence the chains at the left in (18), (19) are bases for their dimensions. We now rewrite these relations taking into account the modulus m.

Let ξ_h^p be the H. C. F. of ζ_h^p and m. We can find two integers β_h^p, x such that

$$\beta_h^p \zeta_h^p + xm = \xi_h^p. \tag{26}$$

To dispose at once of the arithmetical part let us show here and now that β_h^p may be chosen prime to m. Dropping indices for the present let us write the preceding relation in the simplified form

$$\beta\zeta + xm = \xi.$$

If $\zeta = \xi\zeta_1$, $m = \xi m_1$, ζ_1 and m_1 are prime to each other and we have

$$\beta\zeta_1 + xm_1 = 1. \tag{27}$$

This shows also that β and m_1 are prime to each other. Now if β_0, x_0 is a solution of (27),

$$\beta = \beta_0 + tm_1, \qquad x = x_0 - t\zeta_1$$

is one likewise for every t. Let μ be the highest factor of m prime to β. μ is divisible by m_1. Take $t = \dfrac{\mu}{m_1}$ and therefore $\beta = \beta_0 + \mu$. Any prime factor of m divides β_0 or μ. However it cannot divide β else it would have to divide both β_0 and μ. If follows that β is prime to m.

We take then all the numbers β_h^p prime to m. If we rewrite (18) and (19) taking into account m, we must replace the ζ's divisible by m by zero. Since each ζ_h^p divides the following one, when one of them is a multiple of m so are all those that follow it, and similarly of course for the numbers ζ_j^{p-1}. In view of (26), then, the new system assumes the form

$$\left.\begin{array}{lll}
(28\text{a}) & \beta_h^p C_{p+1}^{1h} \to \xi_h^p C_p^{1h}, & h = 1, 2, \cdots, \varrho_p', \\
(28\text{b}) & C_{p+1}^{1h'} \to 0, & h' = \varrho_p' + 1, \cdots, \varrho_p, \\
(28\text{c}) & C_{p+1}^{2i} \to 0, & i = \varrho_p + 1, \cdots, \alpha_{p+1} \\
\hline
(29\text{a}) & C_p^{1h} \to 0, & \\
(29\text{b}) & C_p^{1h'} \to 0, & \\
(30\text{a}) & \beta_j^{p-1} C_p^{2j} \to \xi_j^{p-1} C_{p-1}^{2j}, & j = \varrho_p + 1, \cdots, \varrho_p + \varrho_{p-1}', \\
(30\text{b}) & C_p^{2j'} \to 0, & j' = \varrho_p + \varrho_{p-1}' + 1, \cdots, \varrho_p + \varrho_{p-1}, \\
(30\text{c}) & C_p^{3k} \to 0, & k = \varrho_p + \varrho_{p-1} + 1, \cdots, \alpha_p,
\end{array}\right\} \bmod (L, m).$$

We have $\varrho_p - \varrho'_p = t_p$, the number of torsion coefficients mod L divisible by m.

30. Confining our attention to the dimension p, we see that all p-chains except (30a) are cycles. As for those, since the β's are prime to m, we can find γ, y such that

$$\beta\gamma + ym = 1.$$

Hence (30a) can be replaced by

(31) $$C_p^{2j} \to \gamma_j^{p-1} \xi_j^{p-1} C_{p-1}^{2j} \quad \text{mod}\,(L,\, m).$$

From this and since γ is prime to m, follows

(32) $$\theta_j^{p-1} C_p^{2j} \to 0, \qquad \theta = \frac{m}{\xi},$$

θ being the least positive integer for which this holds. Therefore $\theta\, C$ is a cycle. However,

$$\xi\, \theta\, C = 0 \quad \text{mod}\ (L,\, m)$$

so that the chain is a zero-divisor when $\xi > 1$, a bounding cycle otherwise.

From this we deduce that the base for all p-cycles is composed of the set

$$C^{1h}, \quad C^{1h'}, \quad \theta_j^{p-1} C^{2j}, \quad C^{2j'}, \quad C^{3k}$$

where C^{1h}, $\theta_j^{p-1} C^{2j}$ are the zero-divisors and the others the independent cycles.

The Betti-number is then the number of cycles $C^{1h'}$, $C^{2j'}$, C^{3k}, or

(33) $$R_p(K;\, L,\, m) = R_p\,(K;\, L) + t_{p-1} + t_p.$$

Since there is no η^n matrix, $t_n = 0$, and since there are no torsion coefficients for the dimension zero, (No. 28), $t_0 = 0$.

In particular

(34) $$R_0\,(K;\, L,\, m) = R_0\,(K;\, L).$$

(Regarding the possible definitions of Betti numbers mod m, see A. W. Tucker, Proc. Nat. Acad. 18 (1932), 467-472.)

Therefore the numbers $R_0(K; L, m)$ are susceptible of the same geometric interpretation as the numbers $R_0(K; L)$. Indeed the discussion of No. 28 is directly applicable to them.

31. The torsion coefficients for the dimension p are the invariant factors >1 and $<m$ of the diagonal matrix whose non-zero terms are m and the integers ξ_j^{p-1}, ξ_j^p that are >1. They are calculated as follows: Let e_k be the H. C. F. of all products of k numbers m, ξ^{p-1}, ξ^p. The numbers in question are the integers >1 and $<m$ in the sequence $e_1, \frac{e_2}{e_1}$, etc.

This is as far as it is possible to go with the torsion coefficients. Some simple properties are embodied in the following statements:

I. *If there are no torsion coefficients mod L whatsoever, there are none mod* (L, m).

II. *It is possible to have torsion coefficients for the dimension* p *mod* (L, m) *but not mod* L.

III. *There are no torsion coefficients mod* m *when* m *is a prime. In particular there are none mod* 2.

IV. *The invariants mod* (L, m) *can be calculated once those mod* L *are known.*

V. *When all the Betti numbers* $R(K; L)$ *and* $R(K; L, m)$ *are known the torsion coefficients can be calculated.* (*Alexander* [19]).

The proofs of all but V are implied in our discussion. Regarding V it follows from IV that it is sufficient to show how to calculate the torsion coefficients mod L. Let us designate t_p for m by $t_p(m)$. Since every $t_0 = 0$, (33) gives us the numbers $t_1(m)$, $t_1(m) + t_2(m)$, etc. and hence all the t's. Let μ_1 be the highest integer for which $t_p(\mu_1)$ assumes its highest value. Then $t_p(\mu_1)$ is the number of torsion coefficients for the dimension p and $\mu_1 = \zeta_1^p$. Similarly ζ_2^p is the largest integer m for which $t_p(m)$ assumes its largest value $< t_p(\mu_1)$, etc.

32. **The Euler-Poincaré characteristic.** There is an interesting relation connecting the Betti-numbers of all dimensions. If we multiply each equation (21) or (33) by $(-1)^p$ and sum for p, keeping in mind the special values of R_0, t_0 and R_n, t_n, we find

$$\begin{aligned}\sum_{p=0}^{n}(-1)^p \alpha_p &= \sum_{p=0}^{n}(-1)^p R_p(K; L)\\ &= \sum_{p=0}^{n}(-1)^p R_p(K; L, m).\end{aligned} \tag{35}$$

This is a generalization of the noted relation defining what is called the Euler-Poincaré characteristic of a polyhedron:

$$\alpha_0 - \alpha_1 + \alpha_2 = 2(1-p) \tag{36}$$

where p is the genus of the surface. For a surface bounding a portion of 3-space, p is the number of holes in the solid figure. The figure for which $p = 1$ is the anchor ring. The reader might find it amusing to divide the surface of the anchor ring into triangles and compute this constant by both (35) and (36).

§ 5. Circuits and their Orientation

33. We have already considered oriented simplexes and complexes. In analysis one has frequent occasion to orient spaces and regions. Thus a connected plane region $\mathfrak{R}$ is oriented by means of a small circuit, the *indicatrix* of $\mathfrak{R}$, on which there has been marked a positive direction of advancement. There results a positive direction of rotation, an orientation for other circuits or indicatrices on $\mathfrak{R}$: any two are sensed alike, or are *equivalent*, if they can be deformed into one another over $\mathfrak{R}$ without ever ceasing to be circuits. All indicatrices fall into two non-equivalent classes, and $\mathfrak{R}$ is really oriented by association with one of the classes, which is fixed by means of one of its indicatrices.

The extension to other surfaces suggests itself at once. But here a difficulty appears. When $\mathfrak{R}$ is not a plane region all indicatrices may well be equivalent. The simplest example is the Möbius strip or closed rectangular band with one twist. It is found that an oriented circuit can be deformed into its opposite over the band and therefore the band cannot be oriented.

We thus have to face the possibility that a configuration be non-orientable in the present sense of orientation. A related phenomenon is the fact that the band when on an S_3 is one-sided: an insect describing a closed path on the band may return to its original position upside down. But one-sidedness, two-sidedness point to properties of the band in relation to S_3 (Veblen) whereas *orientability* is an intrinsic property. Therefore we use the now generally accepted term *orientable* introduced by Alexander in place of the older and less accurate "two-sided".

34. We have just introduced orientability by means of continuous deformations of circuits. We have to replace this intuitive device by something that may be described in terms of the incidences of the cells. The correct hint is obtained as follows: Reverting to the plane region $\mathfrak{R}$, it is found that if two indicatrices exterior to one another have a common arc, the latter is a positive arc for one of them, a negative one for the other. Suppose that $\mathfrak{R}$ is convertible into a K_2. Orient any triangle $A^0A^1A^2$ of K_2 so that $A^0A^1+A^1A^2+A^2A^0=F(A^0A^1A^2)$ is a positive circuit, or circuit deformable into the indicatrix. By the preceding remark it follows that whenever two triangles of K_2 have a common side they are oppositely related to it. This is a combinatorial property of the complex which can be stated without reference to deformations. When the triangles of K_2 can be oriented in this fashion, K_2 is called orientable and the scheme can be carried through in two different ways.

A connected region $\mathfrak{R}$ on a surface F which can be turned into a K_2 can be treated in the same manner. $\mathfrak{R}$ is orientable by definition whenever K_2 is. This property, apparently depending upon K_2, is actually topological as will be shown later.

To simplify matters we assume that the surface F is regular over $\mathfrak{R}$. Then every side of the triangles of K_2 will belong to at most two triangles. The sum of those that belong to only one of them is a complex L_1 and $K_2\to 0 \bmod (L_1, 2)$, $F(K_2)\subset L_1$. Thus K_2 is what we shall call a *circuit* mod L_1

or $K_2 - L_1$ a *relative* circuit. It is orientable when and only when the sum, Γ_2, of its 2-cells suitably oriented has no boundary cells on $K_2 - L_1$, i. e. is a cycle mod L_1. This is the most convenient aspect of orientability when it comes to a formal treatment and the extension to any n, which is our next task.

35. **Definition of circuits.** Taking L as a subcomplex of K, we define K as an n-circuit mod L, or $K - L$ as a circuit whenever:

(a) *The chain sum of the n-cells of* $K - L$ *is an n-cycle mod* $(L, 2)$ *which coincides with* $\overline{K - L}$.

(b) $R_n(K; L, 2) = 1$.

The first condition implies that every cell of $K - L$ is an n-cell or a face of an n-cell and every $(n-1)$-cell of it is incident with an even number of n-cells.

From (a) follows that the Betti-number in (b) is at least *one*. Condition (b) in assigning its exact value implies that no subset of the closed n-cells is a cycle mod$(L, 2)$. Its effect is then to make the circuit an irreducible configuration.

When L is of dimension $< n-1$, we have an *absolute circuit* or circuit in the sense of Lennes [1] ($n = 3$) and Veblen [5] (any n). Simple examples of absolute circuits are: (1) an n-sphere, H_n, in which each $(n-1)$-cell is incident with exactly two n-cells; (2) the configuration resulting from pinching a meridian circle of a torus into a closed 1-cell, giving a surface such as that of the accompanying Figure (3) which can obviously be triangulated in such a way that the 1-cell specified above is incident with four 2-cells.—Any n-cell is an n-circuit relative to its boundary; the Möbius band is a 2-circuit relative to its edge, etc.

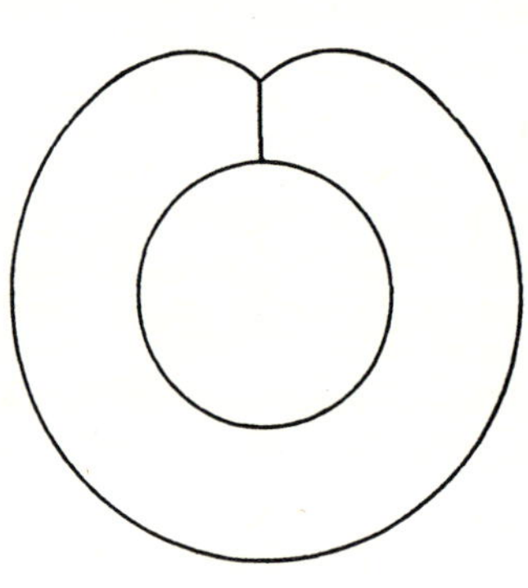

Fig. 3.

All but one of the circuits of our examples have the property that every $(n-1)$-cell of $K - L$ is incident with two

n-cells of it. Whenever this holds the circuit is said to be *simple.* The simple circuit is the same thing as the *pseudo-manifold* of Brouwer [8]. For the applications we have the important

THEOREM. *If every* $(n-1)$*-cell of* $K-L$ *is incident with two of its* n*-cells, and its set of* n*- and* $(n-1)$*-cells is connected,* $K-L$ *is a simple circuit.*

The connectedness here considered may be understood in the following combinatorial sense: Any two cells E'_p, E''_q, $p, q = n-1$ or n, of $K-L$, are end terms of a sequence $\{E^i\}$ of cells of $K-L$, each incident with its subsequent and the dimensions being alternately $n-1, n$.

To prove our assertion we must show that (b) holds. Suppose that $R_n>1$. Let $\Gamma_n = \sum E_n^i$ be the cycle mod$(L, 2)$ sum of all the n-cells of $K-L$, Γ'_n a different cycle; then Γ'_n and the cycle (mod 2) $\Gamma_n + \Gamma'_n$ have no common n-cell. They cannot have in common an $(n-1)$-cell of $K-L$ for it would be incident with two n-cells of each cycle, hence with at least four, contrary to the incidence assumption. Therefore, a sequence such as above between any two of their n-cells is impossible and the connectedness condition of the theorem does not hold. It follows that it implies (b).

While the simple circuit is the circuit commonly found in most applications, the fact that it is not characterised by a Betti number will make the proof of its topological invariance considerably more difficult.

36. Let us consider one of the oriented n-cycles of the circuit

$$\Delta_n = \sum t_i E_n^i.$$

If the t's have a common factor we may divide them by that factor and still have a cycle. Therefore, we may assume that the t's are relatively prime, and if $\alpha_n = 1$ that $t_1 = 1$. In all cases at least one t, say t_1, will be odd. Let t'_i be the least positive residue of t_i mod 2. Then $\sum t'_i E_n^i$ is a cycle mod $(L, 2)$ and therefore every $t'_i = 1$. It follows that in the expression of Δ_n every n-cell of $K-L$ must be present.

Suppose now that there are actual oriented n-cycles mod L and let Δ_n be that cycle whose coefficient t_1 has the least positive value that it assumes for any cycle. By the above remark it follows that the coefficients t_i must be relatively prime, else we could divide by a common factor and have a cycle with a smaller t_1.

Every n-cycle is a multiple of Δ_n. For let any other cycle be

$$\Delta'_n = \sum t'_i E^i_n$$

and call r_1 the H.C.F. of t_1, t'_1, so that there exist integers s, s' such that $st_1 + s't'_1 = r_1$. Then $s\Delta_n + s'\Delta'_n = r_1 E^1_n + r_2 E^2_n + \cdots$ is a non-trivial n-cycle mod L. But $r_1 \leqq t_1$ and in view of our assumption only the equality holds. Therefore t_1 divides t'_1. Let $t'_1 = s'' t_1$. Then

$$\Delta'_n - s''\Delta_n = u_2 E^2_n + \cdots$$

is an n-cycle without E^1_n. But we have shown that this is impossible. Hence that n-cycle is trivial, every $u_i = 0$ and Δ'_n is a multiple of Δ_n. We have therefore:

THEOREM. *The oriented n-cycles on the n-circuit K mod L, if any are present, are all the multiples of a unique n-cycle* Δ_n.

There are now two possibilities:

1. *There are no oriented n-cycles whatever.* Then

$$R_n(K; L) = 0 \text{ and } R_n(K; L, 2) - R_n(K; L) = 1$$

and by (33) there is a single even $(n-1)$-th torsion coefficient.

II. *There exist oriented n-cycles.* Then

$$R_n(K_n; L) = 1 \text{ and } R_n(K_n; L, 2) - R_n(K_n; L) = 0$$

and by (33) all $(n-1)$-th torsion coefficients are odd.

In the second case the circuit is called *orientable*; in the first, *non-orientable.*

There are two alternate necessary and sufficient conditions for orientability of the circuit:

First condition: The n-circuit is orientable when and only when its $(n-1)$-th torsion coefficients are all odd. When it is not orientable there is just one even $(n-1)$-th torsion coefficient.

Second condition: The n circuit is orientable or non-orientable accordingly as

$$R_n(K; L, 2) - R_n(K; L) = 0 \text{ or } 1.$$

37. Granting that the circuit is orientable, we determine Δ_n as follows: Let $E_{n-1}^j, j = 1, 2, \cdots, \alpha_{n-1}$ be the $(n-1)$-cells of $K-L$ and

$$E_n^i \to \sum \eta_{ij} E_{n-1}^j \mod L$$

the associated boundary relations. The t's must satisfy the system

$$\sum t_i \eta_{ij} = 0 \tag{37}$$

which expresses that $\Delta_n \to 0 \mod L$. To any solution in integers t corresponds an n-cycle mod L. We know in advance that there is essentially a unique solution, and the one corresponding to Δ_n is the one for which these integers assume the least absolute value. Of course $-\Delta_n$ is a cycle that may play the same part as Δ_n itself. Which of the two cycles one decides to choose may be specified by stating that any one of the n-cells, say E_n^i, is to have its $t_i > 0$. That n-cell is called the *indicatrix* of the circuit. If we reverse its orientation Δ_n is replaced by $-\Delta_n$. We understand by *K normally oriented* mod L, one or the other of its two oriented subchains $\pm \Delta_n$. Normal orientation is possible in two opposite ways.

38. For simple orientable circuits (37) continues to hold. If E_n^h, E_n^k are the two n-cells incident with E_{n-1}^j, $\eta_{hj} = \pm \eta_{kj} = \pm 1$ and the other coefficients η_{ij} with j as their second suffix are zero. Therefore $t_h = \pm t_k$. Thus when two n-cells have a common boundary $(n-1)$-cell their coefficients t are equal in absolute value. Since the set of n and $(n-1)$-cells is connected, we can pass from any n-cell to any other by crossing only n- or $(n-1)$-cells. Any consecutive n-cells thus crossed must have their coefficients t equal. Therefore, all t's are equal in absolute value. By reorienting the n-cells

if necessary we can also arrange matters so that all the t's are actually equal. But by No. 36 they are relatively prime. Hence their common value is *one* and

$$\Delta_n = \sum E_n^i.$$

Thus in the orientable simple circuit the fundamental n-cycle is the sum of the n-cells properly oriented.

Orientability of the simple circuit merely demands that it be possible to orient the n-cells in such manner that the two incident with any $(n-1)$-cell are oppositely related to it.

39. THEOREM. *An orientable simple circuit has no torsion coefficients for the dimension* $n-1$; *a non-orientable simple circuit has one such coefficient and its value is two.*

We apply the theorem of No. 35. Starting with E_n^1 of the simple circuit $K-L$, it is possible to pass over into a cell E_n^2 having an $(n-1)$-face, E_{n-1}^1, in common with it. We orient E_n^2 so that E_n^1 and E_n^2 are oppositely related to E_{n-1}^1. Then if $C_n^2 = E_n^1 + E_n^2$, the cell E_{n-1}^1 does not belong to $F(C_n^2)$. Similarly we can pass from C_n^2 across some boundary cell E_{n-1}^2 into a new n-cell E_n^3. We orient again the latter in such manner that $E_{n-1}^2 \not\subset F(C_n^3)$, $C_n^3 = E_n^1 + E_n^2 + E_n^3$, etc. Set

$$C_n^i = E_n^1 + \cdots + E_n^i,$$

and designate by E_{n-1}^i the face crossed in passing from C_{n-1}^i to E_n^i. The chains $E_n^1, C_n^2, \cdots, C_n^{\alpha_n}$, constitute a minimum base for their type. Their incidences with the $(n-1)$-cells are represented by a matrix

$$\begin{Vmatrix} 1 & \mathbf{a} \\ 0 & \mathbf{b} \end{Vmatrix}.$$

The invariant factors of this matrix are the same as for the one-rowed matrix $\mathbf{b}$. Therefore, there is at most a single torsion coefficient; the H. C. F. of the elements of $\mathbf{b}$ in case it is >1. The terms of $\mathbf{b}$ are the sums of the terms in the columns of the initial matrix numbered $\alpha_n, \alpha_n+1, \cdots, \alpha_{n-1}$.

Since the circuit is simple these terms are all zero except two equal to ± 1. Therefore, the elements of $\boldsymbol{b}$ are zero or ± 2, and the H. C. F. when $\boldsymbol{b} \neq 0$ is two. We have then two possibilities:

(a) $\boldsymbol{b} = 0$. Then $C_n \rightarrow 0$, i. e. C_n is an oriented cycle on $K - L$ which is then an orientable circuit. In this case there are no torsion coefficients for the dimension $n-1$.

(b) $\boldsymbol{b} \neq 0$. Then the rank of the incidence matrix is α_n, hence $R_n(K; L) = 0$, the circuit is non-orientable and there is a single torsion coefficient for the dimension $n-1$ whose value is two. The theorem is therefore proved.

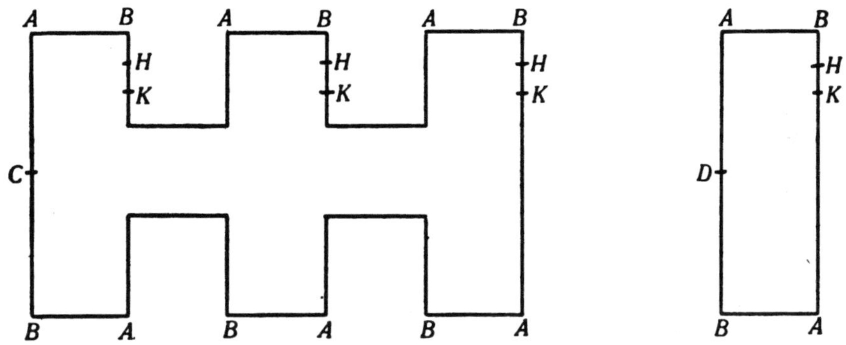

Fig. 4.

40. One would be tempted to surmise that orientable n-circuits are always such that in Δ_n every $t = 1$. It is easy to show by an example that this is false. Consider the adjacent figure. We assume that the 2-circuits K_2 and K_2' of the figure are triangulated although we have not marked the triangles. Furthermore, the arcs AB, HK are assumed to be sides of the triangles. Bring into coincidence all arcs marked AB, and likewise all those marked HK so that the points lettered alike coincide. Call K_1 the one-complex obtained when the arcs AB, HK are removed from the peripheries. Then the new complex L_2 obtained is a two-circuit mod K_1. For the condition (a) of No. 35 is immediately verified. Furthermore, every two-cycle mod $(K_1, 2)$ on it must include K_2 and also K_2',

whereas separately they are not cycles mod K_1, as they have HK for boundary mod $(K_1, 2)$. Hence the only two-cycle mod $(K_1, 2)$ consists of the sum of all two-cells, so that condition (b) loc. cit. holds likewise. On the other hand, K_2, K_2' are separately orientable relatively to their peripheries, and if suitably oriented normally $3K_2' - K_2$ is a two-cycle, so that the circuit is orientable. The two-cycle cannot be the multiple of any other since in its expression each two-cell of the complex K_2 is taken once. Therefore it is the fundamental Δ_2, and we see that in its expression the coefficients of the cells of K_2' are 3 while those of the cells of K_2 are -1. It is scarcely necessary to point out that the present circuit is not simple.

In the following example, communicated to me by A. B. Brown, the circuit is absolute. From one 2-sphere cut p small circular holes; from a seeond q holes; where p and q are odd and unequal. Make the boundaries of the holes coincide so that if an orientation is given to the circle of coincidence, the corresponding circles on either sphere are uniformly oriented. The result is a 2-circuit with the same properties as in the example above. The proof is similar. This example is readily generalized to the case of an n-circuit.

41. In the paper in which he introduced torsion [4] Poincaré associated its presence with certain types of chains. As he defines them they are configurations which generalize the twisted Möbius strip. As Poincaré has it, the presence of torsion for the dimension n is evidence of a sort of internal twist or torsion within K. Hence the name of "variety with torsion" which he applies to K, and the term "torsion coefficient".

We have a *Poincaré chain* if we take a sequence of simplexes $\{\sigma_n^i\}$, $i = 1, \cdots, r$ such that σ^i and σ^{i+1} have a common incident σ_{n-1}. It is orientable whenever the σ's can be so oriented that σ^i, σ^{i+1} are oppositely related to their common face. The Poincaré chain is *open* when $i \neq r$, *closed* otherwise.

All open Poincaré chains are orientable. The Möbius strip is the simplest non-orientable closed Poincaré chain. It is the homeomorph of the following sum of five triangles in S_3:

$$ABC+CBD+CDE+DEA+EAB.$$

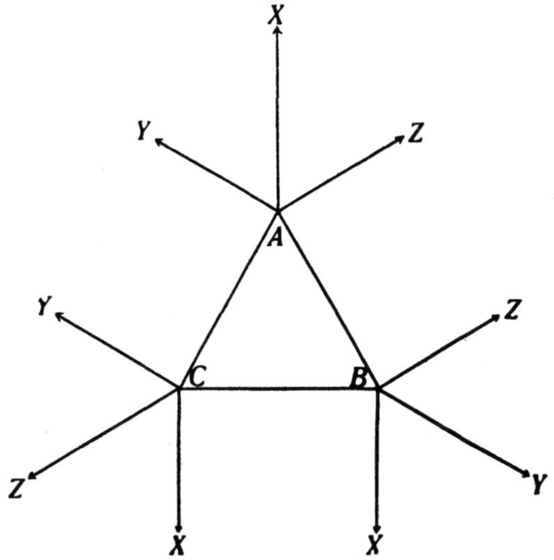

Fig. 5.

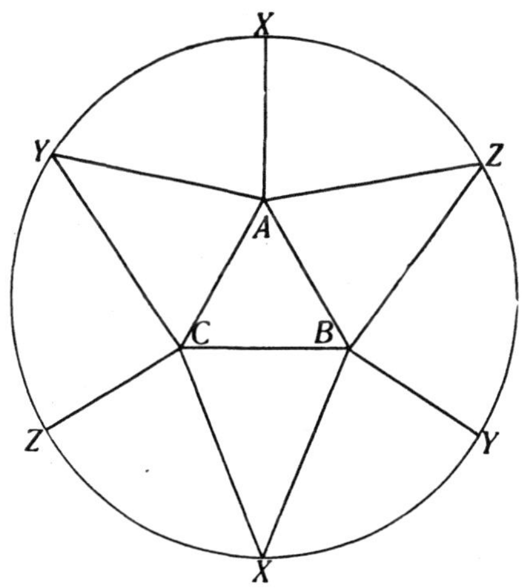

Fig 6.

A simple n-circuit is orientable when and only when all the Poincaré chains composed of its n-cells are orientable.

42. The projective plane is an excellent example of a non-orientable absolute circuit. With the usual conventions of projective geometry on points at infinity we may represent the projective plane as a complex in the Euclidean plane (see the accompanying Figure 5). Or the representation may be made in a finite circular region in S_2 with the convention that diametrically opposite points of the boundary coincide (Fig. 6). In either figure it is evident that the sum of the triangles $XZA+ZAB+ABC+BCX+CXZ$ forms a Möbius strip.

§ 6. Orientation of S_n

43. Various modes of orienting spaces and manifolds have been used in analysis apparently not closely related to our combinatorial procedure. In general they are based upon the properties of the Jacobians of certain analytical transformations of coördinates. We propose to examine in some detail the simplest known procedure for spaces, partly to show the connection with what we have done, also in view of ulterior applications. In many questions regarding complexes one may ultimately reduce the matter of orientation to a single simplex. The analytical machinery to be developed presently is then a convenient tool and simplifies the problem a good deal.

44. There are various modes of attaching orientations to a Euclidean plane π. One of the simplest is to agree that twice the area of any triangle $\sigma_2 = A^0A^1A^2$, with x_{ij} as the coördinates of A^i, is to be in magnitude and sign the determinant

$$|1, x_{i1}, x_{i2}|$$

or else its negative. When the vertices undergo an odd permutation the orientation of σ_2 and likewise the sign of the area are changed. The plane is thus essentially oriented by means of a definite oriented σ_2, its indicatrix, and then the procedure is freed from the special σ_2 chosen in a manner that will be taken up for any S_n. The indicatrix as here defined need not be identified for the present with the indicatrix of No. 37.

45. We will first recall certain well known properties of the collineations of S_n. Let the equations of a collineation be

$$(38) \qquad y_i = \sum a_{ij} x_j + a_{i0}.$$

It is convenient to introduce the variable of homogeneity $x_0 = 1$ for all points of S_n. The equations of the collineation then assume the form

$$(39) \qquad y_i = \sum a_{ij} x_j; \quad a_{00} = 1; \quad a_{0j} = 0, j > 0,$$

the range for both i, j being now from 0 to n. The determinants of the a's in both (38), (39) are equal and $\neq 0$. Let their common value be Δ and when we write $\boldsymbol{a}$ let it be understood that it is the matrix of the coefficients in (39).

The collineation is said to be *direct* when $\Delta > 0$, *inverse* when $\Delta < 0$. The *product* (successive effect) of two collineations whose determinants are Δ, Δ' is a collineation whose determinant is $\Delta\Delta'$. The product is therefore direct if the two are of the same type, inverse otherwise.

If we make a transformation of coördinates with $\boldsymbol{b}$ as the matrix of the coefficients in the equations of the transformation in the form (39), the collineation (39) considered will have new equations of the same type with $\boldsymbol{b}\boldsymbol{a}\boldsymbol{b}^{-1}$ in place of $\boldsymbol{a}$, but $\Delta = |\boldsymbol{a}| = |\boldsymbol{b}\boldsymbol{a}\boldsymbol{b}^{-1}|$ will be unchanged. Hence the *type* of a collineation is invariant under transformation of coördinates.

46: Let now $\sigma_n = A^0 A^1 \cdots A^n$ be a simplex of S_n, with x_{ij} as the homogeneous coördinates of A^i. The determinant

$$|\boldsymbol{x}| = |1, x_{i1}, x_{i2}, \cdots, x_{in}| \neq 0$$

changes its sign with every odd permutation of the vertices. Therefore the orientation of σ_n is uniquely determined by the sign of $|\boldsymbol{x}|$. We designate S_n associated with any σ_n by $+S_n$, associated with its opposite by $-S_n$. The simplex is an indicatrix of the space.

To free the procedure from the special choice of indicatrix we make use of the collineation that carries indicatrices into one another. Let $\sigma'_n = B^0 B^1 \cdots B^n$ with y_{ij} as the homogeneous coördinates of B^i and let us look for a collineation (39) transforming σ_n into σ'_n with vertices in the order named, that is, with A^i going into B^i. We must have

$$y_{ij} = \sum a_{jk} x_{ik},$$

or in matrix form

$$\boldsymbol{y} = \boldsymbol{x} \boldsymbol{a}'.$$

Since $|\boldsymbol{x}| \neq 0$, $\boldsymbol{x}^{-1}$ exists, hence

$$\boldsymbol{a}' = \boldsymbol{x}^{-1} \boldsymbol{y}, \qquad \Delta = |\boldsymbol{y}| \div |\boldsymbol{x}|.$$

Thus the collineation exists, is unique and direct when $|\boldsymbol{x}|$, $|\boldsymbol{y}|$ are of the same sign, inverse otherwise.

The rest is now obvious: We consider σ_n and σ'_n (with vertices named in the order stated) as orienting S_n alike or oppositely, according as the collineation transforming them into one another, with each vertex of the one going into the vertex of same order of the other, is direct or inverse. Analytically the decision rests upon whether $|\boldsymbol{x}|$, $|\boldsymbol{y}|$ are of the same or of opposite sign. It does not matter whether we consider the collineation carrying over σ_n into σ'_n or its inverse.

It is evident that what we have said of the orientation of an S_n can be applied to convex regions of S_n. Thus we can orient such a region $\mathfrak{R}_n$ by associating with it an oriented simplex σ_n lying in it. If σ_n and σ'_n are two oriented simplexes lying in $\mathfrak{R}_n$, it is always possible to find a collineation of S_n taking σ_n into σ'_n. If this collineation is direct we shall say that the orientations of σ_n and σ'_n are *concordant* and that each determines the same orientation for $\mathfrak{R}_n$. Thus there are possible only two distinct orientations of $\mathfrak{R}_n$, which will be designated by $+\mathfrak{R}_n$ and $-\mathfrak{R}_n$.

In practise one often has to compare indicatrices with a common vertex. We can then compare them as regards collineations leaving that vertex invariant, and since the choice of axes is immaterial, let $A^0 = B^0$ be the common vertex and also the origin of coördinates. Then the two determinants whose signs are to be compared are

$$|x_{ij}|, \quad |y_{ij}| \qquad (i, j = 1, 2, \cdots, n)$$

or the determinant of the non-homogeneous coördinates of the faces opposite to the common vertex.

An excellent and much more complete treatment of the orientation of Euclidean or projective spaces is given in Veblen-Young, Projective Geometry, vol. II, Chap. 2.

47. In order to correlate the orientation by means of collineations with the orientation of circuits we need two lemmas.

LEMMA I. *A collineation T of S_n leaving an S_{n-1} invariant induces also a collineation T' on that S_{n-1}. T and T' are of same or opposite types accordingly as T does or does not permute the two regions in which S_n is decomposed by S_{n-1}.*

We may assume that S_{n-1} has for equation $x_n = 0$. Then if (38) are the equations of T we have $a_{nj} = 0$, $j < n$. The points of S_{n-1} undergo a collineation T' whose equations are the first $(n-1)$ of (38). The ratio of the determinants of T and T' is a_{nn}. The regions in question are permuted when and only when $a_{nn} < 0$, that is, when T, T' are of opposite types. This proves Lemma I.

LEMMA II. *Two indicatrices σ_n, σ_n' with a common face σ_{n-1} but no common points are concordant when and only when they are oppositely related to σ_{n-1}.*

We take $\sigma_n = \sigma_{n-1} A_n$, $\sigma_n' = \sigma_{n-1} B_n$. The assumption is that σ_n, σ_n' are on opposite sides of the S_{n-1} of σ_{n-1}. Let T be the collineation turning them into one another, vertex for vertex, and leaving σ_{n-1} invariant. Since T permutes the two regions separated by S_{n-1} and leaves S_{n-1} invariant, it is *inverse* for S_n by Lemma I. Thus σ_n', which bears to σ_{n-1} the same relation as σ_n, is not concordant with the latter as indicatrix of S_n; hence $-\sigma_n'$, oppositely related to σ_{n-1}, is concordant with σ_n, which proves Lemma II.

48. That the orientation of a space by means of collineations and of a circuit by incidence properties of its cells, overlap

whenever the field of application is the same, is completely established by the

THEOREM. *Let K_n be a complex composed of simplexes in S_n, L a subcomplex of it. If $K_n - L$ is a connected region of S_n, then K_n is an orientable simple n-circuit mod L.*

Let E_n be any n-cell of $K_n - L$. If $K_n - L$ contains other points than those of E_n it must contain points of the boundary of E_n, since $K_n - L$ is connected, and these points must include an $(n-1)$-cell incident with E_n and another n-cell of $K_n - L$, since $K_n - L$ is a region of S_n. Hence the points of $K_n - L$ are all on the boundaries of a set of n-cells of $K_n - L$ so situated that each $(n-1)$-cell of $K_n - L$ is on the boundary of exactly two n-cells (since K_n is in S_n, not more than two n-cells can be incident with an $(n-1)$-cell) and the set of n- and $(n-1)$-cells is connected in the sense of No. 35. Hence also, the sum of the n-cells of $K_n - L$ is an n-cycle mod $(L, 2)$. By the proposition of No. 35, $K_n - L$ is an n-circuit mod L.

Let us orient every n-cell of $K_n - L$ concordantly with its S_n. If E_n, E'_n, are two of them with a common boundary E_{n-1}, we can choose concordant indicatrices $A\sigma_{n-1}$, and $\eta B\sigma_{n-1}$ of S_n $(\eta = \pm 1)$, with $A \subset E_n$, $B \subset E'_n$, $\sigma_{n-1} \subset E_{n-1}$. Then $A\sigma_{n-1}$, $\eta \dot{B}\sigma_{n-1}$ are indicatrices of the n-cells. By Lemma II of No. 47, $\eta = -1$, since E_n and E'_n are oppositely related to σ_{n-1} and hence to E_{n-1} itself. Therefore E_{n-1} does not appear in the boundary of $\sum E_n$, which shows that K_n is a cycle mod L. Therefore it is an orientable circuit mod L and the theorem is completely proved.

The desired correlation between the orientations of $K_n - L$ by means of collineations and by incidence relations is established in the course of the proof of the theorem.

§ 7. CONVEX COMPLEXES AND THEIR SUBDIVISION. INVARIANCE OF THE HOMOLOGY CHARACTERS UNDER SUBDIVISION

49. Up to this point we have studied properties of individual complexes. No attempt has been made to relate

different complexes. In this section we shall consider a class of related complexes, those which can be shown to represent homeomorphic point sets by means of a restricted type of *subdivision*, roughly speaking, that is, by cutting up the cells of the given complexes into smaller cells. The two complexes shown by the heavy lines of the accompanying figures, for example, are different representations of the triangle. One mode of demonstrating their equivalence is to subdivide them by the dotted lines. The two complexes are then identical in structure and hence (No. 9) represent homeomorphic point sets. We might also demonstrate this equivalence by subdividing the first figure as indicated and then

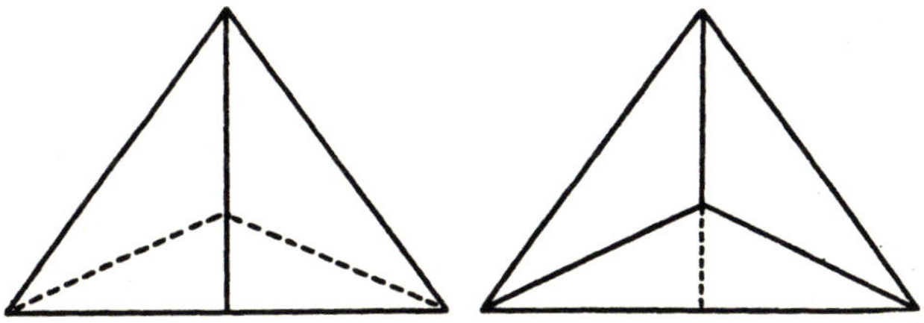

Fig. 7.

combining the triangles at the base to form the original complex of the second figure.

In this example the operations which we have performed can be expressed in terms of the introduction or suppression of a finite number of cells and bounding relations. The only geometric properties of the cells which are explicitly used are those of orientation and bounding and these can be expressed by a purely formal notation subject to operations of arithmetic: addition, subtraction and multiplication. Such a treatment may be termed *purely combinatorial.* A careful examination of the chapter up to this point would show that all intrinsic properties of cells and complexes which have been developed so far have been expressed in this formal manner. Thus, in spite of the fact that our definitions have assigned to cells and complexes properties, such as continuity, infinite divisibility and the like, which belong to their Euclidean prototypes, in determining the Betti numbers, for example, our operations have been of the purely formal type just described. The latter part of § 6 constitutes a digression in which for the moment we reassert the identity of our configurations with certain Euclidean ones and apply the results of the preceding combinatorial work to problems of Euclidean geometry.

50. In the light of these considerations the example just given suggests the following fundamental problem of purely combinatorial analysis situs:

Given two complexes which, regarded as sets of points, represent homeomorphic configurations, is it possible to pass from one to the other by means of a finite number of purely combinatorial operations?

No answer has been given to this question. It is by no means certain that the answer is affirmative. For accounts which deal with problems of this sort the reader is referred to the Encyklopädie article of Dehn-Heegard [2] and to the more recent writings of Weyl [3], Bilz [1], Newman [1, 2, 3, 4], Alexander [21].

We shall not attempt to develop the theory in this manner. Since we are dealing throughout with figures which can be immersed in Euclidean space we shall not hesitate to use the properties of that space when we find it useful to do so. In the present chapter this does not imply the use of continuity, but only the representation of points by sets of numbers. It is consistent, for example, with the restriction of all points to those that have rational coördinates.

51. Before we give an accurate definition of certain important special types of subdivision it is necessary to extend our notions of cell and complex somewhat. If in the definition of simplicial polyhedron (No. 5) we replace *simplex* by *convex polyhedral region* (with the added restriction that no two $(p-1)$-faces of a convex polyhedral R_p lie in the same S_{p-1}), the resulting configuration will be called a *generalized polyhedron*, Π_n. The homeomorph of a generalized polyhedron is the *convex polyhedral complex*, or simply *convex complex*, K_n. Under the homeomorphism taking Π_n into K_n, the images of

the convex polyhedral regions of Π_n are the *convex polyhedral cells* of K_n. As in the case of the corresponding simplicial configurations, we shall use with respect to K_n the terminology which, strictly speaking, belongs to homeomorphic configurations on Π_n. Thus we may speak of cutting K_n by an $S_p (p \leqq n)$, by which we shall mean that we have cut Π_n by an S_p and have then transferred the properties of this intersection in its relation to Π_n to the homeomorphic configuration on K_n in its relation to K_n.

Evidently the convex polyhedral 0-cell and 1-cell are no more general than the simplicial types. The interior of a regular polyhedron in three-space is a simple example of a convex polyhedral 3-cell, and its boundary is a convex 2-complex.

52. A convex polyhedral p-cell is *oriented* by associating with it an oriented p-simplex lying in it. Since it was shown in No. 46 that there are only two distinct orientations of a convex region, we shall distinguish these by $+E_p$ and $-E_p$. We can determine whether a cell E_p is positively or negatively related to an E_{p-1} on its boundary as follows: Orient E_p by means of a σ_p having a face σ_{p-1} in E_{p-1}, and E_{p-1} by means of σ_{p-1}. Then E_p is *positively* or *negatively* related to E_{p-1} according as σ_p is positively or negatively related to σ_{p-1}.

53. The terminology of §§ 1, 2 can now be applied verbatim to convex polyhedral cells. Also the first three properties of No. 15 hold. In place of property IV, however, we must prove that

The boundary of a convex polyhedral p-cell is an absolute $(p-1)$*-cycle.*

Let E^1_{p-1} and E^2_{p-1} be two cells on the boundary of an oriented E_p; let them have in common a boundary E_{p-2}; and let these cells be so oriented that E_p is positively related to E^1_{p-1} and to E^2_{p-1}.

Let σ_{p-2} be any oriented simplex on E_{p-2} which serves as its indicatrix. Let P^1 and P^2 be points of E^1_{p-1} and of E^2_{p-1}, respectively. P^1 and σ_{p-2} determine a σ^1_{p-1} which may be oriented so as to serve as an indicatrix of E^1_{p-1}. Let σ^2_{p-1} be the oriented simplex similarly determined by P^2 and σ_{p-2}.

Finally, let P be a point of E_p and let σ_p^1 and σ_p^2 be concordantly oriented simplexes determined by σ_{p-1}^1 and P and by σ_{p-1}^2 and P, respectively, either of them serving as indicatrix for E_p. These two simplexes have in common a face σ_{p-1}, determined by P and σ_{p-2}; orient σ_{p-1} so that it is positively related to σ_p^1. Orient E_{p-2} so that σ_{p-1} is positively related to σ_{p-2}. Since E_p is convex, σ_p^1 and σ_p^2 lie on opposite sides of σ_{p-1}. Hence by Lemma II of No. 47, σ_p^2 is negatively related to σ_{p-1}. By hypothesis, σ_p^2 is positively related to σ_{p-1}^2. By the proof of IV in No. 15, σ_{p-1}^2 and σ_{p-1} are similarly related to σ_{p-2}, so that σ_{p-1}^2 is positively related to σ_{p-2}. By hypothesis, σ_p^1 is positively related to σ_{p-1} and to σ_{p-1}^1. Again, by the proof of IV in No. 15, σ_{p-1} and σ_{p-1}^1 are oppositely related to σ_{p-2}, so that σ_{p-1}^1 is negatively related to σ_{p-2}. Regarding these now as indicatrices of the respective cells, E_{p-1}^1 and E_{p-1}^2 are seen to be oppositely related to E_{p-2}. Hence $E_p \rightarrow E_{p-1}^1 + E_{p-1}^2 + \cdots \rightarrow -E_{p-2} + \cdots + E_{p-2} + \cdots$, so that E_{p-2} does not appear in the boundary of E_p. Similarly with regard to any $(p-2)$-cell which is on the boundary of exactly two $(p-1)$-cells of the boundary of E_p. Well known theorems of Euclidean geometry assure us that this is the case for every $(p-2)$-cell on the boundary of E_p. (Compare No. 55 below.)

Property V of No. 15 now follows as before from II and this form of IV. From then on everything which has been proved of simplicial cells and complexes depends on those properties which we have just proved to hold for the more general type. Throughout the remainder of this chapter it will be understood that all complexes may be convex.

54. A complex K_n^1 will be said to be a *convex subdivision* of a convex complex K_n if (1) every cell of K_n^1 is a convex polyhedral cell lying wholly in some cell of K_n; (2) every p-cell of K_n is an exact sum of cells of K_n^1 of dimensions $\leqq p$.

The definition imposes certain restrictions on the intuitive idea of cutting up the cells of K_n. For example, it is necessary to extend the process to the boundary of the cell; e. g. if a σ_2 is divided in half by a straight segment the endpoints of the segment must appear as zero-cells on the boundary, which in turn will generally divide the one-cells of the boundary of σ_2 into new one-cells. Again, it is necessary that new cells introduced into a given cell leave the remainder expressed as a set of cells; the introduction of a zero-cell into the interior of a two-cell does not constitute a subdivision of the two-cell since a two-cell minus an interior point is no longer a two-cell.

If K' is a subdivision of K and L is a subcomplex of K, it is at once evident that a certain subcomplex L' of K' is a subdivision of L. We shall speak of L' as the subdivision *induced* on L by K'.

55. We have noted (No. 37) that an n-cell is an orientable n-circuit mod its bounding sphere, and that this sphere is an absolute $(n-1)$-circuit. This property is invariant under convex subdivision, i. e.

If K_n is a convex subdivision of a convex polyhedral n-cell, E_n, with K_{n-1} as the induced subdivision of its boundary sphere, then K_n is a simple orientable n-circuit mod K_{n-1} and K_{n-1} is a simple absolute orientable $(n-1)$-circuit.

The part concerning K_n has already been established in No. 48. Regarding K_{n-1} it can be shown as there that it is a simple absolute circuit having the property that its $(n-2)$-cells are incident with two $(n-1)$-cells. In fact, if we remove from K_{n-1} all cells with a given vertex A, the rest can be projected from A into a convex complex whose cells are on a certain S_{n-1}, and then the incidence properties follow as in No. 48. There remains the question of orientability. Here it is advantageous for the sequel to look a little more closely into the matter.

Let E^i_{n-1} be the boundary cells of E_n with

$$E_n \to \sum \eta_i E^i_{n-1}.$$

This means that there exists on $\bar{E}_n$ a positive indicatrix $A\sigma_{n-1}$ such that $\eta_i \sigma_{n-1}$ is an indicatrix of the convex cell E^i_{n-1} (No. 52).

Designate by $E_n'^h$, $E_{n-1}'^j$ the n- and $(n-1)$-cells of $K_n - K_{n-1}$ and by E_{n-1}^{ik} the cells of K_{n-1} on E_{n-1}^i. The cells $E_{n-1}'^j$ are oriented arbitrarily but $E_n'^h$ and E_{n-1}^{ik} are to be oriented concordantly with E_n and E_{n-1}^i. This implies that when $E_n'^h$ is incident with E_{n-1}^{ik} there is an indicatrix $B\sigma_{n-1}'$ of E_n, concordant with $A\sigma_{n-1}$ above, which is likewise an indicatrix of $E_n'^h$, and such that $\eta_i \sigma_{n-1}'$ is an indicatrix of E_{n-1}^{ik}. It follows that E_{n-1}^{ik} figures in $F(E_n'^h)$ with the coefficient η_i. Since among the cells $E_n'^h$ the only two incident with a given $E_{n-1}'^j$ are oppositely related to it, we have finally

$$\sum_h E_n'^h \to \sum_{i,k} \eta_i \, E_{n-1}^{ik}.$$

The right hand side is then an absolute cycle. Since it is merely the sum of the cells of K_{n-1} suitably oriented, the latter is an orientable circuit, thus proving our theorem.

The result also follows directly by No. 53 and the results at the end of No. 48.

APPLICATION. Let K_n be any convex complex and let us subdivide it into another convex complex K_n'. If E_p' is any cell of K_n' on E_p of K_n, we agree to orient it concordantly with E_p. If E_p' is on a cell of more than p dimensions of K_n we orient it in arbitrary manner. Designate now by E_p^i the cells of K_n and by E_p^{ij} those of K_n' on E_p^i. If the incidence relations for K_n are

$$E_p^i \to \sum \eta_{ir} \, E_{p-1}^r,$$

then according to what has just been established,

$$\sum_j E_p^{ij} \to \sum_{r,s} \eta_{ir} \, E_{p-1}^{rs}.$$

Let then C_p be any subchain of K_n:

$$C_p = \sum_i t_i \, E_p^i \to \sum_{i,r} t_i \, \eta_{ir} \, E_{p-1}^r = C_{p-1}.$$

We associate with C_p the subchain of K_n' defined by

$$C_p' = \sum_{i,j} t_i \, E_p^{ij},$$

obtained by replacing in C_p every p-cell by the sum of the p-cells of K_n' that it carries, and which we define as the *subdivision* of C_p induced by that of K_n. We have at once

$$C_p' \to \sum_{i,r,s} t_i \, \eta_{ir} \, E_{p-1}^{rs} = C_{p-1}',$$

where C_{p-1}' is the subdivision of C_{p-1} in the sense just defined. Since manifestly the subdivision of $\sum t_j C_p^j$ is $\sum t_j C_p'^j$, where $C_p'^j$ is the subdivision of C_p^j, we have finally

THEOREM. *The boundary of the subdivision of a chain is the associated subdivision of its boundary.*

REMARK. We are not at liberty to conclude from this that the combinatorial homology characters are invariant under subdivision. For the subchains of K_n' considered are of a very restricted type—they do not include, for example, the p-chains having p-cells on cells of K_n of more than p dimensions.

56. Certain types of convex subdivision are of especial importance for our purposes. Two of these depend on the obvious fact that the set of points obtained by joining all points of a given convex polyhedral E_p lying in an S_p to a point P external to S_p by straight line segments, forms a convex polyhedral E_{p+1} bounded by the cells of E_p plus cells of dimension $< p+1$ determined in a similar manner by P and the cells of the boundary of E_p.

In a complex K a cell E_p, together with all cells E_{p+q} $(q = 1, 2, \cdots, n-p)$ of K incident with E_p, constitute the *star of* E_p. The *boundary of the star* consists of all those cells of K which are faces of cells of the star but are not in the star.

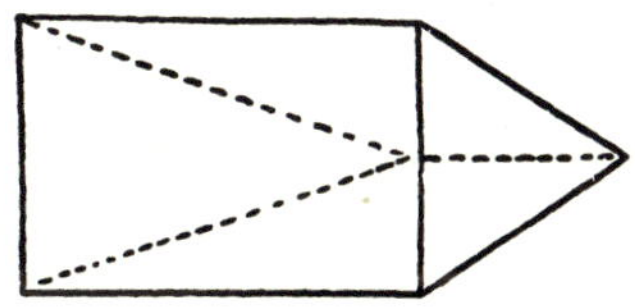

Fig. 8.

57. Elementary subdivision. Let A be any point of any cell E_p of a complex K. Draw segments on the star of E_p joining the point A to all points of the boundary of the star. The segments whose end-points are on a given E_r of the boundary have for sum a convex E_{r+1}. Every point of the star (except A) belongs to exactly one E_{r+1}, and A plus the sum of the E_{r+1}'s coincides with the star. The result of replacing in K the star of E_p by $A+\sum E_{r+1}$ is called an *elementary* subdivision of K (Alexander, [19]). In the simple 2-complex represented by the solid lines of the accompanying figure, the dotted lines indicate a typical elementary subdivision.

58. **Regular subdivision.** Let the cells of K be so ordered that no one precedes a cell of higher dimension, say $E_n^1, \cdots, E_n^{\alpha_n}, E_{n-1}^1, \cdots, E_0^{\alpha_0}$. Form K^1, the elementary subdivision of K using the star of E_n^1. The closed cells $E_n^2, \cdots, E_0^{\alpha_0}$ are left intact by this operation. We can then

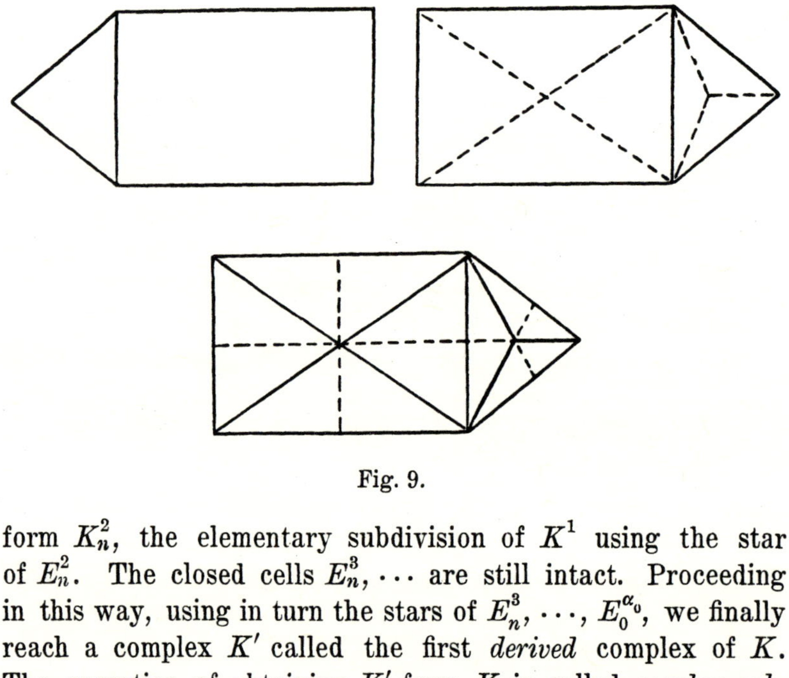

Fig. 9.

form K_n^2, the elementary subdivision of K^1 using the star of E_n^2. The closed cells $E_n^3, \cdots$ are still intact. Proceeding in this way, using in turn the stars of $E_n^3, \cdots, E_0^{\alpha_0}$, we finally reach a complex K' called the first *derived* complex of K. The operation of obtaining K' from K is called *regular subdivision* (Poincaré [3], Veblen [5]). The result of two successive applications of regular subdivision, i. e. the derived complex of the derived complex, will be called the *second derived* complex of K, etc. By "the derived complex" we shall generally understand the first derived.

It is clear that K' is independent of the order of the cells of a given dimension. The accompanying figures show the results on a particular 2-complex of all subdivisions using stars with centers of dimension two and one, successively. Stars whose centers are zero-cells add nothing in this case.

Regular subdivision can also be defined inductively as follows: (1) The regular subdivision of a K_0 is K_0 itself. (2) The regular subdivision of K_p is the complex obtained by introducing new vertices P^i into the corresponding cells E_p^i and replacing (a) the cells of K_p of dimension $<p$ (which form a complex K_{p-1}) by their regular subdivision K'_{p-1}, (b) the cells E_p^i by the cells determined (as in No. 57) by joining P^i to all cells of K'_{p-1} on the boundary of E_p^i (for every p-cell of K_p). The two procedures are easily seen to be equivalent.

One result of regular subdivision is that *it reduces any convex complex to a simplicial one in one step.* Also, successive regular subdivisions of a complex are composed of smaller and smaller cells; in fact we can, by choosing our new vertices properly (say as the centers of gravity of their containing cells), obtain a regular subdivision for which the maximum diameter of its cells is as small as we please.

59. **Subdivision by section.** The fundamental operation is that of replacing the given complex K by the one resulting from the cutting of a single closed p-cell E_p of K by an S_{p-1} of its S_p which meets the cell E_p.

The effect of the operation on K is to replace E_p by two new p-cells, and a new $(p-1)$-cell, which together make up E_p; and similarly to replace every q-cell $(0<q<p)$ on the boundary of the E_p which is cut by S_{p-1}, by a new $(q-1)$-cell and two new q-cells, which together make up the original q-cell. Let us suppose that K' has been derived from K in this manner. We can obviously perform a similar operation on any cell of K' which was also a cell of K; furthermore, if we take the section of a p-cell of K which has been affected by a previous section, this can be expressed as the result of sections on the various cells of K' which make up the original p-cell. The result of any finite succession of sections of cells of K, together with the resulting

sections of new cells in the manner just explained, we shall call a *subdivision of K by section.*

The properties of the Euclidean prototypes of the configurations involved in these operations are evidently sufficient to establish the following properties of subdivision by section: (1) K' is convex. (2) The final K' is independent of the order in which the operations of section take place. (3) If K' is the result of the section of a single E_p of K, it may be expressed as the result of sections performed on cells of dimension $<p$ (all lying on the boundary of E_p) together with the replacement of E_p by two new p-cells and a new $(p-1)$-cell.

60. THEOREM. *The homology characters of a convex complex are unaltered by section.*

It follows from property (3) of the last paragraph in No. 59, that everything reduces down to showing that when any p-cell is replaced by two new p-cells with a common $(p-1)$-face, all cells being convex, the homology characters are unchanged. In this reduced form this proposition is due to Tietze [1].

We may assume that E_p^1 is the altered cell and call E_p', E_p'' the two that take its place oriented concordantly with it, E_{p-1} their common face. Let K be the initial complex, K' the altered complex. The $(p-2)$-faces of E_{p-1} are cells of K and so are the $(p-1)$-faces of E_p' and E_p'' other than E_{p-1}. We label these faces respectively $E_{p-2}^1, \cdots, E_{p-2}^{\alpha}, E_{p-1}^1, \cdots, E_{p-1}^{\beta}, E_{p-1}^{\beta+1}, \cdots, E_{p-1}^{\gamma}$. The $(p-2)$-faces exist only when $p \geqq 1$, but otherwise the lower case $p=1$ is treated like the rest.

To find the variation of the homology characters, if there is any, we shall examine the incidence matrices. The only incidence matrices to be modified are η^p, η^{p-1}, η^{p-2}. Let $\zeta^p, \cdots$ be the corresponding K' matrices. We designate the ranks of η^i, ζ^i by ρ_i, ρ_i', the numbers of i-cells of K and K' by α_i, α_i'. The four types of boundary relations shall be treated simultaneously, it being actually unnecessary to specify the particular type considered.

(a) *Comparison of η^p and ζ^p.* If the fundamental boundary relations for K are

$$E^i_{p+1} \to \sum \eta^p_{ij} E^j_p$$

those for K' are

$$E^i_{p+1} \to \eta^p_{i1}(E'_p + E''_p) + \eta^p_{i2} E^2_p + \cdots.$$

Hence ζ^p is obtained from η^p by mere repetition of the first column. Therefore $\varrho'_p = \varrho_p$ and the invariant factors of η_p and ζ_p are the same.

(b) *Comparison of* η^{p-1} *and* ζ^{p-1}. The cell E'_p is related to E^i_{p-1}, $i \leqq \beta$, like E_p itself and similarly for E''_p and the other faces of E_p, while both E'_p and E''_p are oppositely related to E_{p-1}. Hence we can so orient E_{p-1} that

$$\zeta^{p-1} = \left\| \begin{matrix} 1 & \eta^{p-1}_{11} & \cdots & \eta^{p-1}_{1\beta} & 0 & \cdots & \cdots & \cdots & \cdots \\ -1 & 0 & \cdots & 0 & \eta^{p-1}_{1,\beta+1} & \cdots & \eta^{p-1}_{1,\gamma} & 0 & \cdots \\ 0 & \eta^{p-1}_{21} & \eta^{p-1}_{22} & \cdots & \cdots & \cdots & \cdots & \cdots & \cdots \\ \cdot & \cdot & \cdot & \cdot & \cdot & \cdot & \cdot & \cdot & \cdot \\ 0 & \eta^{p-2}_{\alpha_p 1} & \cdots & \cdots & \cdots & \cdots & \cdots & \cdots & \cdots \end{matrix} \right\|$$

which is equivalent to a matrix

$$\left\| \begin{array}{c|c} 1 & \mathbf{0} \\ \hline \mathbf{0} & \eta^{p-1} \end{array} \right\|.$$

Hence the invariant factors of ζ^{p-1} that are > 1 are the same as for η^{p-1} and $\varrho'_{p-1} = \varrho_{p-1} + 1$.

(c) *Comparison of* η^{p-2} *and* ζ^{p-2}. The only new fundamental boundary relation for the $(p-1)$-cells is that for E_{p-1} or

$$E_{p-1} \to F(E_{p-1}).$$

But from the form of ζ^{p-1} follows

$$E'_p \to E_{p-1} + \eta^{p-1}_{11} E^1_{p-1} + \cdots \sim 0.$$

Hence

$$F(E_{p-1}) \equiv -\eta^{p-1}_{11} F(E^1_{p-1}) - \cdots$$

where we have to write at the end mod $L, \cdots$ in the appropriate cases. Therefore the new fundamental boundary relation is a consequence of the old. Hence again $\varrho'_{p-2} = \varrho_{p-2}$ and the invariant factors of η^{p-2} and ζ^{p-2} are the same.

We are already able to conclude that the torsion coefficients, absolute or mod L, are unchanged. Since also

$$\alpha'_p = \alpha_p + 1, \quad \alpha'_{p-1} = \alpha_{p-1} + 1,$$

formulas (21) and (33) for the Betti numbers of all types show that they are likewise unaltered. From this follows, finally, by reference to No. 31, that the torsion coefficients mod m or mod (L, m) likewise remain the same.

61. We shall say that K_p is *convex on* K_n $(p \leqq n)$ if K_p is on K_n and if there exists a Π_n in an S_r homeomorphic to K_n such that the image of K_p on Π_n is a Π_p.

LEMMA I. *If K_p is a convex complex on K_n there exist subdivisions by section K'_n of K_n and K'_p of K_p such that K'_p is a subcomplex of K'_n.*

Let Π_n on an S_r be such a defining polyhedron for K_n that the image of K_p under the homeomorphism which takes K_n into Π_n is a Π_p. Any S_{r-1} of S_r cutting either or both of Π_n, Π_p, causes a subdivision by section of either or both of Π_n, Π_p; hence, of course, of K_n, K_p. Let $\{S_{r-1}\}$ be a set of hyperplanes of S_r such that every q-cell $(q = 1, 2, \cdots, n)$ of Π_n or Π_p lies on $r—q$ independent hyperplanes of the set. Let the resulting sections of K_n, K_p be K'_n, K'_p. From the mode of selection of the S_{r-1}'s every cell of K'_p arises as the result of the intersection of S_{r-1}'s which subdivide K_n and K_p alike. Hence every cell of K'_p must be a cell of K'_n.

REMARK. From No. 58 follows incidentally that K'_n *may be replaced by a simplicial subdivision.*

LEMMA II. *If K^*_n is a convex subdivision of K_n, both have a common subdivision by section.*

This follows from Lemma I when $K_p = K^*_n$, since K'_n coincides then with K'_p.

62. THEOREM. *If K_n and K'_n are homeomorphic convex complexes and there exists a homeomorphism taking one into the other in such a manner that the homeomorphs of the second*

forms a convex polyhedral complex on the first, then the Betti and torsion numbers of K_n *and* K_n' *are the same.*

Let K_n'' be the homeomorph of K_n' on K_n as in the theorem. By the lemma, there exists a subdivision K_n''' common to K_n and K_n''. By the theorem of No. 60 the Betti and torsion numbers of K_n and K_n'', and hence of K_n', are identical with those of K_n'''.

COROLLARY I. *The homology characters are unaffected by convex, elementary or regular subdivision.*

For elementary and regular subdivision are special cases of convex subdivision.

COROLLARY II. *If* K_n *is an n-circuit mod* L *and a convex subdivision* K_n' *of* K_n *induces on* L *the subdivision* L', *then* K_n' *is an n-circuit mod* L', *and when the first circuit is simple so is the second. Furthermore if* K_n *is orientable mod* L *so is* K_n' *mod* L', *and if* $K_n \rightarrow K_{n-1}$ *then* $K_n' \rightarrow K_{n-1}'$.

This corollary depends upon the fact that condition (a) of No. 35 for the n-circuit is readily verified here, while the other conditions for orientability and for n-circuits are expressible in terms of the Betti numbers. The proof of the part concerning the simple circuit is elementary.

CHAPTER II

TOPOLOGICAL INVARIANCE OF THE HOMOLOGY CHARACTERS. GENERALIZED COMPLEXES

In the preceding chapter there were introduced a series of combinatorial characters of a complex K, or of K and a subcomplex L of K. In the present chapter we shall establish their topological invariance, and as an incidental application the invariance of dimensionality and regionality (Brouwer). The proofs revolve essentially around certain deformation theorems, the first of which was proved by Alexander. Regarding the relative characters, it turns out that when L is a very general subset of K, they are topological invariants of $K-L$ alone, i. e. unchanged under a transformation homeomorphic over $K-L$ but not necessarily over the whole of K.

The theory thus obtained belongs to much more general complexes. How general their class is we do not know as yet. In the study of manifolds however a type essentially due to Alexander appears to be the most natural. It consists of aggregates of circuits which behave like cells only as far as their homology and circuit characters are concerned, and it is with the study of this class of generalized complexes that we conclude the chapter.

§ 1. Topological Homology Characters

1. To establish the invariance of the homology characters of Ch. I we identify them with certain analogous characters that are invariant practically by definition. These characters are defined in terms of continuous single-valued images of the

configurations studied in Ch. I. Following Alexander [2] and Veblen [5] we shall call them *singular* configurations. Explicitly, if g is any configuration, in particular a complex, cell, etc., as considered in Ch. I, a singular g on a point set A, is a continuous single-valued image G of g on A, a "map" of g on A. It is essential to remember that G is not necessarily the homeomorph of g, although *homeomorphism is not excluded.*

EXAMPLES. I. The orthogonal projection of a three-space figure on a plane π of the space is a singular image of the figure. Thus the orthogonal projection of an ordinary sphere is a closed circular region on π, which is to be considered as a singular sphere.

II. The Peano curve is a singular one-cell filling a square. To each point of a certain segment there corresponds a unique point of the square. There is an everywhere dense set on the square of which each point is the image of four points of the segment. Thus we see that the singular image of the simplest possible configuration can be extremely complicated. This is clearly shown by the extensive research of R. L. Moore [1] and his school in the United States, and the Polish school abroad on the subject of the singular one-cell.

Concerning the relationship between a configuration and its singular image very little is known in general and none is needed in the sequel. Some relationship assuredly exists as the following properties indicate. They are given solely to show what sort of restriction one may expect and also to elucidate further the notion of singular configuration.

(a) The singular image of a closed set on a compact space is a closed set.

(b) If g is closed and admits the Borel covering property so does G.

(c) The singular image of a connected set is connected.

2. A $\bar{\sigma}_{n-i}$ may be considered as a singular $\bar{\sigma}_n$ with coincident vertices. A homeomorph of $\bar{\sigma}_{n-i}$ may then also be considered as a singular closed n-cell. We shall call it briefly a *degenerate simplicial cell.* It is interesting to observe that if we write down formally the boundary of σ_{n-i} as if it were a true σ_n, the non-singular σ_{n-1}'s, if any are present, cancel. They can

occur only as a matter of fact when $i=1$. Such a degenerate σ_n can be symbolically represented by $\pm AA\sigma_{n-2}$ and its boundary is $\pm(A\sigma_{n-2}-A\sigma_{n-2})$ plus degenerate cells, so that the latter alone remain.

3. Consider an ordinary simplicial chain

$$c_p = \sum t_i e_p^i$$

with $\{e_q^j\}$ as the set of its cells. Let the complex coincident with $\sum e_q^j$ have for image on A a singular complex with the singular cells E_q^j. We write the symbolic sum

(1) $$C_p = \sum t_i E_p^i,$$

omitting all E's which are degenerate simplicial p-cells. This symbol associated with the set $\{E_q^j\}$ is called a *singular p-chain on A*. The points on the cells are the points of the chain which we name by its symbol: C_p.

Let the boundary relation for the cells e_p^i and for c_p be

$$e_p^i \to \sum \eta_{ij} e_{p-1}^j, \quad c_p \to \sum \eta_j e_{p-1}^j = F(c_p), \quad \eta_j = \sum t_i \eta_{ij}.$$

Then we write down in full analogy

$$E_p^i \to \sum \eta_{ij} E_{p-1}^j,$$

and define the boundary $F(C_p)$ of C_p as the chain

$$F(C_p) = \sum \eta_j E_{p-1}^j$$

with the symbolic boundary relations

$$C_p \to \sum \eta_j E_{p-1}^j, \qquad C_p \to F(C_p).$$

It follows from No. 2 that if E_p^i is a degenerate simplicial cell the sum of its boundary cells is zero, so that

(2) $$F(C_p) = \sum t_i F(E_p^i),$$

the coefficients t_i actually present being the same as in (1). The definition of the p-cycle follows as in Ch. I, No. 14.

4. Let B be any other configuration which may or may not meet A, and let $A \cdot B$ be closed in A. Suppose that C_p has cells on B and if so let us omit them. We then have chains, cycles, boundaries, mod B. Similarly for the chains, cycles, boundaries mod m or (B, m).

For any one of these categories the linear properties of Ch. I, No. 15 are verified and follow from the fact that they hold for the entities whose singular images they are. In fact the terminology loc. cit. can be extended automatically.

The totality of all chains, cycles, boundaries of any particular type (absolute, mod B, etc.) constitutes a class which is topologically definite for the pair A, B. For let A' be a homeomorph of A, with B' a set whose intersection with A' is the image of the intersection of A with B. If, for example, the chain $\sum t_i e_p^i$ has for image on A the cycle Γ_p mod B it will have likewise for image on A' the transform Γ_p' of Γ_p under the homeomorphism between A and A' and Γ_p' is a cycle mod B'. Similarly for the other cases.

Every one of the above classes is a modulus. Again if C_p^i is a singular chain on A, image of the non-singular chain c_p^i, then $\sum t_i C_p^i$ is a chain on A, image of $\sum t_i c_p^i$. Furthermore from (2) follows

$$F(\sum t_i C_p^i) = \sum t_i F(C_p^i) = \text{the image of } \sum t_i F(c_p^i).$$

Hence the boundary relations are again linearly combined like the chains themselves. It follows that the cycles and the bounding cycles constitute moduli. Similarly for the relations mod B, mod m or mod (B, m).

CONCLUSION: *The boundary relations of various types are homeomorphically invariant.*

Let us define the cycle-moduli and the corresponding zero-moduli formally as in Ch. I but in terms of chains on A mod B, mod m or mod (B, m). The associated moduli on A and A' will correspond to one another term for term. It follows that their homology characters are topological invariants. We call these numbers the *topological* Betti and torsion numbers of A, or of A mod B, mod m, or mod (B, m), generically: *topological homology characters.* The same homologies, etc. written for the subchains of a complex are extended to the chains on A.

5. The n-circuit can also be defined in an invariant manner. A is an n-circuit mod B if the following conditions hold: (a) $R_n(A; B, 2) = 1$. It implies that every n-cycle is homologous to a single one, Γ_n, mod $(B, 2)$. (b) Every point of $A - B$ is on any cycle Γ_n mod $(B, 2)$. This implies that no subset of $A - B$ has the same properties. (c) Every $(n+i)$-th Betti number is zero. The circuit is then *orientable* when $R_n(A; B) = 1$, in which case every n-cycle mod $B \sim$ a multiple of a unique Γ_n. The orientation is then possible in two ways. The analogy with the case of complexes is complete.

EXAMPLE: The n-cell is an orientable n-circuit mod its boundary, whether the latter is a complex or not. (See § 5.)

6. The geometric significance of what precedes may be brought out as follows: suppose that $C_{p-1} = F(C_p)$. Given C_{p-1}, usually C_p will not be unique. For example if C_{p-1} is a circumference in S_3, any cone with it as a base will be a suitable C_p. By considering an improper image of the configuration on a plane, we see that the same observation holds for a circumference in the plane. What matters chiefly is then merely whether A carries some $C_p \rightarrow C_{p-1}$.

After we have introduced deformation in the next section we shall be able to give a much more intuitive interpretation of homology.

§ 2. DEFORMATIONS. THE POINCARE GROUP

7. Let there be given a configuration a in some S_{n-1}, itself immersed in an S_n. Impress upon S_{n-1} a parallel translation carrying it into S'_{n-1}, and a into a', denoting by b the cylinder locus of a. Suppose now that we have on a point set G a singular image B of the cylinder b, with A and A' as the corresponding singular images of a and a'. We may

think of it as defining a singular translation of A into A', over G. We will say that A and A' are *homotopically deformable,* or simply *deformable,* into one another, over G. The singular translation of A into A' just described is called a *homotopic deformation,* or *homotopy,* or simply *deformation.*

Let p be any point of a, p' the point of a' into which it is translated. Let P, P', PP' be the images of the points p, p' and the segment pp'. The 1-cell PP' is the *path* of P. When all these paths are of diameters $< \varepsilon$, we have what is called an ε-deformation (Brouwer).

If the position of a at each stage of the parallel translation is homeomorphic to its image on G, we have an *isotopic deformation,* or simply *isotopy.*

EXAMPLES: G is a plane, A a circle, A' a figure 8 curve. There is a homotopic deformation of A into A' On the other hand, if two circles are similarly oriented there is an isotopic deformation of one into the other.

The terms "homotopy", "isotopy" were introduced by Dehn-Heegaard [2].

8. Now let Π_n be a simplicial polyhedron whose faces σ_p^i are cells of a K_n and let its containing space S_{q-1} be itself immersed in an S_q. We impress upon S_{q-1} a parallel translation bringing it into $S'_{q-1} \neq S_{q-1}$, with Π'_n as the image of Π_n.

Let $A^0 A^1 A^2 \cdots$ be the vertices of Π, and $B^0 B^1 B^2 \cdots$ those of Π' into which they go. The locus of $\sigma_p = A^0 A^1 \cdots A^p$ is a convex $(p+1)$-cell $A^0 A^1 \cdots A^p B^0 \cdots B^p$. I say that this closed convex cell is the sum of all closed simplexes $B^0 B^1 \cdots B^i A^i A^{i+1} \cdots A^p$. The statement being obvious for $p = 0$, let us grant it for $p - 1$. If P is any point of the cell, its projection from B^0 onto the boundary is on σ_p, or else on the locus of the face $A^1 \cdots A^p$ since these are the only faces of the $(p+1)$-cell of which B^0 is not a vertex. In the first case $P \subset B^0 \sigma_p$, in the second according to what is assumed, the projection of P is on a closed simplex $B^1 \cdots B^i A^i \cdots A^p = \overline{\sigma}'_p$. Therefore P is on $B^0 \sigma'_p$ which is again of the announced type.

We will define the oriented simplicial components of the convex cell by

$$\sigma_{p+1}^{i} = (-1)^{i-1} B^0 B^1 \cdots B^i A^i \cdots A^p.$$

The sequence of cells σ_{p+1}^{i} is such that any two consecutive cells have a single p-face in common and are oppositely related to it. Therefore $\sum \sigma_{p+1}^{i}$ is merely a normal orientation of the convex cell locus of σ_p. We call that chain the *deformation cell,* $\mathfrak{D}\sigma_p$, of σ_p. The set of all cells σ_{p+1}^{i} plus the cells of Π_n and Π_n' is an $(n+1)$-polyhedron $\mathfrak{D}\Pi_n$, the *deformation polyhedron* of Π_n. Similarly if we have the subchain of Π_n

$$C_p = \sum x_i \sigma_p^i,$$

its deformation chain is by definition the $(p+1)$-chain

$$\mathfrak{D} C_p = \sum x_i \mathfrak{D}\sigma_p^i.$$

We pass now to the boundary relations for the deformation chains. Taking again one cell, say $\sigma_p = A^0 A^1 \cdots A^p = A^0 \sigma_{p-1}$ and the notations being as before we find:

(a) When σ_{p+1}^{i} is not one of the end cells, its boundary consists of cells such as $(-1)^{i-1} B^1 \cdots B^i A^i \cdots A^p$, plus two cells that do not appear in the boundary of $\mathfrak{D}\sigma_p$ (namely the two that do not have both A^i and B^i as vertices). (b) In addition $F(\sigma_{p+1}^{0})$ and $F(\sigma_{p+1}^{p})$ contribute to $F(\mathfrak{D}\sigma_p)$ the respective cells $-\sigma_p$ and σ_p'. Therefore

$$\mathfrak{D}\sigma_p \rightarrow \sigma_p' - \sigma_p - \mathfrak{D}F(\sigma_p).$$

Hence

$$\mathfrak{D} C_p \rightarrow C_p' - C_p - \mathfrak{D}F(C_p). \tag{3}$$

In particular let C_p be a cycle mod L, a subcomplex of Π_n. Then $\mathfrak{D}F(C_p) \subset \mathfrak{D}L$. Hence

$$\mathfrak{D} C_p \rightarrow C_p' - C_p \mod \mathfrak{D}L. \tag{4}$$

As a special case $L = 0$, the chain is an absolute cycle and

$$\mathfrak{D} C_p \rightarrow C_p' - C_p. \tag{5}$$

The analogy with the cylinder and its two bases is quite complete. We can identify one of the bases, C_1, with a one-cycle, and the other C_1', with that cycle displaced along the generators. $\mathfrak{D} C_1$ is the lateral surface decomposed into the cells of a two-circuit and we have the relation (5) between them, as we should expect.

9. All this is now to be translated in terms of singular configurations on the set A. Suppose that we have on A configurations which are the singular images of those just considered. Then the images of $\mathfrak{D} C_p$, etc. are the singular deformation complexes, chains, etc. for the images of C_p, etc.; and we have a homotopic deformation over A from the image of C_p to that of C_p'. In short we extend to the singular behavior on A the terminology that goes with the associated non-singular situation.

Designating now the images themselves by C_p, $\mathfrak{D} C_p$, C_p', etc., relations (3), (4), (5) continue to hold according to our conventions. In particular if $F(C_p) \subset B \subset A$, and if $F(C_p)$ is deformed over B, i. e. if $\mathfrak{D} F(C_p) \subset B$, we shall have from (3)

$$\mathfrak{D} C_p \to C_p' - C_p \mod B \text{ on } A,$$

or

(6) $$C_p' \sim C_p \mod B \text{ on } A.$$

In particular if C_p is an absolute cycle

(7) $$C_p' \sim C_p \text{ on } A.$$

Similarly of course mod m, or mod (B, m). We have then the important

THEOREM. *When a cycle mod B or mod (B, m) is homotopically deformed over A, while its boundary remains on B (is homotopically deformed on B), it remains homologous to itself mod B or mod (B, m) on A. When an absolute cycle, or a cycle mod m, is deformed over A it remains likewise homologous to itself, or to itself mod m, on A.*

10. A very simple and quite useful operation, T, on a chain C_p is equivalent to a deformation. Let $\Gamma_{p-1} = F(C_p)$ be

deformed over A into Γ'_{p-1} with $\mathfrak{D}(\Gamma) \to \Gamma'_{p-1} - \Gamma_{p-1}$ as the deformation chain. T consists in replacing C_p by $C_p + \mathfrak{D}(\Gamma)$ or $T \cdot C_p = C_p + \mathfrak{D}(\Gamma)$.

To show that T is a deformation we first assume C_p simplicial, polyhedral and in S_n, and the deformation of Γ a translation in S_n. Let C'_p be the sum of the cells of C_p with a vertex on Γ, and set $G_{p-1} = F(C') - \Gamma$. Replacing if need be C by a suitable subdivision (the second derived), we may so choose the chain that: (a) $G \neq 0$ and does not meet Γ; (b) through every point μ of C' not on G nor on Γ, there passes a unique segment $M \mu M'$ with M on G, M' on Γ. The construction is exactly the same as will be found in

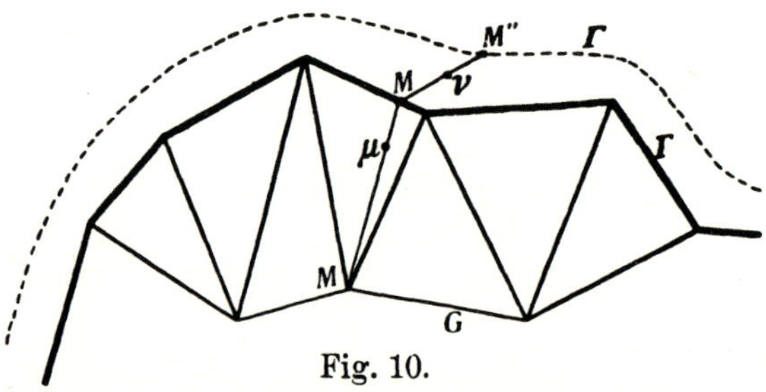

Fig. 10.

No. 21, in connection with the so-called "normal" neighborhoods. In the sense of No. 21, C' is the closure of the C-neighborhood of Γ.

Let now M' be carried by the translation in S_n into M'' on Γ'. We define a deformation of C_p as follows: Every point μ of $\overline{MM'}$ goes over into the point ν of the broken line $MM'M''$ dividing its length in the same ratio as μ divides MM'; points not on the segments remain fixed. The effect on Γ is again to carry it over into Γ', and as for C_p, it is turned into a (singular) chain having $C_p + \mathfrak{D}(\Gamma)$ for subdivision, which proves our assertion in the present instance.

If we consider chains on A, there is no formal change in the treatment, the paths and cells being merely singular, hence the final result is the same.

11. At the end of the preceding section we asserted that deformations would yield a very good intuitive interpretation of homology. It rests on the

THEOREM. *In order that $\Gamma_p \sim 0$, on A in any one of the four senses of the homology, it is necessary and sufficient that by adding to it a sum of p-cells identically zero, or each taken m times in the modular case, or also cells on B in the relative cases, it may be expressed as an algebraic sum of cycles, each of which is deformable over A into a point, or into a point counted zero times when $p = 0$.*

This theorem gives a sharp intuitive content to the concept of homology.

When the condition is fulfilled we have a Γ'_p whose cells must be taken zero times according to our conventions for the chains on A. Therefore (7) gives $\Gamma_p \sim 0$.

Conversely when $\Gamma_p \sim 0$, say mod B,

$$A \supset C_{p+1} = \sum t_i E^i_{p+1} \rightarrow \sum t_i F(E^i_{p+1}) = \Gamma_p, \text{ mod } B.$$

Now $F(E^i_{p+1})$ can be deformed into a point on E^i_{p+1}, hence the chain $\sum t_i F(E^i_{p+1})$ is a sum of cycles each deformable into a point on A. But at worst it differs from Γ_p by a sum of cells identically zero or by cells on B, or by m times some chain in the modular case. This completes the proof of the theorem.

In the light of the preceding theorem it is plain that the operation on chains that primarily matters in the homology theory is what we may call *homological deformation:* homotopic deformation combined with the addition at any time of a sum of p-cells $\equiv 0$, or $\equiv 0$ mod m, or else on B, as the case may be. We have thus four types of homological deformations which may be distinguished when need be by the terms: absolute, mod m, mod B, or mod (B, m). The above theorem can then be stated more briefly: *In order that $\Gamma_p \sim 0$ it is necessary and sufficient that it be homologically deformable (in the proper sense) into a set of points, each taken zero times when $p = 0$.*

12. The Poincaré group of a set. This group introduced by Poincaré [2,7] depends in an essential manner upon homotopy. It may be said to mark the place where homological deformations become inadequate.

We start with a set A which we take to be *arc-wise* connected: any two points can be joined by an arc on A. Let P be a fixed point of A. The elements of the group G are the oriented arcs with end points coincident with P, singular arcs being allowed. The product $\delta\gamma$ of two elements δ, γ of G, consists of the oriented arc obtained by describing γ followed by δ. The identity consists of the elements that may be

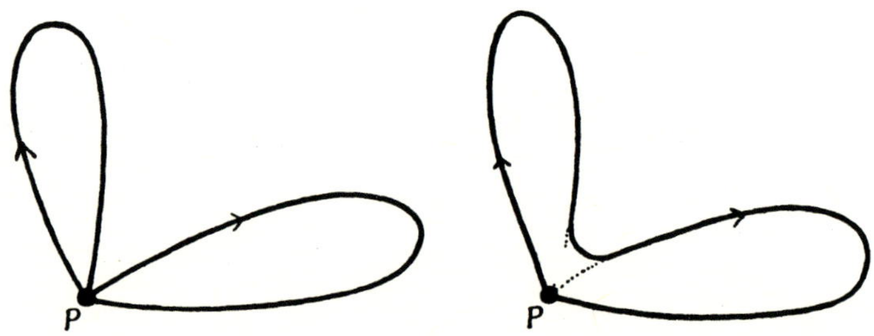

Fig. 11.

shrunk into P over A by a homotopy leaving the end points fixed. Every element γ has an inverse γ^{-1}, which corresponds to the same arc with its orientation reversed. Therefore the aggregate G is a group. If P is replaced by any other point Q of A, G is merely replaced by a *similar* group G', that is by a group abstractly identical with G. It is that abstract group which is the Poincaré group of A. This group is a topological invariant of A.

In the simplest cases: cells, spheres, Euclidean spaces, G is the identity. For the anchor ring G is a commutative infinite group generated by two independent elements γ, δ. For the surface of the two-sided disk with p holes — Clifford's normal form for a two dimensional orientable manifold of genus p — the group is infinite and generated by $2p$ operations

γ_i, δ_i, $i = 1, 2, \cdots, p$, (Riemann's retrosections), between which there is essentially a unique relation

$$\Pi \gamma_i \delta_i \gamma_i^{-1} \delta_i^{-1} = 1.$$

When $p = 1$, or the anchor ring, this reduces to $\gamma \delta = \delta \gamma$, so that G is commutative, as asserted.

With G we may abstractly associate a new group G_0 whose elements correspond term for term to those of G, satisfy all the similar identical relations and in addition are commutative. G_0 is nothing more nor less than the group whose elements are the one-cycles of A, the product of two elements being the sum of the corresponding cycles, the identity the cycles ~ 0 on A. In short, if homotopy is replaced by homological deformation G reduces to G_0. To sense the difference let us take on the Clifford disk of genus $p > 1$ the element $\gamma_1 \delta_1 \gamma_1^{-1} \delta_1^{-1}$. It is not an identical element of G although it represents an identical element of G_0. And in fact it cannot be shrunk into P as demanded for the identity of G, but can evidently be shrunk into it by a homological deformation, since it consists of pairs of opposite arcs.

It is clear then that G is a more accurate character of A than G_0. One of the aims of Poincaré was in fact to use G in order to discriminate between manifolds with the same group G_0, i. e. more generally with the same homology characters. He did obtain [7] a three-manifold M_3 with the homology characters of the sphere but whose G is not the identity, so that M_3 is not an H_3. However since then Alexander [4] has shown by an example that two three-manifolds may have the same homology characters and the same Poincaré group without being homeomorphic.

Recently the Poincaré group has been the object of highly interesting investigations, notably in connection with the theory of knots by Alexander [10, 20, 23] and Reidemeister [5,6], and with the isolation of fixed points of certain surface transformations by Nielsen [1, 2, 3] and other manifolds by Hopf [5]. The group appears also in disguise in Morse's investigations [1, 2] on geodesics of open and closed surfaces.

§ 3. The Fundamental Deformation Theorem for Complexes

13. The theorem which we have in view forms the connecting link between the combinatorial and topological numbers of a complex. It was first proved for manifolds by Alexander [2], then extended to complexes by Veblen [5].

The general idea of the theorem is intuitively simple enough and will be made clear by an example. Suppose that we have on a K_2 (Fig. 12) a singular K_1 represented by the dotted lines of the figure, its vertices being the points enclosed in small circles. Then our theorem asserts essentially that a certain subdivision of it can be deformed into a subcomplex

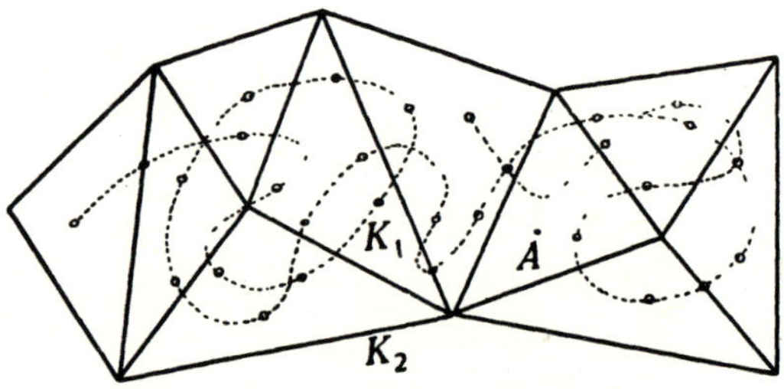

Fig. 12.

of K_2, a statement whose truth seems fairly obvious in the example given. For we have only to project the points of K_1 onto the boundaries of the 2-cells on which they lie from suitably chosen points interior to the 2-cells, such as the point A of the figure, and then "smooth out" the resulting deformation of K_1 on the 0- and 1-cells of K_2, introducing new vertices where necessary, so that each 1-cell of the final deformation of K_1 covers exactly one 1-cell of K_2. Of course the points A must not lie on K_1. But in the fact that K_1 may very well fill the whole of a 2-cell of K_2, as shown by Peano curves, for example, lies a real difficulty which renders the proof not wholly elementary.

14. As before we designate the basic complex by K, its dimension being n, while L is a subcomplex of K.

Before proceeding with our proof we shall make a few preliminary remarks regarding a singular K_p ($p \lesseqqgtr n$) on K. Let Π_p be the polyhedron of which it is the image. A subdivision Π'_p of Π_p will have for image a subdivision K'_p of K_p. By repeated subdivision Π_p can be reduced to a new simplicial polyhedron whose cells are as small as we please. We shall call *mesh* of K the maximum diameter of its cells. Since each subdivision of Π_p has for image a subdivision of K_p and since the correspondence between the two is continuous, we can find a K'_p whose mesh is less than any preassigned ε.

Any point P of Π'_p is on a definite cell E_q of Π'_p whose image E'_q is a cell of K'_p carrying the image P' of P. The point P had definite barycentric coördinates relative to E_q. These are also taken as the barycentric coördinates of P' with respect to E'_q. Since P' may belong to several cells of K'_p it may have several sets of barycentric coördinates, one set for each of the cells.

It is convenient to define a particular type of one-valued continuous transformation of a simplex $\bar{\sigma}_p\,(A^0, A^1, \cdots, A^p)$ into a simplex $\bar{\sigma}_q\,(B^0, B^1, \cdots, B^q)$, $p \geqq q$. Let each vertex of σ_p correspond to a single vertex of σ_q, in such a way that each vertex B^s is the correspondent of at least one vertex A^r. The coördinates $x_0, \cdots, x_p$ of a point P of $\bar{\sigma}_p$ may be regarded as a set of weights with sum unity, each attached to the corresponding vertex, as x_r to A^r, and so selected that P is the center of gravity of the resulting system. If now we attach to each vertex B^s of $\bar{\sigma}_q$ all weights x_r formerly belonging to those vertices A^r which correspond to B^s, the resulting system of weights on the vertices of $\bar{\sigma}_q$ will form the barycentric coördinates of a single point Q of $\bar{\sigma}_q$. The transformation of $\bar{\sigma}_p$ into $\bar{\sigma}_q$ thus defined for each point of $\bar{\sigma}_p$ is evidently single-valued and continuous. We shall call it a *barycentric transformation* (Brouwer [3]). The $p+1$ weights attached to the $q+1$ vertices of $\bar{\sigma}_q$ may be regarded as *singular barycentric coördinates* of $\bar{\sigma}_q$ when $p > q$.

A barycentric transformation of $\bar{\sigma}_p$ into $\bar{\sigma}_q$ evidently determines a barycentric transformation of any face of $\bar{\sigma}_p$ into the face of $\bar{\sigma}_q$ determined by the correspondence of vertices. Hence we may extend the transformation to simplicial complexes and define a barycentric transformation (or deformation) of a simplicial complex K as one in which each cell of K is transformed barycentrically. Note that any continuous transformation of K which is defined by means of a barycentric deformation of each *closed* cell is a barycentric deformation.

15. THE FUNDAMENTAL DEFORMATION THEOREM. *Every chain or complex on K has a subdivision homotopically deformable over K into a subcomplex or subchain of K in such manner that no point leaves the closure of the cell of K that carries it.*

The proof being identical for chains and complexes we will take a K_p. We shall show by induction on n that a K_p whose mesh is less than a certain ε has the required property. Since every K_p possesses a subdivision whose mesh $< \varepsilon$ this proves the theorem. As it is obvious for $n = 0$, we can assume that it holds for the dimension $n - 1$, in particular for the complex K_{n-1} obtained from K by the removal of its n-cells, the suitable value of ε being ε'.

Select a point P^i on the cell E_n^i of K and on that cell draw a sphere H_{n-1}^i of center P^i. Let G be a point set exterior to all the spheres and G' its rectilinear projection from the P's onto K_{n-1}. The points on K_{n-1} remain fixed and those on E_n^i are projected from P^i alone onto the boundary of the cell. For a given diameter ξ of G, the diameter of G' has an upper bound η which tends to zero with ξ. Hence ξ can be so chosen that $\eta < \varepsilon'$. We take ε less than ε', ξ and $d(\sum H^i;\ K_{n-1})$.

Let then K_p' be the sum of all closed cells of K_p which meet K_{n-1}. Under the circumstances no point P^i is on K_p'. Hence the cells of K_p' have well defined projections on K_{n-1} and these projections have for sum a complex of mesh ε'. This complex is by assumption barycentrically deformable into a subcomplex K_p'' of K_{n-1} in such manner that its points do not leave their carrying closed cells of K_{n-1}.

Each cell of $K_p - K_p'$, in particular each vertex of it, is on a unique n-cell of K. Displace each such vertex into some vertex of that n-cell. Now if E_q is any cell of $K_p - K'_p$, lying say on E^i_n, each of its vertices, whether in $K_p - K'_p$ or in K'_p, lies on E^i_n. The deformation of K'_p into K''_p plus the displacement of the vertices of $K_p - K'_p$ has established a displacement of each vertex of E_q into a vertex of E_n. Hence we can deform the cell E_q barycentrically over E_n into the face of E_n determined by the displacement of vertices.

Whatever then the cell E_q of K_p we have established a barycentric correspondence between its points and those of a certain cell of K, which is a true or a degenerate σ_q. The point pairs P, Q, associated by the correspondence are always on a closed cell of K and hence the segment PQ is uniquely determined. The set of all these segments determines a deformation complex $\mathfrak{D}K_p$ for a deformation of K_p into a subcomplex of K. This proves the theorem.

REMARKS. I. If K_p has a subcomplex K_q in common with K we may choose the barycentric deformation such as to leave K_q invariant. All that is necessary is to carry out the subdivision of K_p in such a manner as to leave K_q untouched. The restriction as to the diameter of the cells is to be applied to the cells having no cell of K_q for a face, while if a cell has such a cell for a face, its points must not be farther than η from it. Similarly for C_p in place of K_p.

II. It is important to bear in mind that *the dimension p of K_p or C_p has been in no sense restricted.* It may for example very well exceed the dimension of K.

§ 4. INVARIANCE OF THE HOMOLOGY CHARACTERS

16. Let C_p be a non singular chain and C'_p a subdivision of it. A translation of every vertex of C'_p to one of the vertices of the cell of C_p that carries it defines, as in the proof of the deformation theorem, a barycentric deformation $\mathfrak{D}$ of C'_p into a subchain C''_p of C_p. I say that $C''_p = C_p$. This being true for $p = 0$ we can assume it for $p-1$ and prove it for p. Furthermore due to the linearity of the chain symbols and the fact that every p-cell of C'_p is deformed into a multiple (possibly zero times) of the carrying p-cell of C_p,

it is sufficient to prove that the property in question holds for $C_p = E_p$ a single p-cell.

This being the case, $F(C'_p)$ is on $F(E_p)$ and remains on it throughout $\mathfrak{D}$. Also since the property holds on $F(E_p)$, $F(C''_p) = F(E_p)$. But $C''_p = t(E_p)$. Hence $F(C''_p) = tF(E_p)$, and therefore $t = 1$, which proves our assertion.

Since the behavior on a singular chain is the singular picture of the behavior on the ordinary chain, and since $F(C'_p) \subset F(C_p)$ throughout $\mathfrak{D}$ we have the

THEOREM. *A cycle of any one of our types is homologous on itself to its own subdivisions, homology being understood in each case in the appropriate sense.*

17. Combining this with the deformation theorem we have the

THEOREM. *The combinatorial and topological homology characters of the same type are equal.*

The deformation theorem implies that any p-cycle Γ_p, of one of our four types, has a subdivision deformable into a subcycle Γ'_p of the same type, the boundary on L remaining on L. Hence by that theorem, the one just proved and the theorem of No. 9, $\Gamma_p \sim \Gamma'_p$ in the appropriate sense.

Let several cycles Γ^i_p be thus reduced to subcycles Γ'^i_p and suppose that

$$\sum t_i \Gamma^i_p \sim 0 \mod L \quad \text{or} \quad (L, m).$$

We shall also have

$$\sum t_i \Gamma'^i_p \sim 0 \mod L \quad \text{or} \quad (L, m).$$

Therefore

$$K \supset C_{p+1} \rightarrow \sum t_i \Gamma'^i_p \mod L \quad \text{or} \quad (L, m).$$

From the deformation theorem applied to C_{p+1} it follows that C_{p+1} may be replaced by a subchain with the same boundary, i. e. we may assume it to be a subchain. Therefore the homology between the cycles Γ' may be taken in the combinatorial sense, i. e. as in Ch. I.

CONCLUSION. Let homologies between cycles be understood in the topological sense, those between subcycles in the combinatorial sense. Then there is a one-to-one homology-preserving correspondence between the families of homologous cycles of each of the two kinds. This proves our theorem.

An immediate consequence is the

FUNDAMENTAL INVARIANCE THEOREM. *The combinatorial homology characters of a complex are topological invariants.*

COROLLARY. *The topological homology characters of a complex are finite.*

Incidentally we have proved the very useful

THEOREM. *If a subchain of K is ~ 0 in one of the four possible senses of the homology (absolute, mod L, etc.), it bounds in the same sense a subchain of K.*

In regard to the invariance proofs see Alexander [2, 19], Veblen [5]. The proof given here is essentially Veblen's and closely related to Alexander's first proof. A second proof by Alexander is based upon a condition for the homeomorphism of two complexes and will be recalled in Ch. 6, § 3.

18. The invariance of the combinatorial n-circuit properties is again obtained by identifying them with the corresponding topological properties (No. 5).

Suppose then that K is a combinatorial n-circuit mod L. The problem is to show that it is also a topological n-circuit. The Betti number and base conditions are satisfied by what precedes. We have a unique cycle Γ_n mod $(L, 2)$, subchain of K, and we merely have to show that if $\Gamma'_n \sim \Gamma_n$ mod $(L, 2)$ then every point of $K-L$ is also on Γ'_n.

Suppose that there exists a point P of $K-L$ not on Γ'_n. We may replace complexes and chains by subdivisions of arbitrarily small mesh. In particular let the mesh for the subdivision K' of K be less than $d(P, \Gamma'_n)$. Then a suitable subdivision of Γ'_n will have no point on at least one $\bar{E}_n$ of K' and yet come under the deformation theorem. In the linear expression for the corresponding reduced subcycle Γ^*_n of K' the cell E_n will not appear at all, hence K' is not an n-circuit. As the n-circuit has been shown to be invariant under subdivision we have here a contradiction. Hence every point of $K-L$ is on Γ'_n and *the combinatorial n-circuit is also a topological one.*

The invariance of combinatorial orientability depends only upon the Betti-numbers and therefore follows from No. 17. For the same reason it coincides with topological orientability.

§ 5. Invariance under Partial Homeomorphism

19. We have already observed that it is *a priori* probable that the characters of $A \bmod B$ are not so much topological invariants of the pair of sets A, B as of their difference $A-B$. We propose to substantiate this for the following very general case: $A-B$ is open, $\overline{A-B}$ a complex, $L=\overline{A-B}-(A-B)$ is *locally connected*, an important property to be defined below, introduced by Alexander.

The term "partial homeomorphism" refers to the fact that we are considering a homeomorphic transformation, not of the whole of K into, say K', but of its open subset $K-L$ into a similar subset, $K'-L'$, of K'.

20. We first consider certain properties of neighborhoods.

Let A be a compact metric space and B, C, closed subsets of it. If $d(B,C)=0$, the sets meet one another. For the set of all pairs of points of which one is on B, the other on C is also closed. Therefore there exists an actual pair whose points are at a distance apart less than any $\varepsilon>0$, and therefore coincide.

It follows that if N is a neighborhood of B, $d(B, A-N)>0$. For otherwise B and $A-N$, both closed, would meet and so would N and $A-N$, which is absurd.

The number $d(P, A-N)$, P a point on the boundary of B, has a maximum ε actually reached for some P since the boundary of B is closed. It is called the *width* of the neighborhood N. The distances between points of B and points of $A-N$ are therefore included between two positive numbers.

21. Let L be a closed set on the complex K, N the set of all cells of K whose closures meet L. The complement $K-N$ of N is the set of all closed cells which do not meet L. It is therefore a complex and in particular a closed subset of K. It follows that N is open and as $N\supset L$ it

is a neighborhood of L on K. This type of neighborhood, which recurs frequently in the sequel, shall be called the *K-neighborhood of L*. If K' is a simplicial subdivision of K whose mesh is η, the K'-neighborhood of L is of width $<\eta$.

When L is a subcomplex of K, its K-neighborhood, N^L, consists of all cells of K having vertices in common with L. There are now two possibilities:

(a) Whenever a cell has all its vertices on L, it is a cell of L. The neighborhood N^L will then be called *normal*. In that case if E_p is any cell of $N^L - L$ we have $E_p = \sigma_q \sigma_{p-q-1}$ with σ_q on L, σ_{p-q-1} on $K - L$ and $q < p$. A normal neighborhood has the following noteworthy and very useful property: Through every point R of E_p there passes a unique segment PQ with P on σ_q and Q on σ_{p-q-1}. It is the segment that *projects* R on L.

(b) There are cells of $N^L - L$ with all their vertices on L. Then the first derived K' of K will be as in (a), and the associated neighborhood N'^L will be normal.

In short either K or its first derived will yield a normal neighborhood with its convenient system of projecting lines. That is quite satisfactory in practise for in general all that is needed is a convenient K^0-neighborhood, where K^0 is some subdivision of K.

Incidentally we can consider a convex K as falling under (b).

22. In deriving the results of the present section it is not necessary to restrict L to a non-singular complex on K. The proofs hold indeed just as readily when L is allowed to be an arbitrary closed subset of K having *local connectedness*. A compact metric set $\mathfrak{R}$ is locally connected whenever for every ϵ there is an η such that every sphere H_p on $\mathfrak{R}$ whose diameter $<\eta$ bounds a cell E_{p+1} equally on $\mathfrak{R}$, whose diameter $<\epsilon$, where both H and E are allowed to be singular.

For $p = 0$ local connectedness was first introduced by Pia Nalli. One might impose connectedness for all values $p \leqq q$, thus obtaining *local q-connectedness*. *The different orders of local connectedness are independent in the following sense. Given any finite set of integers $r, s, \cdots, t$ all positive or* *zero, it is possible to construct a closed point set in* S_{t+1} *having local connectedness of all orders lower than the smallest* $r, s, \cdots, t$ *and no local connectedness of any of the latter orders.*[a]

The following proof is by A. B. Brown. If we can show how to construct a closed point set having all orders of local connectedness except some assigned one, say q, the required point set may be had by taking the sum of point sets for which this integer is $r, s, \cdots, t$, and which do not meet one another. If $q = 0$ a set of isolated points having a single limit point, plus that limit point, may be taken. For $q > 0$, we take two line segments L and L', with end points PA and PA', making an angle of two degrees in S_{t+1}, and choose some S_{q+1} in S_{t+1}, containing L and L'. Mark the sequence of points $B^1 = A$, B^2, B^3, $\cdots$, each half way from its predecessor on L towards P. Corresponding to B^j is taken the q-sphere H_q^j in S_{q+1} with center at B^j and tangent to L'. No two of the spheres H^j have a point in common. The required point set consists of L' and the spheres H_q^j.

REMARK. *Local connectedness of any one of the above types is a topologically invariantive property.*

23. THEOREM. *Every complex is locally connected.*

Given K and $\epsilon > 0$, subdivide the complex into K' of mesh $\delta < \frac{1}{2}\epsilon$. There exists an $\eta < 2\delta < \epsilon$ such that any subset of K whose diameter $< \eta$ is on the star of some vertex of K'. For otherwise, whatever α, there would exist pairs of points on two closed cells without a common vertex, yet not more than α apart. Since K is closed, we could find a limiting pair consisting of two coincident points on two closed cells without a common vertex, which is absurd.

If $d(H_p) < \eta$, H_p is carried by the star of some vertex A of K. The segments from A to H_p generate an E_{p+1} on the star, hence of diameter $< 2\delta < \epsilon$, bounded by H_p, which proves the theorem. It is scarcely necessary to point out that $\eta \leqq \epsilon$.

24. THE EXTENDED DEFORMATION THEOREM. *Let L be a subset of K, both being locally connected and closed. Then for*

[a] See Addenda.

every ε there exists an η such that any singular K_p on K whose points are not farther than η from L is ε-deformable onto L on K.

For $p = 0$ this is a direct consequence of the Borel Theorem and the local connectedness of K. Grant it for the dimension $p-1$ and for a certain $\varepsilon' < \varepsilon$; let η be the η of the theorem corresponding to that dimension. Let also K_{p-1} be the complex left on removing the p-cells from K_p. We will assume the points of K_{p-1} within a distance η' from L so that it becomes ε'-deformable into K'_{p-1} on L over K.

Let E_p be any cell of K_p, H_{p-1} its boundary sphere, $\mathfrak{D}(H_{p-1})$ its deformation complex, associated with the deformation of K_{p-1} (§ 2). That deformation turns it into a sphere H'_{p-1}, subcomplex of K'_{p-1}.

Now $d(H')$ tends to zero with ε' and with the mesh of K_{p-1} or, if we wish, with ξ, the mesh of K_p. We can replace K_p by a subdivision of mesh as small as we please, i. e. dispose of the magnitude of ξ. Since L is locally connected, we can so choose ε' and ξ that H'_{p-1} bounds on L an E'_p whose diameter $< \zeta$ assigned, and that furthermore this be the case for all spheres analogous to H'_{p-1}.

Now $E_p + E'_p + \mathfrak{D}$ is a singular H_p image of the boundary of the deformation complex of an $\bar{E}_p$, and $d(H_p)$ tends to zero with ε', ξ, ζ. Since K is locally connnected we can so choose these three quantities that H_p bounds on K a cell $\mathfrak{E}_{p+1}$ whose diameter $< \varepsilon$, and that, furthermore, this be the case for all spheres analogous to H_p. The set of all cells $\mathfrak{E}$ plus the deformation complex for K_{p-1} constitutes a deformation complex for K_p into K'_{p-1} plus the sum of all cells E'_p on L and the corresponding deformation of K_p onto L is ε. Our construction has resulted in assigning an upper bound to ε'. We choose as η the corresponding η'. When the points of K_p are not farther than this η from L the preceding construction can be carried through. The theorem is therefore proved.

25. Remarks. I. When K is a complex, η can be chosen independent of p. For designate the η of the theorem by

$\eta(\varepsilon, p)$. We will take for fixed η the least of the numbers $\frac{1}{2}\varepsilon$ and $\frac{1}{2}\eta$ $(\frac{1}{2}\varepsilon, q)$, $0 \leqq q \leqq n$. By the fundamental deformation theorem any K_p whatever is η-deformable into a subcomplex K_q of a subdivision of K. If the points of K_p are not farther than η from L, those of K_q will be not farther than $\eta(\frac{1}{2}\varepsilon, q)$ from L. Hence K_q will be $\frac{1}{2}\varepsilon$-deformable onto L and K_p ε-deformable onto L.

II. K_p may be replaced everywhere by a C_p.

III. If K_p or C_p have cells on L they may be kept fixed throughout the deformation.

IV. If C_p is a cycle mod L or mod(L, m) behaving as in the theorem, it is $\sim 0 \bmod L$ or mod(L, m) on K.

V. An equivalent form of the theorem is as follows: *For every N'^L there exists $N^L \subset N'^L$ such that every $K_p \subset N$ is deformable onto L over N'.*

This theorem which we first proved in [14], may be considered as an extension of Alexander's deformation theorem for manifolds [2], to whose proof it is also very closely related. Alexander's theorem for complexes is obtained if we take for L what is left when the n-cells of K are removed. The special property of manifolds utilized by Alexander, that every point Q has an n-cell for neighborhood, is replaced by the local connectedness of K.

26. Our deformation theorem will enable us to replace the study of cycles mod L by that of cycles mod some N. In order to carry out this idea take three neighborhoods $N \subset N' \subset N''$ of L such that N, N' and N', N'' are related like N, N' in the modified statement of the theorem. Let now Γ_p be any p-cycle mod L. We associate with it a p-cycle mod N, say Δ_p, *not meeting* L, obtained by removing from Γ_p, or from some subdivision of it, a p-chain on N' whose boundary is on N. We obtain, for example, a Δ_p by subdividing Γ_p into a cycle of mesh $< d(L, K-N)$ and removing the cells whose closure meets L. Other modes of obtaining a Δ_p are however by no means excluded.

It follows from the deformation theorem that any Δ_p mod N, not meeting L, is associated with some Γ_p mod L. For $F(\Delta_p)$

is deformable onto L on N' and its deformation complex plus Δ_p constitutes precisely a Γ_p mod L.

We have the following basic properties:

(a) *If* Δ_p^h *is associated with* Γ_p^h *then* $\sum t_h \Delta_p^h$ *is associated with* $\sum t_h \Gamma_p^h$.

(b) *The following homologies are equivalent*

$$\Gamma_p \sim 0 \bmod L, \qquad \Delta_p \sim 0 \bmod N'.$$

We may replace Γ_p by the subdivision of which Δ_p is part, hence assume $\Gamma_p = \Delta_p + C_p$, $C_p \subset N'$. If then $C_{p+1} \to \Gamma_p$ we have $C_{p+1} \to \Delta_p \bmod N'$, hence the second homology is a consequence of the first.

When the second holds there exists a $C_{p+1} \to \Delta_p + C_p'$, $C_p' \subset N'$. Hence $\Gamma_p + C_p' - C_p \sim 0$. But a chain ~ 0 is a cycle. Hence $F(C_p - C_p') = F(\Gamma_p) \subset L$, or $C_p - C_p'$ is a cycle mod L on N'. By the corollary of the deformation theorem as modified in No. 25 and applied to the pair of neighborhoods N', N'', $C_p - C_p' \sim 0 \bmod L$. Therefore also $\Gamma_p \sim 0 \bmod L$.

27. *Let now K be once more an n-complex.* We can choose for one of our neighborhoods sets whose closures are subcomplexes of a subdivision K' of K. Hence there exists a base for the p-cycles mod N consisting in fact of subchains of K'. As a consequence for $p > n$, $\Delta_p \sim 0$. From any cycle of that base we may remove the cells whose closure meets L and then we have a base $\{\Delta_p^h\}$, $h = 1, 2, \cdots, r$, whose elements are of the type of the Δ_p considered above. Let Γ_p^h be the Γ_p associated with Δ_p^h. Then given any Γ_p and the associated Δ_p we have

$$\Delta_p \sim \sum t_h \Delta_p^h \bmod N, \text{ hence } \bmod N'.$$

Therefore by (a) and (b) combined

$$\Gamma_p \sim \sum t_h \Gamma_p^h \bmod L.$$

Hence the set $\{\Gamma_p^h\}$ is a base for the p-cycles mod L. It is clear that mod L or mod N, $\cdots$ can be replaced throughout by mod (L, m) or mod $(N, m) \cdots$. Therefore we have proved

THEOREM. *When K is a complex and L a locally connected subset of K there exists a base for all cycles mod L or mod (L, m). Hence there exist torsion coefficients and all homology characters mod L or mod (L, m) are finite. For any dimension $> n$ there are no torsion coefficients and the Betti numbers are zero.*

28. It is now a simple matter to establish invariance under partial homeomorphism, whether K is a complex or otherwise.

Let then $\mathfrak{K}$, $\mathfrak{L}$, be a pair analogous to K, L with $\mathfrak{K}-\mathfrak{L}$ the homeomorph of $K-L$ under a transformation T. The image of $K-N^L$ is a closed subset of $\mathfrak{K}$, which does not meet $\mathfrak{L}$. Its complement $\mathfrak{N}^{\mathfrak{L}} \supset \mathfrak{L}$ and is open, hence it is a neighborhood of $\mathfrak{L}$. In other words to N^L corresponds $\mathfrak{N}^{\mathfrak{L}} = \mathfrak{K} - T(K-N^L)$. Conversely, given $\mathfrak{N}^{\mathfrak{L}}$, $N^L = K - T^{-1}(\mathfrak{K}-\mathfrak{N}^{\mathfrak{L}})$ is a neighborhood of L on K. In short, the complements of the neighborhoods are transformed into one another. It is also clear that if N, N' on K correspond to $\mathfrak{N}$, $\mathfrak{N}'$ on $\mathfrak{K}$ then $N \subset N'$ implies $\mathfrak{N} \subset \mathfrak{N}'$ and conversely.

Let now N, N' be related as in No. 26. Then we are not assured that $\mathfrak{N}$, $\mathfrak{N}'$ will be so related on $\mathfrak{K}$. However, there exists an $\mathfrak{N}^*$ related to $\mathfrak{N}'$ like N to N'. Let its image on K (in a transparent sense) be N^*. Then $N \cdot N^*$, $\mathfrak{N} \cdot \mathfrak{N}^*$ are again neighborhoods, images of one another, and respectively related to N', $\mathfrak{N}'$ in the desired manner. In other words, given N', we can always choose N related to it as in No. 26, such that their images $\mathfrak{N}'$, $\mathfrak{N}$ will be similarly related on $\mathfrak{K}$.

It follows that we can find a triple N, N', N'' such as in No. 26, whose images $\mathfrak{N}$, $\mathfrak{N}'$, $\mathfrak{N}''$ are similarly related. Once this is acquired, the correspondence between the chains Δ_p on K with their images on $\mathfrak{K}$, is one-one with the homologies mod N' and mod $\mathfrak{N}'$ corresponding to one another. The bases and their fundamental homologies must then likewise correspond to one another. Hence the homology characters are the same.

29. We still have to consider the n-circuit and orientability. If the Betti number conditions for one or the other properties are verified as regards one of the two sets $K-L$, $\mathfrak{K}-\mathfrak{L}$, they are as regards the other. We merely have to show

that if every n-cycle mod $(L, 2)$ on K carries every point of $K-L$ the same holds for $\mathfrak{K}$ and $\mathfrak{L}$. This however follows at once from the fact that the n-cycles on the two sets correspond to one another in such manner that except for neighborhoods of $L, \mathfrak{L}$ as narrow as we please, they are homeomorphic to one another under T. Summing up then we have

THEOREM. *Let $K, L, \mathfrak{K}, \mathfrak{L}$, be locally connected and closed and let $K-L$ and $\mathfrak{K}-\mathfrak{L}$ be homeomorphic. Then their homology characters and their circuit properties are the same.*

REMARK. When K is an orientable n-circuit mod L we have a base cycle Γ_n for all the n-cycles mod L (not merely mod $(L, 2)$). In place of Γ_n we may also take $-\Gamma_n$ and to orient K mod L is equivalent to choosing one of the two opposite cycles. The choice is topologically invariant in this sense: If Γ_n' is any other base cycle then $+\Gamma_n'$ or $-\Gamma_n' \sim \Gamma_n$ mod L. Of the two opposite cycles $\pm\Gamma_n'$ we must choose the only one for which the homology is true and it is unique.

§ 6. INVARIANCE OF DIMENSIONALITY, REGIONALITY AND THE SIMPLE CIRCUIT

30. In this section we have gathered some invariantive properties of which the most striking are those pertaining to spatial regions due to Brouwer. One of our purposes in taking them up here is to show their intimate connection with the invariance of the homology characters.

THEOREM I. *The dimension n of K is the highest integer such that there exist locally connected subsets L of K for which $R_n(K; L) \neq 0$.*

The notations being as in § 5, any Δ_p can be reduced to $\Delta_p' \sim \Delta_p$ mod N' and subchain of a subdivision of K. When $p > n$, Δ_p' is a degenerate p-cycle, hence ~ 0. Therefore $R_p(K; L) = 0$ for $p > n$. On the other hand let E_n be an n-cell of K, and let $L = K - E_n$. We have here $\alpha_n = 1$, $\varrho_n = 0$, these numbers being as in Ch. I, No. 27, hence

$R_n(K; L) = 1$ and in fact $R_n(K; L, m) = 1$ whatever L. This proves the theorem.

THEOREM II. *The dimension n of K is a topological invariant.*

This follows from Theorem I plus the fact that the L's have topologically invariant character.

31. As a preliminary to the proof of an important lemma we shall make certain simple observations. Let C_p be a non-singular chain on our usual n-complex K. Since C_p and $F(C_p)$ coincide with complexes we can take them as our pair K, L of § 5. The analogues N^F, N'^F of the neighborhoods figuring in V of No. 25 may then be taken as neighborhoods on our present fundamental complex K and not merely on C, for it is only the width of N that matters.

Observe next that in the equivalence of homologies in No. 26 it is actually immaterial whether Δ meets L or not. Since $\sim 0 \bmod N$ is stronger than $\bmod N'$, we have this result: There exists an N^F on K such that if a cycle $\bmod F$ on C, is $\sim 0 \bmod N$, it is also $\sim 0 \bmod F$.

We shall designate by T^ξ a barycentric ξ-deformation by means of which we reduce a chain or complex on K to a subchain or subcomplex of a subdivision whose mesh is ξ.

LEMMA I. *For every non singular C_p on K there exists an η such that $T^\xi C_p \neq 0$ when $\xi < \eta$.*

Let N be the neighborhood of $F(C_p)$ on K considered above and 2ε the width of N. We take $\eta < \varepsilon$ and $<$ the η of No. 24 corresponding to ε and to $L = C_p$. Suppose that $T^\xi C_p = 0$ for $\xi < \eta$ and call D_{p+1} the deformation chain of C_p under T. Since its points are not farther than η from C_p it can be ε-deformed into D'_{p+1} on C_p without displacing any point of its subchain C_p (No. 25). The associated deformations impressed upon F never make it leave N. Hence $D'_{p+1} \rightarrow -C_p \bmod N$, $C_p \sim 0 \bmod N$, hence $\bmod F$ on ts elf which is impossible since it would have to bound a proper $(p+1)$-subchain of itself (No. 17) and none exists. This proves the lemma.

32. COROLLARIES. I. *For every non-singular K_p on $\bar{K}$ there exists an η such that no $T^{\xi} K_p$, $\xi < \eta$, is a complex of less than p dimensions.*

For if this were untrue the lemma would fail to apply to the C_p sum of the p-cells of K_p.

II. *A non-singular C_p or K_p cannot be on K_n when $p > n$.*

For every $T^{\xi} C_p = 0$ and every $T^{\xi} K_p$ is of dimension $< p$.

III. *A non-singular cell cannot be on one of lower dimension.*

For if $E_p \subset E_n$, E_p carries a K_p, which is a closed p-cell, on a K_n which is a closed n-cell, and this is impossible when $p > n$.

REMARK. Lemma I and Corollary I are special cases of a very general theorem on the deformation of closed sets due to Alexandroff [10], his proof being quite unlike ours.

33. INVARIANCE OF DIMENSIONALITY. THEOREM. *When two spatial regions $\mathfrak{R}$, $\mathfrak{R}'$, are homeomorphic their dimensions are the same. Brouwer* [6].

Let $\mathfrak{R} \subset S_n$, $\mathfrak{R}' \subset S_{n'}$. We may assume $\mathfrak{R}$ and $\mathfrak{R}'$ bounded, since in any case they have homeomorphic subsets that are bounded regions. Then $S_n \supset \sigma_n \supset \mathfrak{R}$. Hence there is an n-cell on $\mathfrak{R}'$, i. e. on $S_{n'}$ and therefore on some $\sigma_{n'}$ of $S_{n'}$. Hence $n' \geqq n$. Similarly $n' \leqq n$, hence $n' = n$.

We see thus that Brouwer's theorem is a quite rapid corollary of the deformation theorems.

34. LEMMA II. *The situation being as in Lemma I, let E_p be a cell of C_p, N a neighborhood of $\bar{E}_p$ on K. There exists an η such that $T^{\xi} C_p$, has p-cells on N when $\xi < \eta$.*

The proof is exactly as in Lemma I, with $F(C_p)$ replaced by $C_p - t E_p$, where t is the coefficient of E_p in C_p, and $\eta < d(E, K - N)$.

COROLLARY. *Lemma II holds with E_p replaced by any non-boundary point Q of C_p.*

Subdivide C_p into C_p' with Q as a vertex, and with a p-cell having Q for vertex lying on N^Q. The corollary follows then by applying Lemma II to that p-cell.

35. THEOREM. *If $\bar{E}_n \subset E_n'$, E_n is an open set on E_n'.*

We may assume that E'_n is a σ_n, and that $F(E_n)$ is an $(n-1)$-sphere, since every point of E_n is on an n-cell which is on E_n and whose boundary is such a sphere. Taking any point Q on E_n, we choose as N^Q the interior of a Euclidean sphere of center Q and radius $< \frac{1}{2} d(Q, F(E_n))$. Subdivide σ_n into K whose mesh ξ is as small as we please and in any case less than the radius of N, and then reduce E_n by an appropriate T to a subchain C_n of K. From the last Corollary we learn that for suitably small ξ, C_n will have n-cells on N. Let one of them be E^1_n. On the other hand, due to the choice of ξ, $F(C_n)$ does not meet N. Let E^2_n be any other cell of C_n on N. We can join two points of E^1_n and E^2_n by an arc l on N crossing only n- and $(n-1)$-cells of K. Since every $(n-1)$-cell thus crossed is a non-boundary cell of C_n and is incident with two n-cells of K to which it is oppositely related, if one of the n-cells appears in C_n so does the other. Consequently all n-cells thus crossed belong to C_n and in particular so does E^2_n. Thus every closed n-cell on N is a cell of C_n. Hence C_n carries every point of N at a distance $> \xi$ from the boundary of N. Since $C_n = T^\xi E_n$, every point of N is at a distance $< 2\xi$ from E_n. But N is fixed and ξ arbitrarily small, hence $N \subset E_n$, which proves the theorem.

36. Invariance of Regionality. Theorem. *If $\mathfrak{R}$ is a region of S_n and $\mathfrak{R}'$ its homeomorph on S'_n, then $\mathfrak{R}'$ is likewise a region of S'_n. Brouwer* [8,11].

This is an immediate corollary of the preceding theorem: Since $\mathfrak{R}'$ is the homeomorph of $\mathfrak{R}$, there is a bounded non-singular E_n on $\mathfrak{R}'$ through any point Q of $\mathfrak{R}'$. As $S'_n \supset \sigma_n \supset E_n$, E_n is an open set on S'_n. As it is on $\mathfrak{R}'$ and $\supset Q$, this proves the theorem.

Remark. A number of proofs of Brouwer's theorems are in existence besides his own. We may mention those of Lebesgue [2] noteworthy owing to their importance in the dimensionality theory, Alexander [8], E. Sperner [1], Knaster-Kuratowski-Masurkiewicz [1]. Regarding the proofs for $n = 2$, see Kerekjartó [10].

37. INVARIANCE OF THE SIMPLE CIRCUIT. THEOREM. *If $K-L$, $K'-L'$ are partial homeomorphs and one of them is a simple circuit, so is the other.*

We designate the image on $K'-L'$ of any configuration of $K-L$ by accenting its symbol. Let G be the sum of all cells of $K-L$ whose dimensions $< n-1$ and E'_{n-1} any $(n-1)$-cell of $K'-L'$. We already know that $K'-L'$ is a circuit (No. 29) and it remains to be shown that E'_{n-1} is incident with two n-cells of $K'-L'$.

In the first place $E'_{n-1} \not\subset G'$, else $E_{n-1} \subset G$, which contradicts No. 32 Corollary II. Let then Q' be a point of E'_{n-1} not on G'. Q is on an n-cell on $K-L$ which is an N^Q. For when it is on an n-cell of the circuit that is obvious. When it is not on such a cell, it is on an E_{n-1} and that E_{n-1} plus the two incident n-cells form a configuration homeomorphic to one, $\mathfrak{R}$, of same structure on S_n. But $\mathfrak{R}$ is a convex polyhedral region of S_n, hence an n-cell, and so is the configuration on $K-L$.

Passing then to $K'-L'$, Q' is on an n-cell, on $K'-L'$ which is an $N^{Q'}$ for $K'-L'$. It follows that every n-cell through Q' on $N^{Q'}$ is likewise a neighborhood for Q' (No. 35). But if E'_{n-1}, the cell of $K'-L'$ that carries Q', is incident with more than two n-cells of $K'-L'$, $N^{Q'}$ carries an E_n on only two of the n-cells of $K'-L'$ just mentioned and hence assuredly not a neighborhood of Q'. It follows that E'_{n-1} is incident with only two n-cells of $K'-L'$.

38. **Invariant properties of cells and spheres.** We list here a number of properties of cells and spheres which are rapid consequences of some of the theorems in this and the preceding section.

I. *The dimension n of E_n or H_n is a topological invariant.*

For we can take as representative of two homeomorphic cells E_n, $E_{n'}$, simplexes σ_n, $\sigma_{n'}$. Since these are regions of spaces S_n, $S_{n'}$, we must have $n = n'$.

As for H_n, the theorem follows from the similar one for K_n.

II. *Every H_n is an absolute orientable simple circuit without torsion coefficients whose Betti numbers are*

$$R_p(H_n) = \begin{cases} 0 & \text{for } p \neq 0, n, \\ 1 & \text{for } p = 0, n. \end{cases}$$

For $F(\sigma_n)$ has these properties and since they are invariant they also belong to H_n.

III. *Let* $E_n = K - L$, *where* K, L *are as in* § 5. *Then* E_n *is an absolute orientable circuit mod* L *without torsion coefficients, the Betti numbers being*

$$R_p(E_n; L) = \begin{cases} 0 & \text{for } p < n, \\ 1 & \text{for } p = n. \end{cases}$$

For these properties are invariant and belong to σ_n.

§ 7. Extension of Boundary Relations

39. Let K be a circuit mod its subcomplex K_{n-1} and K_{n-1} a circuit mod its subcomplex L. Denote by Γ_n, Γ_{n-1} cycles of K mod K_{n-1} and of K_{n-1} mod L which form bases for their types.

$F(\Gamma_n)$ is an $(n-1)$-cycle whose $(n-1)$-cells are all cells of K_{n-1}. We can write

$$F(\Gamma_n) = C_{n-1} + C'_{n-1}$$

where C_{n-1} is a subchain of K_{n-1} and C'_{n-1} is a subchain of L. Since $F(\Gamma_n)$ has no boundary, $F(C_{n-1}) = -F(C'_{n-1})$. Hence C_{n-1} is a cycle mod L. But every such cycle, subchain of K_{n-1}, is a multiple of Γ_{n-1} mod L. We may assume that Γ_{n-1} has no $(n-1)$-cell on L since these cells may be neglected. Then $C_{n-1} = \eta \Gamma_{n-1}$. Finally

$$\Gamma_n \rightarrow \eta \Gamma_{n-1} \quad \text{mod } L.$$

The integer η *has an important property of invariance.* For let us break up the point set K into the cells of a new complex K^* in such manner that the various complexes and chains become individually similar elements designated by the same letters starred. Thus K_{n-1} has its point set decomposed into the cells of a new complex K^*_{n-1}, etc. We assume Γ^*_n, Γ^*_{n-1}

oriented like the associated non-starred chains in accordance with No. 29. We shall have

$$\Gamma_n^* \sim \Gamma_n \bmod K_{n-1}, \qquad \Gamma_{n-1} \sim \Gamma_{n-1}^* \bmod L,$$
$$\Gamma_n^* \to \eta^* \, \Gamma_{n-1}^* \bmod L.$$

Let us subdivide the starred complexes so as to make it possible to apply the deformation theorem, then deform them into subcomplexes of K. Denote the new complexes by the same symbols as the old. The above relations will continue to hold. This time Γ_n^*, subchain of K, being homologous to $\Gamma_n \bmod K_{n-1}$, coincides with that cycle and similarly for Γ_{n-1}^* and Γ_{n-1}. Therefore

$$\Gamma_n \to \eta^* \, \Gamma_{n-1} \bmod L.$$

Since the coefficient of every cell of Γ_{n-1}, and hence that of Γ_{n-1} itself, in such a boundary relation is unique, $\eta^* = \eta$. Therefore, η depends not so much upon the complexes K, K_{n-1}, L as upon their total configuration.

40. What we have just brought out can be extended at once to the decomposition of K_{n-1} into q circuits $\bmod L$, say $K_{n-1}^1, \cdots, K_{n-1}^q$. If Γ_{n-1}^i is a base cycle for K_{n-1}^i we find

$$\Gamma_n \to \sum \eta_i \, \Gamma_{n-1}^i \bmod L,$$

the η's being invariant in the same sense as above.

A decomposition such as the above is always possible; for example, by taking as K^i's the closed $(n-1)$-cells of K_{n-1}.

41. The most interesting case is when $K - K_{n-1}$ and the sets $K_{n-1}^i - L$ are all cells. We designate the corresponding cycles by E_n, E_{n-1}^i. Each is a single circuit mod its boundary. Thus E_n consists of a sum of σ_n's two and only two of them incident with any non-boundary σ_{n-1} and similarly for the cells E_{n-1}^i. The boundary relations now take the usual form

$$E_n \to \sum \eta_i \, E_{n-1}^i$$

with the important modification, however, that the η's need not be ± 1 or 0, but may assume arbitrary integral values.

Furthermore η_i may well be zero even when E_{n-1}^i is on the boundary of E_n.

EXAMPLE. Consider the customary decomposition of the anchor ring into a rectangular surface with coincident opposite sides. We have here an E_2, the surface of the rectangle, and a pair of one-cells, E_1^1, E_1^2, on its boundary. However each one-cell appears in $F(E_2)$ twice, with opposite orientations, hence

$$E_2 \to 0$$

that is $\eta_1 = \eta_2 = 0$.

If two of the opposite sides of the initial rectangle are matched in such manner as to have a non-orientable Möbius strip we shall have

$$E_2 \to E_1^1 + E_1^2 + 2\,E_1^3$$

where E_1^3 is the cell with which the two sides coincide, and E_1^1, E_1^2, are the free sides of the rectangle, or borders of the strip with one point out.

Let K be the actual homeomorph of a $\bar{\sigma}_n$, with $K_{n-1} = H_{n-1}$ as the homeomorph of the boundary sphere, and $K - K_{n-1} = E_n$ as the oriented cell. We have here

$$E_n \to \eta\, H_{n-1}.$$

Since $\eta = \pm 1$ for the simplex, its value is the same here, or

$$E_n \to \pm H_{n-1}.$$

When the sign is $+$, cell and sphere are positively related, they are negatively related otherwise. If E_{n-1}^i are the cells of a decomposition of H_{n-1} and

$$H_{n-1} = \sum \eta_i\, E_{n-1}^i, \qquad \eta_i = \pm 1,$$

we have

$$E_n \to \sum \eta_i\, E_{n-1}^i$$

and E_n, E_{n-1}^i are positively or negatively related accordingly as $\eta_i = +1$ or -1. Here of course the cells E_{n-1}^i need not be simplexes.

§ 8. Combinatorial Complexes

42. The preceding results point to the possibility of extending the scope of our combinatorial machinery. To return again to the anchor ring, when our initial rectangle is turned into the ring we have a system consisting of the cells E_0, E_1^1, E_1^2, E_2, with the boundary relations

$$E_0 \to 0, \ E_1^1 \to 0, \ E_1^2 \to 0, \ E_2 \to 0,$$

whose simplicity is striking. It would be adequate in every way were we assured that all homology relations on the ring are reducible to homology relations between subchains of the above four cells. In short what is wanted is some analogue of Alexander's deformation theorem for the system. Furthermore the generalized aggregate of cells should be susceptible of simplicial subdivision, since we can make but little progress otherwise.

What is the broadest class of cellular aggregates having these properties? No answer can be provided to the question at the present time. The best that we are able to do, and this covers all practical needs at the present stage of development of the science, is to provide a degree of generality sufficient for a well rounded out theory of manifolds.

43. **Combinatorial elements (Alexander).** Our first move is to extend the notions of cell and sphere. To allow the boundary relations of § 7 free play the new elements must be circuits. But that is not sufficient and would not enable us, for example, to get a suitable extended deformation theorem for chains on the complex. The correct hint is found in a far reaching remark made to us several years ago (1924) by Alexander and followed up recently with considerable success by van Kampen [3]. Another important point in this connection is that given a simplicial K we can ascertain by a purely combinatorial (finite) procedure whether it possesses the homology characters of cells or spheres. On the contrary to determine whether it is a true cell or sphere is a transcendental and as yet unsolved problem. In the

whole of combinatorial topology scarcely any other properties of cells and spheres were used or needed than the values of their homology characters. As an illustration we may refer to the developments of § 7. This indicates that we may go far without having occasion to call upon the deeper properties of cells or spheres. Thus since the whole structure deals exclusively with homology and circuit characters, homotopy and isotopy are in the nature of luxuries, and homological deformations amply sufficient. This suggests the following definitions:

Combinatorial n-cell. For $n=0$ it consists of a single point. For $n>0$ it is a relative simplicial circuit $K-L$ whose homology characters mod L are as for the n-simplex. It has no torsion coefficients mod L and

$$R_p(K;L)=\begin{cases}0 \text{ for } p \neq n,\\ 1 \text{ for } p=n.\end{cases}$$

This implies in particular that $K-L$ is an orientable circuit, but not, for example, that it is a simple circuit.

Combinatorial n-sphere. For $n=0$ it consists of two points. For $n>0$ it is a simplicial absolute circuit K, with the homology characters of the n-sphere. There are no torsion coefficients and

$$R_p(K)=\begin{cases}0 \text{ for } p \neq 0, n,\\ 1 \text{ for } p=0, n.\end{cases}$$

The implication here also is that K is an absolute orientable circuit, not necessarily simple.

Ordinary cells and spheres are also combinatorial, and conversely, for $n<3$. Poinaré [7], however, has given an example of a combinatorial H_3 which is not an ordinary sphere and from this we can construct any number of others (No. 48).

Since the homology characters and circuit properties are both subdivision and topological invariants, combinatorial cells and spheres are configurations invariant topologically and under mere subdivision.

Combinatorial n-complex. An aggregate K of a finite number of non intersecting combinatorial cells of various dimensions up to and including n, such that: (a) The closure of any k-cell of the set consists of that cell plus cells of the set whose dimension $<k$; (b) K possesses a simplicial subdivision.

In other words K is merely an ordinary simplicial complex whose simplexes have been regrouped into new aggregates, the combinatorial cells of K. Observe that (a) implies that the boundary of every cell of K is a sum of cells of K.

Combinatorial chains. The definition is as usual except that the cells are now combinatorial. Let C_p be a combinatorial p-subchain of K, E_p^i the p-cells of the latter, K' a simplicial subdivision of K, E_p^{ij} the cells of the subdivision induced by it on E_p^i. The cell E_p^i, oriented, stands as in No. 41 for a certain chain $\sum x_{ij} E_p^{ij}$ and

$$C_p = \sum y_i E_p^i$$

stands for the chain $\sum y_i x_{ij} E_p^{ij} = C'_p$. The chain C'_p is the simplicial subdivision induced on C_p by K'.

It is scarcely necessary to point out that combinatorial complexes include all types previously examined.

We may also introduce combinatorial elements mod m, mod L or mod (L, m), i. e. as above except with homology characters mod m, etc. In particular the elements mod 2 are sufficient in every question where orientation may be disregarded.

44. The formal extension of the homology theory on the combinatorial side is based on § 7. To prove the topological invariance of the new combinatorial characters we need a deformation theorem. In its statement there will occur a new operation on C_p, that of *cell annexion,* which is merely the inverse of subdivision. More precisely C_p is said to be derived from C'_p by cell annexion whenever C'_p is a subdivision of C_p. This is, for example, the case for the chains so designated in the preceding number.

45. Deformation Theorem for Combinatorial Complexes. *Every chain on K is reducible to a subchain of K by the*

combined operations of subdivision, cell annexion, homological deformation. Throughout the deformation no point leaves its closed cell on K.

The theorem being obvious for a K_0, and for a C_0 on K_n, we will prove it by induction. Let us assume it proved for K_{n-i}, $i = 1, 2, \cdots, n$, and all p; and also proved for K_n and C_{p-i}, $i = 1, 2, \cdots, p$. The induction will be complete if we can prove it for K_n and C_p.

a) $p < n$. Let K_{n-1} denote the complex composed of all cells of K_n of dimensionality lower than n. Now $F(C_p)$ can be deformed onto K_{n-1}, by hypothesis, since the dimensionality of $F(C_p)$ is less than p. It has been proved that the chain obtained by adding the deformation chain of $F(C_p)$ to C_p can be obtained by deforming C_p on the sum (No. 10). The boundary of the new C_p will then be on K_{n-1}. To save notation let us suppose that $F(C_p)$ is on K_{n-1}.

By assumption C_p and K_n have simplicial subdivisions. Hence by the deformation theorem for simplicial complexes, C_p can be deformed into a subchain C_p' of a subdivision K_n' of K_n, in such a manner that $F(C_p)$ remains on K_{n-1}.

Let L^i be the point-set boundary of a given n-cell E_n^i of K_n, C_p^i the chain obtained by removing from C_p' the terms corresponding to cells not on E_n^i. Then C_p^i is a cycle mod L^i since $F(C_p')$ is on K_{n-1}. Therefore

$$C_p^i \sim 0 \bmod L^i \text{ on } \bar{E}_n^i,$$

as follows from the definition of combinatorial cell. Hence there exists a simplicial chain D_p^i on L^i such that

$$C_p^i \sim D_p^i \text{ on } \bar{E}_n^i.$$

It has been proved that under these circumstances C_p^i can be deformed homologically into the chain D_p^i (No. 11).

Now this process, applied to every n-cell of K_n, will convert C_p' into a simplicial chain on K_{n-1}. This chain can be deformed into a subchain of K_{n-1}, by hypothesis of the induction, for $(n-1)$-complexes. Therefore we have reduced C_p to a subchain of K_n.

b) $p \geqq n$. The reduction to a subchain of K_n is as before, hence for $p > n$, C_p is reduced to zero. When $p = n$ we find this time that C_n^i is a multiple of the unique fundamental cycle on E_n^i. As it is precisely that cycle which we designate by E_n^i (No. 41), we have

$$C_n^i \sim t_i E_n^i \mod L^i.$$

But since L^i is of dimension $< n$, and C_n^i is a subchain of the simplicial subdivision induced by K_n' on E_n^i, C_n^i coincides with $t_i E_n^i$ and C_n' with $\sum t_i E_n^i$, so that the initial chain has been reduced to a subchain of K_n, even for $p = n$.

In the reduction to a subchain of K_n' no point leaves its closed cell on K_n', hence *a fortiori* not its closed cell on K_n itself. The subsequent operations involve only deformations on K_{n-1} or else on the cells E_n^i, both of the proper type. Therefore the deformations never displace any point from its closed cell on K_n and the proof of the theorem is complete.

From this theorem follows at once as before the *invariance of the various combinatorial homology characters computed as if one were dealing with ordinary cells.*

46. **Further properties of combinatorial elements.** In the sequel the only combinatorial cells encountered will have combinatorial spheres, of one dimension less, for boundaries. A simple property of such cells is that in their definition absolute invariants may replace relative invariants except that now $R_0(K) = 1$, $R_n(K) = 0$. The circuit conditions are of course as before. For let Γ_r be an absolute cycle with $r \neq 0$. Under our definition $\Gamma_r \sim \Gamma_r' \subset H_{n-1}$ for $r \neq n$. If $r < n-1, \Gamma_r' \sim 0$ on H_{n-1}, hence $\Gamma_r \sim 0$ on $\bar{E}_n$. For $r = n-1, \Gamma_r' = tH_{n-1} = tF(E_n) \sim 0$. Since E_n is a circuit mod L every n-cycle on it $\sim$ a multiple of a chain whose boundary is $H_{n-1} \neq 0$. The reduced chain can only be an absolute cycle if it is $\equiv 0$, hence the initial absolute n-cycle ~ 0. Thus $\Gamma_r \sim 0$ for every $r > 0$, the case $r > n$

being automatically disposed of, since we are on an n-complex, so that the absolute homology characters are zero for $r \neq 0$.

Conversely let every absolute $\Gamma_r \sim 0$ on $\bar{E}_n$ for every $r \neq 0$. If Δ_r is a cycle mod H_{n-1} and $\gamma_{r-1} = F(\Delta_r)$, we have for $r \neq n$, $\gamma_{r-1} \sim 0$ on H_{n-1}. This is true even when $r-1 = 0$, for γ_{r-1} will be of type $\sum t_i A^i$, where the A's are points and $\sum t_i = 0$. If A is a fixed point on the sphere and l^i an arc whose end points are A, A^i,

$$\sum t_i l^i \rightarrow \sum t_i (A^i - A) = \sum t_i A^i \sim 0 \text{ on } H_{n-1}.$$

Thus for every $r \neq 0$ or n, $H_{n-1} \supset C_r \rightarrow \gamma_{r-1}$ and $\Delta_r - C_r \rightarrow 0$. This absolute cycle of dimension $\neq 0$ is by assumption ~ 0 on $\bar{E}_n$, hence $\Delta_r - C_r \sim 0$ on $\bar{E}_n$, $\Delta_r \sim 0$ mod H_{n-1} and the relative numbers mentioned in the first definition are actually zero except that $R_n\ (K;\ L) = 1$ from the orientable circuit conditions. This proves the equivalence of the two definitions.

47. **The joins of complexes.** There exists a noteworthy procedure for deriving new cells and spheres from given cells or spheres. We consider it at length as it occurs frequently later. Thus from Poincaré's combinatorial sphere (No. 43) we shall be able to deduce other combinatorial elements that are not homeomorphic to ordinary elements.

Let first K, K' be two simplicial complexes, for the present identified with their representations in some S_n and so chosen that segments with end points on K and K' do not meet unless they coincide. If A, A' are two configurations on K, K' any homeomorph of the locus of all closed segments PP', $P \subset A$, $P' \subset A'$, will be called the *join* of A, A' and designated by (A, A'). Thus from two configurations we obtain a third by a very simple construction first utilized in topology by Poincaré [4], later also by Newman [1]. There is a close relationship with the *products* of configurations studied at length in Ch. V.

The join (σ_i, σ_j') of two closed oriented simplexes $\bar{\sigma}_i$, $\bar{\sigma}_j'$, is the closed $(i+j+1)$-simplex that we have previously designated by $\overline{\sigma_i \sigma_j'}$. If σ_i, σ_j' are oriented as written the same is to be true by definition for their join.

The join of two subchains of K, K':

$$C_p = \sum x_i \sigma_p^i, \qquad C'_q = \sum y_j \sigma_q'^j$$

is defined as the $(p+q+1)$-chain

$$(C_p, C'_q) = \sum x_i y_j (\sigma_p^i, \sigma_q'^j).$$

It is easy to verify that when p and q are both >0

$$(8) \qquad (\sigma_p, \sigma'_q) \rightarrow (F(\sigma_p), \sigma'_q) + (-1)^{p+1} (\sigma_p, F(\sigma'_q)).$$

From this follows:

$$(9) \qquad (C_p, C'_q) \rightarrow (F(C_p), C'_q) + (-1)^{p+1} (C_p, F(C'_q)).$$

These boundary relations hold even for p or q zero provided that we make the conventions

$$(F(\sigma_0), \sigma'_q) = \sigma'_q; \qquad (\sigma_p, F(\sigma'_0)) = \sigma_p.$$

In other words if A is a point

$$(9') \qquad \begin{cases} (C_p, A) \rightarrow (F(C_p), A) + (-1)^{p+1} C_p, \\ (A, C'_q) \rightarrow C'_q - (A, F(C'_q)). \end{cases}$$

48. **The joins of closed cells or spheres.** We will show that they are closed cells or spheres. More precisely let for the present H_p stand for a combinatorial sphere, E_p for a combinatorial cell whose boundary is an H_{p-1}. Then symbolically

$$(10) \qquad \begin{aligned} (H_p, \bar{E}'_q) &= (\bar{E}_p, H'_q) = (\bar{E}_p, \bar{E}'_q) \\ &= \bar{E}_{p+q+1}; \quad (H_p, H'_q) = H_{p+q+1}. \end{aligned}$$

The proof will rest on the following

THEOREM. *Let $\{\Gamma_h^i\}$, $\{\Gamma_k'^j\}$ be the minimum bases for the absolute cycles of two complexes K, K' without torsion coefficients. The cycles $(\Gamma_h^i, \Gamma_k'^j)$, $h+k=r-1$, constitute a minimum base for the r-cycles of (K, K') mod $(K+K')$.*

The proof of this proposition is word for word the same as that of an important similar theorem treated in Ch. V Nos. 11, 13, with (9) taking the place of formula (6) of Ch. V in the proof. There is then no necessity for giving it explicitly here.

The application to (10) is immediate. It will be sufficient to consider one of the joins, say (H_p, H_q'). The minimum bases for the relative cycles of all dimensions will be composed of the cycles (H_p, H_q'), (A, H_q'), (H_p, A'), (A, A'), where A, A' are vertices of H_p, H_q'. Of these chains (H_p, H_q') is an absolute cycle, and the others are not. An absolute cycle Γ_r, $r > 0$, of (H, H') will be a linear combination of the above chains plus subchains of H and H'. Now we have by (9')

$$(H_p, A') \rightarrow (-1)^{p+1} H_p \sim 0; \quad (A, H_q') \rightarrow H_q' \sim 0.$$

If we compute $F(\Gamma_r)$ by means of (9) or (9'), express the fact that it is $\equiv 0$, and make use of the preceding homologies, we find

$$r = p + q + 1, \quad \Gamma_{p+q+1} \sim t(H_p, H_q').$$

Hence the multiples of the absolute cycle (H_p, H_q') are the only non-bounding cycles of dimension $\neq 0$. Since (H_p, H_q') is connected, its $R_0 = 1$. Therefore the homology characters of (H_p, H_q') are as for an H_{p+q+1}. By construction it is the sum of its closed $(p+q+1)$-cells, hence it is an absolute circuit and therefore a sphere.

49. There remains the question of boundaries. Referring to (9) or else directly we find that, orientations disregarded:

I. When E_q' has for boundary an H_{q-1}', the cell $(H_p, \bar{E}_q')$ has for boundary the sphere (H_p, H_{q-1}').

II. When E_p, E_q' have for boundaries H_{p-1}, H_{q-1}', that of their join is

$$(H_{p-1}, \bar{E}_q') + (\bar{E}_p, H_{q-1}') = H_{p+q}, \tag{11}$$

a symbolic relation analogous to (10) and of interest for its own sake. The left hand side consists of two combinatorial cells with a common boundary sphere. To prove (11) it is therefore sufficient to show that if E_r, E_r' are combinatorial cells with a common boundary combinatorial sphere H_{r-1},

(12) $$E_r + E'_r + H_{r-1} = H_r,$$

a combinatorial sphere. The fundamental incidence relations for H_r are those of H_{r-1} plus

$$E_r \to H_{r-1}, \qquad E'_r \to H_{r-1}.$$

Hence H_{r-1}, the only cycle of dimension >0, $\not\sim 0$ on itself, is ~ 0 on H_r. In addition H_r possesses a unique base subcycle $E_r - E'_r$ of the maximum dimension. Hence its homology characters are those of the sphere. As condition (a) of Ch. I No. 35 for the circuit, is immediately verified, H_r is a sphere, and hence both (11) and (12) are proved. This also completes the proof of (10).

Remark. It is possible to invert in a sense the result just proved when one of the joined elements is a point A. Namely if (A, K) is a closed E_{p+1} whose boundary is K, K is an H_p. For if σ is a generic cell of K, the incidence relations for the cells (A, σ) of the join are the same as for the cells σ of K. Referring then to the definitions of combinatorial cells and spheres the required result follows.

It is scarcely necessary to point out that all these proofs are greatly simplified when one of the elements joined is an ordinary cell or sphere.

50. **Normal cells and complexes.** The new complexes that we have introduced are actually too general for the theory of manifolds. In that theory no advance can be made unless K is susceptible of being regularly subdivided—a property which, so far as we know, combinatorial complexes do not necessarily possess.

The minimal type that manifolds demand is the *combinatorial normal complex*, alone considered in the sequel. It is a combinatorial K such that every one of its cells is the join of a point and a sphere. Cells of that type are said to be *normal*. They are essentially clusters of simplexes with a common vertex. K has a simplicial first derived K' obtained directly by the second construction of Ch. I No. 58 which may be applied here without modification.

By simplicial subdivision of K we shall always understand a simplicial subdivision of its first derived K'.

In the future we simplify the language by omitting, so far as possible, the word "combinatorial" for cells, complexes, manifolds, chains.

51. An interesting type of ordinary complex belongs to the class just introduced. It is the type considered exclusively by Veblen in his Colloquium Lectures [5] and defined as a closed aggregate of non-intersecting cells whose boundaries are spheres (an H_{p-1} for p-cells). Thus simplicial subdivision is not imposed *a priori*. It can be proved at once however that *such a K possesses a simplicial regular subdivision K'*. The construction of K' is carried out by induction as for the simplicial type (Ch. I, No. 58). The only difference is that the rays drawn, say on E_p, from the new vertex on it, to the points of its boundary H_{p-1}, are images of segments on the basic σ_p corresponding to E_p under the definition of K. It follows that K is a special type of combinatorial complex.

We can attach to a polyhedral image of K' a straightness and a metric for K. A segment is then uniquely determined whenever its end points are on the same closed simplex of K'.

Another interesting property which follows from the existence of K', is that *K is uniquely determined when the incidences of its cells are specified*—or if we prefer, when its boundary matrices mod 2 are known. For then those of K' are uniquely determined, hence if two K's have the same structure, so have their K''s, the latter are homeomorphic (Ch. I, No. 9) and so are the given complexes.

CHAPTER III

MANIFOLDS AND THEIR DUALITY THEOREMS

The complexes which have been considered in the two preceding chapters were not specialized in any manner. If we take however, say a polyhedral region $\mathfrak{R}$ of S_n and decompose it into simplexes so as to obtain a K_n, we find that the stars of all cells of K interior to $\mathfrak{R}$ are themselves cells. A similar observation holds for Riemann surfaces and their generalizations (in certain important cases). This suggests that complexes, or parts of complexes, whose structures are more or less of the above type, deserve special attention. They are the *manifolds* (absolute or relative) to which we largely confine our attention henceforth.

The present Chapter is strictly combinatorial in scope, and the true topological questions raised will be taken up in Chapter VII. We first make a thorough study of the general theory of manifolds, then take up their duality theory. This is a live question and it may not be amiss to recall its present status (1929). There were at first in existence three unrelated groups of relations:

(a) Poincaré's formulas for the non-modular homology characters of absolute manifolds (manifolds without boundary) [2, 3, 4] and their modular extensions by Veblen-Alexander [22] and Alexander [19].

(b) Alexander's duality relation [8] between the Betti numbers mod 2 of $L \subset H_n$ and those of $H_n - L$.

(c) A duality relation for characters similar to the Betti numbers but referring to the dependence between the cycles of a manifold and those of its boundary, Lefschetz [11].

Various extensions have been given in recent years by Alexandroff [6, 7, 8, 10], Pontrjagin [1], Frankl [1], van Kampen [1], Morse [4], Lefschetz [12, 14]. In particular, we established the important fact that the two apparently unrelated relations of types (a), (b) for Betti numbers are actually both special cases of a third between absolute and relative invariants. We continue and carry here to completion this work of unification leading to very comprehensive duality theorems. For Betti numbers and the more general characters mentioned in (c) we have now two basic duality relations, (7), (20), of which all known duality relations are special cases: groups (a), (b) fall both under (7), group (c) under (20).

In method we cleave as nearly as possible to the scheme used by Poincaré, Veblen and Alexander for proving relations (a). The chief weapon is a generalized dual complex recently introduced by ourselves [15] and likewise considered by W. A. Mayer [1].

Following Alexander, and more recently van Kampen [3], we have combinatorial cells and spheres throughout, which alone makes it possible to prove the topological invariance of manifolds (van Kampen [3]). Except for that all our results will hold for ordinary cells and spheres; the passage from one type to the other requires but minor modifications in some proofs.

§ 1. The Structure of the Stars of a Normal Complex

1. Taking a normal n-complex K (Ch. II, No. 50), let us first point out a simple but most useful property of its cells: *If the star of E_p includes an E_q it includes cells of all intermediary dimensions from $p+1$ up to q.* This is true for $n=1$, suppose that it holds for $n-1$. Remove the n-cells from K, leaving a K_{n-1}. The only case to be considered is that of an E_p of K_{n-1}, with $p<n-1$, on the boundary H_{n-1} of an E_n of K. Our affirmation will evidently follow if we show that E_p is on the boundary of some E_{n-1} of H_{n-1}, hence of K. But that is a consequence of the fact that H_{n-1} is an abso-

lute $(n-1)$-circuit, hence all its points are on the sum of its closed $(n-1)$-cells.

2. The first derived complex K' of K will play a most important part in the sequel. It is a simplicial complex with a single vertex on each cell of K (Ch. II, No. 50). The cell being designated generically by E_p, we shall denote the vertex of K' on it by A^p. In the rare cases where there are several p-cells, E_p^i, $i = 1, 2, \cdots$, we can designate the corresponding A's by A^{ip}.

From the fact that K is normal plus an elementary induction based on the second construction given for K' in Ch. I, No. 58, we find, that the cells of K' are represented without omission or duplication by all symbols $A^p A^q \cdots A^r$, $p < q \cdots < r$, such that the carrier cells $E_p, E_q, \cdots, E_r$ are mutually incident. The last one, E_r, carries the cell of K' corresponding to the above symbol.

3. From the cell symbol it follows that if a cell of K' has just one vertex on E_p, that vertex can only be A^p and the cell symbol must be of type $A^p \cdots A^r$, hence in particular its dimension $\leqq r - p$. There exists such a cell for every sequence $E_p, E_q, \cdots, E_r$ of mutually incident cells with E_p as its first member. From No. 1, we learn in fact that there always exists a sequence $E_p, E_{p+1}, \cdots, E_r$ for any cell E_r of the star of E_p. Therefore if π is the maximum dimension of any cell of the star, $\pi - p$ is that of any cell $A^p \cdots A^r$ of K with the sole vertex A^p on $\bar{E}_p$, and this maximum is actually reached. The sum of all these cells is an open subcomplex of K' meeting E_p at a single point, which must be A^p. This open $(\pi - p)$-subcomplex is called *transverse complex* of E_p and designated by $Tv(E_p)$, or $Tv(E_p)_K$ if we wish to indicate the complex K relatively to which it is taken. The sum of the cells of $\overline{Tv}$ which do not have A^p for vertex is a $K_{\pi-p-1}$ called the *linked complex* of E_p and designated by $Lk(E_p)$ or $Lk(E_p)_K$. *The closure of the transverse complex is the join of A^p and the linked complex.*

When we change the vertices of K' on the carrying cells of K, Lk and Tv are changed but their structure is not.

Hence their topological characteristics depend solely upon the star of E_p.

The case of most importance is when Lk is a sphere and Tv a cell. We return to it later.

The situation is clarified by examining what happens for example when K consists of a set of $\bar{\sigma}_3$'s of S_3 having a common edge PQ of which they constitute a complete neighborhood (closed) relatively to S_3. It is then easily verified that Lk is a simple closed polygonal line surrounding PQ, i. e. *linking* the *line* PQ, while Tv is a polyhedral 2-cell meeting PQ at a single point, or 2-cell *transverse* to the one-cell PQ. This justifies also the terms "linked, transverse" applied to the complexes.

4. *When K is simplicial and no two of its simplexes have the same vertices, it is not necessary to have recourse to the derived complex K'.* Let in fact E_q be a cell of the star of E_p and E_{q-p-1} its face opposite E_p. The two sets of vertices A^q, A^{q-p-1}, of K' are in one-to-one correspondence. Let $A^q \cdots A^s$ be a cell of K' determined by the vertices on the mutually incident cells $E_q, \cdots, E_s$ of the star E_p. Then the cells $E_{q-p-1}, \cdots, E_{s-p-1}$, opposite to E_p on them, constitute a mutually incident set. Therefore $A^{q-p-1} \cdots A^{s-p-1}$ is a cell of K' of the same dimension as $A^q \cdots A^s$. Hence the subcomplex of K' whose vertices are the points A^{q-p-1} has the same structure as the one whose vertices are the points A^q, i. e. as the linked complex. It is however merely a regular subdivision of the subcomplex K^*_{n-p-1} of K, sum of the cells E_{q-p-1}, or maximal subcomplex of the closed star of E_p having no vertex in common with E_p.

There is an interesting application of what we have just found to the case when K is not simplicial. The first derived K' is then of the type just considered. Let E'_p be a cell of K' on an E_p of K, hence a cell of type $A^0 A^1 \cdots A^p$. If we form the complex K^*_{n-p-1} with respect to E'_p and K', we find that it is exactly the linked complex of E_p relatively to K. By the preceding paragraph the first derived complex of K^*_{n-p-1} has the same structure as the linked complex of E'_p relative to K', or: *The linked complex of E'_p relatively to K' has the same structure as a first derived complex of the linked complex of E_p relatively to K.*

Returning to the simplicial case, we can prove, exactly as above, that the transverse complex has the structure of a subdivision of the maximal subcomplex of the closed star having just one given vertex in common with E_p.

The same results can be obtained by a more geometric procedure. If P is any point of $Lk(E_p)$ there is a unique σ_{p+1} on K with P as vertex and E_p as its face. The notations being as above, if P is on E_q, the space of the simplex meets E_{q-p-1}, hence K^*_{n-p-1}, at a unique point P'. The correspondence between P and P' is precisely a homeomorphism between Lk and K^*_{n-p-1}. The join of a fixed vertex of E_p and K^* generates the above substitute for the closure of the transverse complex, their homeomorphism being obvious.

§ 2. Definition of Manifolds and their General Properties

5. In the definitions to follow E, H stand for combinatorial cells and spheres, K is our usual combinatorial normal n-complex while L is a subcomplex of K. As we are solely concerned with $K-L$ we shall assume that $K = \overline{K-L}$, i. e. that L has no cells that are not cells of $\overline{K-L}$, a convention to which we adhere throughout the present section. Another useful convention will be to introduce configurations of dimension -1: K_{-1}, H_{-1} etc. They all stand for the null set. Their purpose is to simplify the recurrence in the definitions. We may say, for example, if E_n is a cell of K, $Lk(E_n)$ is a K_{-1} or an H_{-1}. The part here played by the null set is wholly analogous to that which it plays in the definition of dimensionality by Urysohn-Menger. Another instance of the same thing occurs later (Ch. IV, § 5).

The generic notation for an n-manifold will be M_n.

Definitions. I. According to our conventions an absolute M_{-1} is merely the null set. $K-L$ is an M_n whenever for every cell E_p of $K-L$:

(a) The boundary sphere of E_p is an absolute M_{p-1}.

(b) $Lk(E_p)$ is an H_{n-p-1}.

The M_n is *open* or *relative* when $L \neq 0$, *absolute* when $L = 0$. In the first case L is its *boundary.*

II. E_p is a *regular* cell of K whenever all cells of its star (and hence E_p itself) satisfy both conditions (a), (b).

III. The boundary L of the open manifold $K-L$ is *regular* whenever all its cells satisfy (a) and

(c) $Lk(E_p) = E_{n-p-1} + H_{n-p-2}$

where H consists of all closed $(n-p-2)$-simplexes of $Lk(E_p)$ incident with a single $(n-p-1)$-simplex of $Lk(E_p)$.

IV. A cell E_p is *semi-regular* whenever it satisfies (a) and (c) and so do some cells of its star, the rest being regular. The star consists then of regular and semi-regular cells, among the latter being E_p and among the former the n-cells.

According to Ch. II, Nos. 48, 49, (b) is equivalent to

(b′) $Tv(E_p)$ is an $(n-p)$-cell whose boundary is a sphere.

It may likewise be shown by considerations analogous to those of Ch. II, No. 49 that (c) ist equivalent to

(c′) $Tv(E_p) = E_{n-p} + E_{n-p-1}$ where E_{n-p-1} is on the boundary of E_{n-p} and consists of all $(n-p-1)$-simplexes of $\bar{E}_{n-p}$ incident with a single $(n-p)$-simplex of E_{n-p} and having no vertex on E_p.

The $(n-p)$-cell attached to every regular E_p by virtue of (b′) is the *dual* of E_p and designated by E^*_{n-p}. These dual cells are of the greatest importance for the sequel.

Referring to No. 3 we see that the regular and semi-regular cell properties are properties of the stars of $K-L$ independent of the choice of K'.

A further classification of manifolds is possible by specifying that the cells and spheres in the definition are to be understood mod m or mod (L, m). We then have manifolds mod m or mod (L, m). In particular, *whenever no questions of orientation arise, cells and spheres may be understood mod 2.*

6. The absolute M_1 consists of a finite number of non-intersecting one-spheres (Jordan curves), the relative M_1 of a certain number of one-cells (Jordan arcs). The sum of their end

points is then L. For $n = 0, 1, 2$, there is no difference between modular or non-modular manifolds. The absolute (orientable) M_2 is merely the homeomorph of a Riemann surface. A more interesting and practical example is that of a σ_{n+1} or a convex polyhedral E_{n+1}. We will prove the

THEOREM. *The convex polyhedral cell E_{n+1} is an open M_{n+1} with its boundary H_n as regular manifold boundary. H_n itself is an absolute M_n.*

The proof, elementary for a σ, is somewhat more complicated for the general convex polyhedral cell. Let E_p be one of its faces, S_{n-p+1} a subspace of the S_{n+1} of E_{n+1}, meeting E_p at a single point A^p. We can construct the first derived K' of E_{n+1} with A^p as a vertex and with its vertices on the cells of the star of E_p all on S_{n-p+1}. In particular, the vertex A^{n+1} on E_{n+1} itself will also be on S_{n-p+1}. As a consequence, the linked and transverse complexes of E_p relatively to $\bar{E}_{n+1}$ or H_n will be the same as those of A^p relatively to their sections by S_{n-p+1}. These sections form a convex E_{n-p+1} and its boundary, with A^p as a vertex. Therefore we have reduced the situation to one of same type but with $p = 0$.

To simplify matters we will then simply go back to the initial situation with E_p replaced by $E_0 = A$. Now if H_n' is the first derived of H_n induced by K', $Tv(A)_H$ is merely the star of A relatively to H_n', and $Lk(A)_H$ is the boundary of Tv. Project $Tv + Lk$ from A^{n+1} onto an S_n meeting the straight line $A^{n+1}A$ at the single point A. The projection of the sum is the closed star of a vertex of a K_n covering a region of S_n. Such a closed star is readily shown to be a normal E_n. Hence the projection of Lk is a sphere and so is Lk itself, since its structure is the same as that of its projection.

Going back to the initial cell E_p it follows that $Lk(E_p)_H$ is an H_{n-p-1} and $Lk(E_p)_{\bar{E}_{n+1}}$, the join of $Lk(E_p)_H$ and A^{n+1} is then a normal cell. Since the cells are all convex, they fulfill condition (a). According to what precedes, every cell of H_n is then regular for H_n, semi-regular for $\bar{E}_{n+1}$. This proves the theorem.

COROLLARY. *For a simplicial or convex $K-L$ the manifold conditions are reduced to (b), (the linked complexes are all spheres).*

This makes it tempting to limit the whole theory to the simplicial or convex types. Unfortunately, as we have already observed, the duals of such complexes are not even convex in general.

7. Various definitions of manifolds have been given equivalent to demanding that the stars be ordinary cells (Poincaré [2], Brouwer [7], Lefschetz [5,6]), but there is as yet no proof of the topological invariance of such manifolds. Veblen [6] defines an absolute manifold by the invariant requirement that every point possess a neighborhood which is an n-cell. However no way is known of reducing this condition to a property depending upon the mutual relations of the cells. All these difficulties disappear when, following Alexander and more recently van Kampen [3], we base the definitions, as above, on the combinatorial elements. In particular topological invariance of the combinatorial manifolds has been proved by van Kampen [3].

In what may be called the strict combinatorial theory, i. e. with cells and spheres derived from σ, $F(\sigma)$, by a series of elementary subdivisions or their inverses, the definition of manifolds assumes an exceptionally simple form, as shown recently by Alexander [21]. For in that theory regularity has to be imposed on the vertices alone and follows then for the other cells.

8. **Elementary properties of regular or semi-regular cells.** From the relation between the star of E_p and $Lk(E_p)$ plus the properties of the joins of cells and spheres (Ch. II, Nos. 48, 49) we find:

I. *When E_p is $\left\{\begin{matrix}\text{regular}\\ \text{semi-regular}\end{matrix}\right\}$ its star is an $\left\{\begin{matrix}E_n\\ E_n \text{ plus an}\end{matrix}\right.$ E_{n-1} of its boundary and the boundary of E_n is an H_{n-1}.*

Since[a] $Lk(E_p)$, $p<n$, is a K_{n-p-1} the highest dimension of the cells incident with E_p is n, that is its star includes an n-cell. Therefore

II. *When K defines an absolute M_n, it is the sum of its closed n-cells. Similarly for the open M_n the sum being then* $\overline{K-L}$.

When E_{n-1} is regular Lk is a zero-sphere, i. e. two points, each corresponding to an incident n-cell. When it is semi-regular the complex is a zero-cell or one point and there is just one incident n-cell.

Let now $p < n-1$, with E_p regular or semi-regular. Any σ_{n-p-2} of Lk has a symbol $A^{p+1} \ldots A^{q-1} A^{q+1} \ldots A^n$, the cells $E_p, E_{p+1}, \cdots$ being mutually incident. If $q < n$, E_{q-1} is a cell of the boundary of E_{q+1} which is an absolute M_q; hence, by the property just brought out, E_{q-1} is a face of two cells E'_q, E''_q of that M_q. If the vertices A'^q, A''^q of K' on the two cells are inserted in the right place in the symbol of σ_{n-p-2} we have two σ_{n-p-1}'s of Lk incident with it. If $q = n$ there may or may not be two cells E'_n, E''_n incident with E_{n-1}, and Lk is an absolute circuit when and only when the first circumstance always occurs, that is, when and only when every E_{n-1} of the star of E_p is regular. Therefore,

III. *A regular $(n-1)$-cell is incident with two n-cells, and when E_p is regular all the incident $(n-1)$-cells have that property.*

IV. *A semi-regular $(n-1)$-cell is incident with a single n-cell. When E_p, $p < n-1$, is semi-regular it is a face of such an $(n-1)$-cell, and conversely.*

V. *Whenever all cells of K are regular or semi-regular, the sum of the semi-regular cells is a subcomplex L of K, and* $\overline{K-L}$ *is an M_n with the regular boundary L.*

When $Lk(E_p) = E_{n-p-1} + H_{n-p-2}$, a cell plus a sphere (its boundary), the $(n-p-2)$-cells on H are characterized by being incident with a single $(n-p-1)$-cell of Lk (No. 5c). Any one of them is then necessarily of type $A^{p+1} \ldots A^{n-1}$, with E_{n-1} a cell of the star of E_p and on the boundary of a single n-cell of K. Hence H is the sum of the closed $(n-p-2)$-cells of the linked complex that are on L itself. H is therefore $Lk(E_p)_L$, hence:

VI. *When K defines an M_n with regular boundary, the latter is an absolute M_{n-1}.*

When E_p is regular or semi-regular its boundary H_{p-1} is an M_{p-1}. Hence if E_q is a face of E_p, $Lk(E_q)_H$ is an H_{p-q-2}. It follows that $Lk(E_q)_{\bar{E}_p}$ which is the join of H_{p-q-2} and the vertex of K' on E_p, is a normal $\bar{E}_{p-q-1}$. Therefore E_q is semi-regular for E_p. Since E_p is regular for E_p we have

VII. *A regular or semi-regular E_p is an M_p having its boundary sphere for regular boundary.*

9. **Invariance under subdivision. (Combinatorial invariance.)** THEOREM. *If E_p of K is regular or semi-regular so is any cell of a simplicial subdivision of K which is on E_p, relatively to its complex.*

The only simplicial subdivisions of a normal K that we undertake to consider are its first derived K' or simplicial subdivisions of K'. Therefore we have to consider the effect of passing from K to K', then from K' to any simplicial subdivision, or which is the same from a simplicial K to any simplicial subdivision.

Since an n-simplex of a K_n is regular and since for an M_1 the theorem is obvious we may operate by double induction on p and n. We assume then the theorem correct up to $n-1$, and also for n with cells of dimension $>p$, where $p<n$, and prove it for p and n.

Let first E'_p of K' be on the regular cell E_p of K. Since $Lk(E'_p)_{K'}$ has the structure of a regular subdivision of $Lk(E_p)_K$, and since the conditions for spheres are invariant under subdivision, $Lk(E'_p)_{K'}$ is a sphere like the other linked complex. Moreover if E'_p is a face of E'_q, E'_q must be on E_r, $r>p$, of E_p. Hence E'_q is regular by the hypotheses of the induction so that E'_p, which is a σ_p, fulfills all conditions for regularity.

10. We will now consider a cell E'_q, $q<p$, of K' on E_p, and use induction on q. We assume then that for dimensions $>q$ (and $\leqq p$) the cells are regular. We have then $E'_q = A^r A^{r+1} \cdots A^s A^t \cdots A^p$, where $E_r, \cdots, E_p$ are mutually incident. In the sequence $r, \cdots, p$, there may be several gaps: we have explicitly indicated one between s and t. The star of E'_q is the sum of the closed n-cells:

$$(A^0 \ldots) A^r A^{r+1} \ldots A^s (A^{s+1} \ldots A^{t-1}) A^t \ldots A^p (A^{p+1} \ldots A^n),$$

the parentheses being filled with vertices in all possible ways, such that the $n+1$ cells of K corresponding to the total set are all mutually incident. Then $Lk(E'_q)_{K'}$ has the structure of a simplicial subdivision of the sum of all closed $(n-q-1)$-simplexes obtained by writing down *all* vertices in parentheses. Let K_{n-p-1} be that sum. It is obtained as follows: (a) Construct the complex K^i sum of all simplexes obtained by filling in the i-th parenthesis, for example all simplexes $(A^{s+1} \ldots A^{t-1})$. (b) Take the join (K^1, K^2) then the join of that complex and K^3 etc. To show that the ultimate complex, and hence that $Lk(E'_q)_{K'}$ is a sphere it is sufficient to show that every K^i is a sphere. As for the last it is merely $Lk(E_p)_K$, which is a sphere since E_p is a regular cell of K. Consider now any K^i, for example the sum of all cells $(A^{s+1} \ldots A^{t-1})$. Let H_{t-1} be the boundary sphere of E_t, or rather the complex into which it is decomposed by the cells of K' on it. Then K^i is the homeomorph of $Lk(E_s)_{H_{t-1}}$ (its substitute in the sense of No. 4 for a simplicial complex). Since H_{t-1} is an M_{t-1}, K^i is a sphere.

We have then shown that $Lk(E'_q)_{K'}$ is a sphere. Now if E'_r of K' is a cell of the star of E'_q, it is on E_p or else on some cell of the star of E_p. Hence by the various hypotheses governing our induction, E'_r is regular. It follows as before that E_p is regular.

11. To complete the proof we have to establish invariance when we pass from any simplicial K to any simplicial subdivision K^0. Actually we shall prove the same result for the case where K, K^0 are convex.

Let E^0_q of K^0 be on E_p of K. The star of E^0_q relatively to K^0 is made up of cells that are on those of the star of E_p relatively to K, i. e. of cells that are subdivisions of regular cells of K. Since the cells of K^0 are convex, referring to No. 6 we find that we merely have to show that $Lk(E^0_q)_{K^0}$ is an H_{n-q-1}. Now we have shown (Ch. I, No. 61) that K and K^0 have a common subdivision by section K^1, obtained by successive sections of cells one at a time. We shall then compare K^1 with K and K^0, and this will yield the required result.

Let us first assume that K^1 is obtained from K by section of the single cell E_p. The operation consists in introducing a new convex E'_{p-1} on E_p, whose faces are faces of E_p; E'_{p-1} disconnects E_p into two new convex cells E'_p, E''_p and E_p is replaced by $E'_p + E''_p + E'_{p-1}$. The first derived K'^1 of K^1 is to have the same vertices as K' on all cells common to K and K^1. The effect on the linked complexes is as follows: (α) For the cells common to K and K^1, Lk is either unchanged, or if it is, the new Lk has the same structure as the old (for cells that are faces of E_p but not of E'_{p-1}), or else a subdivision having the structure of a subdivision of the old (for the faces of E'_{p-1}). (β) Since the stars of E'_p, E''_p as to K^1 and of E_p as to K coincide except for the p-cells themselves, their Lk's are the same. (γ) As for E'_{p-1}, if A', A'' are the new vertices of K'^1 on E'_p, E''_p we verify directly, with $Lk = Lk(E_p)$,

$$Lk(E'_{p-1})_{K^1} = (A', Lk) + (A'', Lk),$$

the two terms at the right having only Lk as common subcomplex. In particular if E_p is regular it follows from Ch. II, Nos. 48, 49, that $Lk(E'_{p-1})_{K^1}$ is an H_{n-p}.

Let now K^1 be the same as before, and E_p regular. As we take repeated sections it follows from (α), (β), (γ) that only spheres appear among the Lk's of the cells on E_p. Hence the Lk's of the cells of K^1 on E_p are all spheres.

Compare now K^0 with K^1. The cell E^0_q being as before a cell of K^0 on E_p, the repeated application of (α), (β) as between K^0 and its sections, shows that the Lk of some q-cell of K^1 on E^0_q, and hence on E_p, will have the structure of a subdivision of $Lk(E^0_q)$. Hence by the above, the latter is a sphere. This completes the proof of our theorem for regularity.

The treatment of semi-regular cells differs from the preceding merely in insignificant details and need not be taken up here, the more so since it has no application in the sequel.

REMARK. From this point on wherever convenient we shall be free to replace K by any simplicial subdivision, or which is the same, to assume that it is already simplicial. It will be observed that as much follows already from invariance under regular subdivision, without the discussion of the present number.

12. **Orientation.** It follows from No. 8, II and III, that when $M = K - L$ is connected it is a simple circuit. When M consists of $q > 1$ connected sets the subcomplex of K corresponding to each is a simple circuit.

M is *orientable* whenever each of the q connected subsets is orientable mod L, *non-orientable* otherwise. It is *normally* oriented whenever each circuit has been so oriented that its fundamental n-cycle is $\pm$ the sum of its oriented n-cells, or (which is the same thing) in such fashion that the sum of the n-cells is a chain whose boundary consists only of cells of L. That chain is temporarily designated also by M_n.

Normal orientation in the orientable case is possible in 2^q ways. Each mode of normally orienting M_n is determined by assigning an *indicatrix* on every one of the q circuits, that is an oriented simplex on one cell of some simplicial subdivision of the circuit, or rather of the closure of the sum of its cells, which is an n-complex. This being done, we can determine whether any particular oriented σ_n on K, which is on a simplicial subdivision, is oriented concordantly with K, or oppositely to it: We simply compare its orientation with that of the carrying n-cell of K.

If M_n is orientable and its boundary L is regular, it is likewise orientable. In that case $F(M_n)$ is merely the latter normally oriented.

This follows at once from the fact that $F(M_n)$ is a chain without boundary, the sum of the $(n-1)$-cells of L each taken with a coefficient ± 1.

13. **The duals of the regular or semi-regular cells.** Let the cell E_p of K be regular. We have seen (No. 5) that $Tv(E_p) = E^*_{n-p}$, a combinatorial cell, the dual of E_p. E^*_{n-p} is

the sum of all simplexes $A^p A^q \cdots A^s$, such that $E_p, E_q, \cdots, E_s$ are mutually incident cells of K. The boundary of E^*_{n-p} is a sphere—the linked complex $Lk(E_p)$.

Let $\sigma = A^q \cdots A^s$ be an arbitrary cell of Lk. It is uniquely determined by the mutually incident cells $E_p, E_q \cdots, E_s$, and conversely σ determines that set uniquely. Therefore Lk is the sum of cells such as σ. On the other hand σ is also a component of the uniquely determined dual cell E^*_{n-q} of E_q, and all components σ of such dual cells are cells of Lk. Therefore *Lk is the sum of the duals of the cells of the star of $E_p (\neq E_p)$*. The components of any two such duals are distinct—hence no two of them meet. Their sum is closed, since it is the complex Lk. Therefore the E^*_{n-q}'s are the cells of a complex $Lk^*(E_p)$. Since each cell of the latter is an exact sum of cells of Lk (such as σ), it has Lk as subdivision. This subdivision is in fact regular, for Lk has one and only one vertex on each cell of Lk^*. The boundary of E^*_{n-q} is $Lk(E_q)$, i. e. an $(n-q-1)$-sphere. Therefore each closed cell of the new complex is the join of a point and a sphere and hence that new complex is normal.

The situation is essentially the same when E_p is semi-regular. The closed dual $\overline{Tv} = \overline{\mathfrak{E}}_{n-p}$ is not a cell however but the join of a point with a cell and its boundary sphere, hence $\mathfrak{E}$ is the sum of two cells E'^*_{n-p} and E''^*_{n-p-1}, where the first is on $K-L$ and the second on L. If E_q is regular E^*_{n-q} is again a cell on the linked complex, but the linked complex is the sum of these cells plus the cells E'^*, E''^*, corresponding to the semi-regular cells of the star of E_p. There results once more an Lk^* whose regular subdivision is Lk.

14. THEOREM I. *The linked complex of a* $\left\{\begin{matrix}\textit{regular}\\ \textit{semi-regular}\end{matrix}\right\}$ E_p *is an* $\left\{\begin{matrix}\textit{absolute } M_{n-p-1}\\ M_{n-p-1} \textit{ with regular boundary}\end{matrix}\right\}$ *and this property is equivalent with condition* $\left\{\begin{matrix}b\\ c\end{matrix}\right\}$ *of No. 5 for* $\left\{\begin{matrix}\textit{regular}\\ \textit{semi-regular}\end{matrix}\right\}$ *cells, as applied to the cells of the star of E_p other than E_p itself.*

THEOREM II. *Invariance under the inverse of subdivision (cell annexion):* E_p *is* $\begin{Bmatrix} \textit{regular} \\ \textit{semi-regular} \end{Bmatrix}$ *whenever the cells of a subdivision on its star are all* $\begin{cases} \textit{regular} \\ \textit{regular or semi-regular with some} \end{cases}$ *of the latter type on* E_p *itself.*

These two propositions will be proved simultaneously. Their verification both for $n = 1$ and that of Th. I for $p = n - 1$ is trivial. We assume then that both hold for dimensions $< n$, also that Th. I holds for n and q-cells $(q > p)$ and prove them for n and p.

If K is replaced by its first derived K', and E'_p is a cell of K' on E_p, $Lk\,(E'_p)_{K'}$ has the structure of a first derived of $Lk\,(E_p)_K$. Since we may assume $p < n$, to prove Th. I, under the hypotheses of the induction for Th. II, we may replace the second linked complex by the first. In essence this means that to establish Th. I we may take K simplicial.

Under the circumstances the cells of K' that are on $Lk\,(E_p)$ and meet E^*_{n-q} of Lk^* (the complex thus designated in No. 13) at a single point, necessarily A^q, are all of type $A^r \cdots A^s A^q$, the cells $E_p, E_r, \cdots, E_s, E_q$ being mutually incident. The sum of the cells $A^r \cdots A^s$ is the linked complex of the face E_p of the simplex E_q relatively to its boundary H_{q-1} and therefore they constitute a $(q-p-2)$-sphere. But the sum in question is also the linked complex of E^*_{n-q} relatively to Lk^*. Hence the linked complexes of the cells of Lk^* relative to Lk^* are all spheres. The boundaries of these cells have for regular subdivisions the complexes $Lk(E_q)_K$, where E_q is a cell of the star of E_p (here $\neq E_p$) and hence they are absolute manifolds under the hypotheses of the induction. Thus all conditions for regularity are fulfilled and Lk is an absolute M_{n-p-1}. Conversely when that is the case, and since Th. II holds for Lk under the hypotheses of the induction, Lk^* is also an absolute M_{n-p-1}. Therefore the boundaries of its cells are spheres, and as they are the linked complexes of the cells of the star of E_p, condition (b) of No. 5 is fulfilled for that star. This completes the proof of Th. I for regular cells.

15. Regarding Th. II we first take for subdivision of K its first derived K' and $p = n$. Let A^n be a vertex of K' on E_n of K. Then $Lk(A^n)_{K'}$ has for structure that of a first derived of the complex into which K' decomposes $F(E_n)$. Since A^n must be regular that linked complex is an absolute M_{n-1} by Th. I, and therefore under the hypotheses of the induction this is also true for $F(E_n)$, hence, by II of No. 5, E_n is regular. When we pass from a subdivision of K' back to K' in case $p = n$, the preceding result holds *a priori* since K' is simplicial. Therefore Th. II is proved for $p = n$ and regularity.

Let us now take $p < n$ and suppose that Th. II (for regularity) has been proved for all cells of dimension $> p$. Then under our hypotheses all cells of the star of E_p are regular. We take again K' as the subdivision. If E'_p is a cell of K' on E_p, $Lk(E'_p)_{K'}$ has the structure of a first derived of $Lk(E_p)_K$. Hence the latter, like the former, is an $(n-p-1)$-sphere, so that condition Ib, of No. 5 for regularity is verified.

Let now $E'_{n-p} = A^p A^{p+1} \cdots A^n$ be a cell of K', component of the dual of E_p. The n-cells of its star are of type $(A^0 \cdots A^{p-1}) A^p \cdots A^n$, where the parenthesis represents a component of $F(E_p)$. Hence $F(E_p)$ has for first derived a complex whose structure is the same as for $Lk(E'_{n-p})_{K'}$. Since E'_{n-p} is on the star of E_p, it is a regular cell of K'. Therefore the linked complex just named is an absolute manifold. From our hypothesis follows then that $F(E_p)$ is also an absolute manifold. Thus Ia of No. 5 is likewise fulfilled and E_p is regular.

We have thus proved Th. II for regularity when we pass from K' to K. Let now K^0 be another simplicial subdivision of K. Since it is also a subdivision of K', it is sufficient to prove our theorem for annexion from K^0 to K', hence for annexion from any simplicial subdivision K^0 of a simplicial K back to K. Under the hypotheses of the induction all that is necessary is to show that $Lk(E_p)$ is a sphere. But as an incident result in No. 11, we have shown that if E^0_p of K^0 is on E_p, $Lk(E^0_p)_{K^0}$ and $Lk(E_p)_K$ have subdivisions whose structure is the same. Since the first is a sphere this is likewise the case for the second. This completes the proof of Th. II for regularity.

The modifications for semi-regularity are again wholly secondary and need not be taken up separately.

Combining the theorem of No. 9 and Th. II of the present number we have:

THEOREM III. *In order that* $K - L$ *be a manifold it is necessary and sufficient that some subdivision of it be a manifold.*

An alternative definition of manifolds. The preceding result suggests the following procedure for defining the manifold $K - L$: Introduce regular cells for a simplicial K by means of the conditions of No. 5, of which Ia is naturally satisfied. There follows the definition of the manifold for the simplicial case. When $K - L$ is not simplicial, it is to be an M by definition whenever $K' - L$ is an $M(K' =$ the first derived of K as usual).

16. **The dual complex.** Let E_p be a cell of the manifold $K - L$. Any cell E_q of its star is regular and has boundary points *not* on L. Since L is a closed set E_q has points not on it, and since L is an exact sum of cells of K, E_q is a cell of $K - L$. Thus the total star of E_p is on $K - L$.

Since E_p and all cells of its star are regular they all have duals which are cells, and since E^*_{n-p} (the dual cell of E_p) is on the star, it is also on $K - L$. Furthermore

$$\bar{E}^*_{n-p} = E^*_{n-p} + K^*_{n-p-1},$$

where the last complex is the sum of the duals of the cells of the star $\neq E_p$.

Consider the sum K^* of all duals of cells of $K - L$. It has the following properties:

(a) *No two of its cells meet.* For let them be E^*_{n-p} and E^*_{n-q}, $q \geqq p$. They are respectively on the stars of E_p and E_q of K. Hence they can meet only if E_q belongs to the star of E_p. But in that case E^*_{n-q} belongs to $F(E^*_{n-p})$ and hence does not meet E^*_{n-p} itself.

(b) *Every cell of* $F(E^*_{n-p})$ *is a cell of* K^*. For that boundary is precisely K^*_{n-p-1}, the sum of the duals of the cells of the star of E_p other than E_p.

(c) *The boundary of every cell of* K^* *is a sphere which is an absolute manifold.* For the first derived of K^*_{n-p-1} is $Lk(E_p)$, which is both a sphere and an absolute manifold since E_p is regular. It follows that K^*_{n-p-1} is likewise a sphere and also an absolute manifold (No. 14, Th. II).

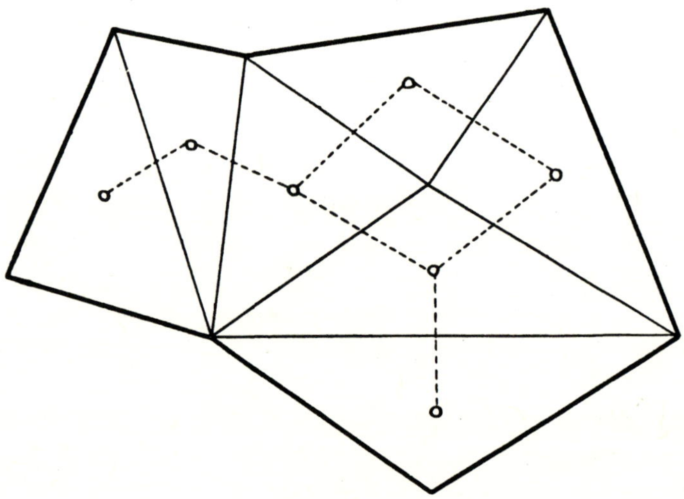

Fig. 13.

Since E^*_{n-p} is the join of a point and its boundary sphere, it is normal. Hence the preceding properties imply

(d) K^* *is a closed normal complex.*

This closed complex is called the *dual* of the open complex $K-L$. Fig. 13 illustrates the case $n = 2$, with the perimeter of the polygon as L.

Thus the duality relationship takes place here between the *open* complex $K-L$ and the ordinary or *closed* complex K^*. Exception must be made of course when $L = 0$ and K is an absolute M_n. Then both are closed and K^* becomes Poincaré's dual, [3], which he called "polyèdre réciproque". The derivation of the duality relations and a good part of the intersection theory to be taken up in Ch. IV rest on the properties of dual complexes.

17. The justification of the term "dual" is found in a certain symmetry between $K-L$ and K^* relatively to K'. For any vertex, A^{ip}, of $K-L$ is the unique vertex of K' on the cell E^{*i}_{n-p} of K^*. Designate it then by $A^{*i, n-p}$. As a result each cell of K' on K^* will have a new symbol consisting of A^*'s in the reverse order from that of the corresponding A's. Now a component cell of the cell E_p of $K-L$ was of type $A^r \cdots A^p$. Its new symbol will be $A^{*n-p} \cdots A^{*n-r}$. This shows clearly that E_p is the dual of its dual $\overset{*}{E}_{n-p}$ relatively to K'. *Thus the cells of $K-L$ and K^* are one another's duals relatively to K'.* The cells of the latter complex whose sum is K^* are the cells of a first derived complex K'^* of K^*. The symmetrical relationship is thus made apparent.

18. **The neighborhood of L when $K-L$ is a manifold.** Let N be the K'-neighborhood of L. It follows from the properties of the cells of K' that if one of them has all its vertices on L it is on L. Therefore N is a normal neighborhood (Ch. II, No. 21). We denote its boundary by Φ, and as we shall see, N and Φ are in a remarkable relation to $K-L$ and K^*.

Any cell of K^* is a sum of cells of K' of type $\sigma = A^p \cdots A^q$, where if $E_p, \cdots, E_q$ are the associated mutually incident cells, they are all cells of $K-L$. $\bar{\sigma}$ being on the star of E_p, is then wholly on $K-L$. Conversely if $\bar{\sigma}$ is on $K-L$, σ is a component of a $Tv(E_p)$ where E_p is a cell of $K-L$ and hence regular. Therefore $Tv = \overset{*}{E}_{n-p}$, and hence σ is a component of the cell $\overset{*}{E}_{n-p}$ of K^*. Therefore $K-K^*$ consists of all cells of K' with a vertex on L, or $K-K^* = N$.

Let now σ be a cell of Φ. It must be a face of a cell of K' with a single vertex on L. Let A^r be that vertex, E_r the corresponding cell of K. E_r must be a cell of L, since its point A^r is on L. It is also incident with the cells $E_p, \cdots, E_q$. Since E_p is not on L, E_r can only be a boundary cell of E_p, $r < p$, and the cell of which σ is a face, is $A^r A^p \cdots A^q$.

This implies that E_p, to whose dual E^*_{n-p}, the cell σ belongs, has a vertex on L. Thus σ is a cell of K', component of the dual of a cell of $K-L$ having a vertex on L. Conversely if σ is of this nature, and if A^0 is a vertex of E_p on L, σ is a face of the cell $A^0 A^p \cdots A^q$ of K' having the single vertex A^0 on L and σ as its opposite face. Hence σ belongs to Φ.

Combining our results we have:

THEOREM. *The K'-neighborhood of L coincides with $K-K^*$. Its boundary is the sum of the duals of the cells of $K-L$ that have at least one vertex on L.*

REMARK. Since the manifold conditions are invariant under subdivision, when we consider neighborhoods we are free to replace at any time K by a subdivision. We shall take advantage of this by assuming wherever need be that K is simplicial and that the K-neighborhood of L is normal.

19. THEOREM. *When $K^*-\Phi$ is not vacuous it is an M_n.*

The assumption $K^*-\Phi \neq 0$ is equivalent to assuming that K has at least one vertex not on L. This being granted we take a maximal connected subset of $K-L$ having at least one vertex and all that is necessary is to show that the part of K^* on that subset behaves as asserted. Practically this means that we merely need to prove the theorem under the additional restrictions that $K-L$ is connected and has at least one vertex.

Let $K'^* = K'-N =$ the first derived of K^* induced by K'. By Th. II of No. 14 it is sufficient to show that K'^* is an M_n with the boundary Φ.

Now if E'_p is a cell of $K'^*-\Phi$ so is every cell E'_q of its star. For it cannot be a cell of Φ, since Φ is closed and with E'_q every cell of its boundary, hence also E'_p, must be a cell of Φ. Since $K'-L$ is a manifold E'_p is a regular cell of K'. As a consequence of the preceding property however it is likewise a regular cell of K'^*. This proves our theorem.

20. Suppose now that K is simplicial and that the K-neighborhood of L is normal. Then it is possible to proceed a step

further. For let Φ^* be the subcomplex of K^* coincident with Φ, or the decomposition of the point-set Φ into a sum of cells of K^*. I say that now *K^* has Φ^* for regular boundary.* For let E^*_{n-p} be a cell of Φ^*, E_p its dual, A^p the vertex of K' on it. E_p is a simplex of $K—L$ with some, but not all of its vertices on L. Hence $E_p = \sigma_q \sigma_{p-q-1}$, where σ_q has no vertices on L and σ_{p-q-1} is on L. Let H_{p-1} be the boundary of E_p as decomposed into cells by K'. It is the first derived of the sum of the faces of E_p. The complex $Lk(E^*_{n-p})_{K^*}$ is merely the sum of the cells of H that have no vertex on σ_{p-q-1}. The structure of Lk is actually independent of the choice of vertices of K' on E_p. Let us choose those not on σ_q or σ_{p-q-1}, on one and the same S_{p-1} of the space of E_p. Then Lk is merely one half of the section of H by that space, which is manifestly the homeomorph of a closed hemisphere of S_p, that is, a closed E_{p-1} with a sphere as boundary. Since we already know that $F(E^*)$ is a manifold and a sphere, the result just obtained together with an elementary recurrence is sufficient to show that E^*_{n-p} is semi-regular and proves our assertion.

§ 3. Duality Relations for the Homology Characters

21. In the present section we propose to derive certain duality formulas for an orientable manifold $K—L$. The treatment will apply directly to the characters mod m, the cells and spheres being then likewise taken mod m. To simplify notations we omit all mention of mod m, and write henceforth in general $R_p(K)$, $R_p(K; L)$ for $R_p(K; m)$ or $R_p(K; L, m)$ as well. In addition when $m = 2$ all orientability conditions may be omitted both as regards $K—L$ and the cells or spheres. All formulas to be derived will hold in this very general sense. In some special cases we shall have occasion to deal with Betti numbers mod 2; the corresponding relations and statements will hold only for that special modulus.

22. **A fundamental matrix equation.** The duality relations will follow quite readily from a fundamental relation,

to be established presently, between the incidence matrices of $K-L$ and of its dual K^*.

We take $K—L$ normally oriented by means of its decomposition into n-simplexes by the first derived K'. This orients in a definite way the n-cells of $K—L$. The cells of dimension $<n$ are oriented in arbitrary fashion. We shall assign orientations to the dual cells and then the required matrix equation will follow as a result of comparing the incidence relations.

Let E_p be a cell of $K—L$, E^*_{n-p} its dual, the latter oriented in some as yet unspecified way. Let $\sigma_p = A^0 A^1 \cdots A^{p-1} A^p = \sigma_{p-1} A^p$ be a (simplicial) cell of K' on E_p, and $\sigma_{n-p} = A^p A^{p+1} \cdots A^n = A^p \sigma_{n-p-1}$, a (simplicial) cell of K' on E^*_{n-p}. The vertices $A^0, \cdots, A^n$ are those of a cell $\sigma_n = \sigma_{p-1} A^p \sigma_{n-p-1}$, of K' on a cell E_n of $K—L$.

Since E_p, E^*_{n-p}, E_n are orientable circuits they can be normally oriented by means of their subdivisions by K'. Assume this done in such a way that these orientations are concordant with the given orientations. Then we may write $\beta_p \sigma_p$, $\beta_{n-p} \sigma_{n-p}$, $\beta_n \sigma_n$ as indicatrices of E_p, E^*_{n-p}, E_n (and hence of $K—L$), respectively. The chief point in introducing the β's ($=\pm 1$) is that they enable us to take care at one stroke of the exceptional cases $p = 0$ or n, the formulation being thus completely general.

We propose to show that the product of the β's obtained as above from a given E_p, is independent of the choice of the cells σ_p, σ_{n-p}, σ_n. In other words, *for a given cell E_p of $K—L$, the number $\beta_p \beta_{n-p} \beta_n$ depends solely on the orientations of E_p and its dual.*

23. Consider first the effect of changing σ_p, if it is possible (i. e. if $p>0$), and leaving σ_{n-p} fixed. Since E_p is an open M_p and a circuit, it is a simple circuit. It follows that any two p-cells σ_p, σ'_p, of K' on E_p are the end cells of a Poincaré-chain consisting of p- and $(p-1)$-cells of K' on E_p (Ch. I, No. 41). Therefore we merely have to prove that the product $\beta_p \beta_{n-p} \beta_n$ is unchanged on passing from a p-cell of the Poincaré-chain to one that follows it, i. e. from any σ_p to a σ'_p

having a σ_{p-1}, but no interior points, in common with it. Since we deal with a normally oriented simple circuit the two cells, when oriented as indicatrices of E_p, are oppositely related to their common $(p-1)$-face. Let A^q, $q<p$, be the sole vertex of σ_p that is not also a vertex of σ'_p. We have then, say, the two indicatrices

$$\beta_p \sigma_p = \beta_p A^0 A^1 \ldots A^{q-1} A^q A^{q+1} \ldots A^p,$$
$$\beta'_p \sigma'_p = \beta'_p A^0 A^1 \ldots A^{q-1} A'^q A^{q+1} \ldots A^p,$$

and therefore by the remark just made $\beta'_p = -\beta_p$.

Now the two simplexes $\sigma_p \sigma_{n-p-1}$ and $\sigma'_p \sigma_{n-p-1}$ are n-simplexes of the manifold $K'-L$ having a common $(n-1)$-face. Since they belong to the same connected subset of $K'-L$, when oriented as indicatrices of $K'-L$, they must be oppositely related to that face. Therefore, β'_n being the analogue of β_n for the second simplex, we have as above $\beta'_n = -\beta_n$. Finally our operation does not alter β_{n-p}, i.e. $\beta'_{n-p} = \beta_{n-p}$. Therefore

$$\beta_p \beta_{n-p} \beta_n = \beta'_p \beta'_{n-p} \beta'_n.$$

In the same way we can show that this number is independent of the choice of σ_{n-p}, and hence of σ_p and σ_{n-p}. Any change in σ_n can be expressed as the result of changes in σ_p, σ_{n-p}, so that the proof is complete.

24. The number $\beta_p \beta_{n-p} \beta_n = \pm 1$ is what we shall designate later (Ch. IV) as the *Kronecker index* of E_p and E^*_{n-p}. We leave aside for the present all questions pertaining to it, contenting ourselves with the little that we need for the immediate sequel. Since this number is fixed any two β's will determine the third. Granting that $K-L$ and E_p, given configurations, are oriented *ad libitum*, we have assigned values for β_n and β_p. The value of the product will change with the orientation of E^*_{n-p}, and conversely, when the product is known, an indicatrix is determined for the cell which is thus oriented. *We fix the orientation of E^*_{n-p} by imposing the condition*

$$\beta_p \beta_{n-p} \beta_n = 1. \tag{1}$$

We have thus assigned a definite orientation to all dual cells.

25. When $p = 0$ or n it is convenient to impose the further condition that whichever of the cells E_p, E^*_{n-p} is an n-cell shall be oriented concordantly with the orientation of $K—L$. In the first case we then have both E^*_{n-p} and E_n oriented by $\beta_n A^0 A^1 \cdots A^n$ so that $\beta_o = \beta_n \beta_{n-o} = \beta_n^2 = 1$, i. e. every vertex of K is taken with a plus sign. In like manner the case $p = n$ yields the result that the vertices of K^* must be taken with the plus sign.

In the preceding treatment *the orientability of* $K—L$ *has played a vital role.* For it demands that the star of E_p be orientable. Hence in order to be able to apply it throughout $K—L$ the cells must be normally orientable.

In the non-orientable case there remains merely the obvious statement that E_p and E^*_{n-p} intersect at a single point.

26. We have seen (No. 16) that the cells on the boundary of E^*_{n-p} are the duals of the cells of the star of $E_p (\neq E_p)$. Therefore the incidence relations of K^* are completely determined if we know what they are for a pair E^*_{n-p}, E^*_{n-p-1}, duals of two incident cells E_p, E_{p+1} of $K—L$.

Let then E_p appear in the boundary of E_{p+1} with the coefficient $\eta = \pm 1$ or zero. From what we have just said E^*_{n-p-1} will appear in that of E^*_{n-p} with the coefficient $\pm\eta$ and our present problem is to determine the choice of sign. In so doing we need only consider the case where E_p and E_{p+1} are incident. The answer is obtained by comparing indicatrices which is most easily done by means of a table. Setting

$$\sigma_{p-1} = A^0 A^1 \cdots A^{p-1}, \qquad \sigma_{n-p-2} = A^{p+2} \cdots A^n$$

we have:

Cells	Indicatrices
E_{p+1}	$\beta_{p+1} \sigma_{p-1} A^p A^{p+1}$
E_p	$(-1)^{p+1} \eta \beta_{p+1} \sigma_{p-1} A^p$
$K—L$	$\beta_n \sigma_{p-1} A^p A^{p+1} \sigma_{n-p-2}$
E^*_{n-p-1}	$\beta_{p+1} \beta_n A^{p+1} \sigma_{n-p-2}$
E^*_{n-p}	$(-1)^{p+1} \eta \beta_{p+1} \beta_n A^p A^{p+1} \sigma_{n-p-2}$

The orientations of the dual cells are obtained by the rule (1) of No. 24. The last column shows that in the expression for the boundary of E^*_{n-p} we must affect E^*_{n-p-1} with the sign $(-1)^{p+1}\eta$, since

$$\eta\, \beta_{p+1}\, \beta_n\, A^p\, A^{p+1}\, \sigma_{n-p-2} \rightarrow \eta\, \beta_{p+1}\, \beta_n\, A^{p+1}\, \sigma_{n-p-2} + \cdots.$$

27. Let E^i_p $(i = 1, 2, \cdots, \alpha_p)$ be the set of p-cells of $K - L$ and denote by E^{*i}_{n-p} the dual of E^i_p. Let the incidence relations for the $(p+1)$- and p-cells of $K - L$ be

$$E^i_{p+1} \rightarrow \sum \eta^p_{ij}\, E^j_p \mod L.$$

We infer from our discussion that the (absolute) incidence relations for the $(n-p)$- and $(n-p-1)$-cells of K^*, the dual of $K - L$, are

$$E^{*i}_{n-p} \rightarrow (-1)^{p+1} \sum \eta^p_{ji}\, E^{*j}_{n-p-1}.$$

Therefore if η^{*q} designates the incidence matrix for the $(q+1)$- and q-cells of K^* we have the *fundamental matrix equation*

(2) $$\eta^{*n-p-1} = (-1)^{p+1} (\eta^p)'.$$

28. **Duality relations for absolute manifolds.** Let K define an absolute M_n, that is, let L be vacuous so that K^* coincides as a point-set with K. Both are then merely different defining complexes of the same manifold. Hence their combinatorial Betti and torsion numbers are the same. Denote by ϱ_p, ϱ^*_p, the ranks of η^p, η^{*p} and by α_p, α^*_p, the numbers of p-cells of the respective complexes. Then (Ch. I, No. 27):

$$R_p(M) = \alpha_p - \varrho_{p-1} - \varrho_p; \quad R_{n-p}(M) = \alpha^*_{n-p} - \varrho^*_{n-p-1} - \varrho^*_{n-p}.$$

But in view of the matrix equation (2), $\varrho_p = \varrho^*_{n-p-1}$, and we know that $\alpha_p = \alpha^*_{n-p}$, a fact also implied by (2). Hence

(3) $$R_p(M) = R_{n-p}(M).$$

From the same matrix equation follows for the torsion coefficients

(4) $$\theta^p_i(M) = \theta^{n-p-1}_i(M);$$

for η^p and η^{*n-p-1} have the same invariant factors. We have therefore (Poincaré [2, 3, 4], Veblen-Alexander [22], Veblen [5]):

DUALITY THEOREM FOR AN ABSOLUTE M_n: *The Betti numbers are the same for* p *and* $n-p$, *and likewise the torsion coefficients for* p *and* $n-p-1$.

COROLLARY. *The Euler-Poincaré characteristic of an absolute* M_{2n+1} *is zero.* (*Poincaré* [2]).

29. **Fundamental duality relations for the homology characters of open manifolds.** We return to our manifold $K-L$ with L subcomplex of K, and consider a closed subcomplex L^1 of L, setting $L^2 = L - L^1 =$ an open subcomplex of L. The duality relations that we shall derive will be between the characters of $K-L^1$ mod L^2 and of $K-L^2$ mod L^1. As we shall see *they include all duality relations between the homology characters as special cases.* For the present we are dealing exclusively with the characters of "polyhedral" configurations. The extension to the general case will be found in Ch. VII, § 3.

We have just stated that we have polyhedral configurations throughout, whereas at the outset we appear to have restricted the discussion to K and its subcomplexes. The two statements are easily reconciled: The situation is in no wise affected when K is arbitrarily subdivided, nor are the characters that we shall consider, changed. When the L's are polyhedral, we can find a subdivision K^0 of K such that the L's have subdivisions induced by K^0 related as above, and all that is necessary is to replace K by K^0.

We take advantage of the possibility of passing from K to K^0, and choose for K^0 a simplicial complex such that the associated neighborhoods of L and L^1 are normal. From now on we write K for K^0, and L, L^1, L^2 for the subcomplexes of K^0 coincident with them. It means, practically, that we assume K simplicial, the L's subcomplexes of K, and the K-neighborhoods N, N^1, of L, L^1, normal.

30. Let Φ be the sum of the duals of all cells of $N^1 - L$. If E_p is any cell of $N^1 - L$ so is every cell of its star. But

the duals of the cells of the star $\mp E_p$ are the boundary cells of E^*_{n-p}, the dual of E_p. Hence when E^* is a cell of Φ so is every cell of $F(E^*)$. Therefore Φ is a closed subcomplex of K^*, the dual of $K-L$.

By construction then the cells of $K^*-\Phi$ and those of $K-L-N^1$ are the duals of one another. Let $\mathfrak{F}$ be the sum of the cells of L (or L^2) that have no vertices on L^1 and let $\mathfrak{K} = K - N^1$. Both $\mathfrak{F}$ and $\mathfrak{K}$ are closed complexes, $\mathfrak{F}$ is a subcomplex of $\mathfrak{K}$ and $\mathfrak{K} - \mathfrak{F} = K - L - N^1$.

Now the cells of the open complexes $\mathfrak{K} - \mathfrak{F}$ and $K^* - \Phi$ are the duals of one another. Furthermore when E_p of K, is a cell of $\mathfrak{K} - \mathfrak{F}$ so is any cell E_{p+1} of its star. Their duals E^*_{n-p}, E^*_{n-p-1} are then incident cells of $K^*-\Phi$ and *vice versa.* Hence the fundamental matrix equation (2) and the reasoning of No. 29 are applicable as between the boundary relations of $\mathfrak{K}$ mod $\mathfrak{F}$ and those of K^* mod Φ. They yield here

$$R_p(\mathfrak{K};\ \mathfrak{F}) = R_{n-p}(K^*;\ \Phi), \tag{5}$$

$$\theta_i^p(\mathfrak{K};\ \mathfrak{F}) = \theta_i^{n-p-1}(K^*;\ \Phi). \tag{6}$$

31. Let C_p be any chain on $K-L^1$. Since N^1 is regular we can draw the usual projecting segments through all points of $N^1 - L^1$ and by sliding the points of C_p on N^1 along them, we deform C_p into a chain C'_p on $\mathfrak{K}$. Since the projecting segments of points on any cell of K are on that cell, the deformation leaves on L^2 all points of C_p that it may carry. We then take C'_p and by means of the fundamental deformation theorem for simplicial complexes, applied to $\mathfrak{K}$, we reduce C'_p to a subchain C''_p. This reduction leaves likewise on L^2 the points of the chain that it may carry. Thus on passing from C to C'' the points on L^2 have remained on L^2. Therefore if C is a cycle of $K-L^1$ mod L^2, C'' is a cycle of $\mathfrak{K}$ mod $\mathfrak{F}$, since $\mathfrak{F}$ consists of the cells common to $\mathfrak{K}$ and L^2. Finally if $C_p \sim 0$ on $K-L^1$ mod L^2, the same reasoning applied to the C_{p+1} which it bounds mod L^2, shows that C''_p bounds a subchain of $\mathfrak{K}$ mod $\mathfrak{F}$. If follows as in Ch. II, § 4, that *the homology characters of* $K-L^1$ *mod* L^2 *and* $\mathfrak{K}$ *mod* $\mathfrak{F}$ *are the same.*

32. Designating as usual by K' the first derived of K, let N' be the K'-neighborhood of L and N'^1 the sum of the cells of N' with vertices on L^1 but not on L^2. Since N' is normal any $C_p \subset K-L^2$ can be deformed into a subchain of $K-N'+N'^1 = K^*+N'^1$, (No. 18). This deformation is of the usual kind, along the projecting segments in the K'-neighborhood of $\mathfrak{F}$. It follows then exactly as above that the homology characters of $K-L^2$ mod L^1 and of $K^*+N'^1$ mod L^1 are the same.

But $K^*+N'^1 = \mathfrak{K}^*$ is a closed complex $\supset L^1$, with N'^1 as the $\mathfrak{K}^*$-neighborhood of L^1. Let us compare $\mathfrak{K}^*$ and L^1 in the same way as K, L in Ch. II, § 5. Since N'^1 is normal the neighborhoods analogous to N, N', N'', loc. cit., can all be taken coincident with N'^1. The cycles analogous to Δ loc. cit. are the cycles of K^* mod Φ. Therefore here the homology characters of K^* mod Φ and $\mathfrak{K}^*$ mod L^1 are equal. But the latter, as already shown, are the same as those of $K-L^2$ mod L^1. Hence finally *the homology characters of $K-L^2$ mod L^1 and of K^* mod Φ are the same.*

33. Comparing the conclusions of the last three numbers we have then ultimately

$$R_p(K-L^1;\ L^2) = R_{n-p}(K-L^2;\ L^1), \tag{7}$$

$$\theta_i^p(K-L^1;\ L^2) = \theta_i^{n-p-1}(K-L^2;\ L^1). \tag{8}$$

This proves the

FUNDAMENTAL DUALITY THEOREM. *Let K, L, L^1 be closed complexes, the last two subcomplexes of their predecessors and let $K-L$ be a manifold. If $L^2 = L-L^1$ the duality theorem for absolute manifolds holds between the homology characters of $K-L^1$ mod L^2 and those of $K-L^2$ mod L^1.*

34. **Important special cases.** I. Let $L^1 = L$, $L^2 = 0$. Then we have

$$R_p(K-L) = R_{n-p}(K;\ L), \tag{9}$$

$$\theta_i^p(K-L) = \theta_i^{n-p-1}(K;\ L). \tag{10}$$

This shows that *the duality theorem holds between the absolute characters of $K-L$ and those of K mod L (L a subcomplex* of K, *$K-L$ a manifold).*

II. For $L = 0 = L^1 = L^2$ we have an absolute manifold and (7), (8) reduce to (3), (4) as they should.

III. *Alexander's duality relation.* The situation being as in I, let this time $L \neq 0$ and $K = H_n$, a sphere. Assuming first $n > 1$ and $1 < p < n$, let Γ_p be a cycle of H mod L and γ_{p-1} its boundary, so that $\gamma_{p-1} \sim 0$ on H. I say now that the two homologies

$$\text{(11)} \qquad \Gamma_p \sim 0 \bmod L \text{ on } H,$$

$$\text{(12)} \qquad \gamma_{p-1} \sim 0 \text{ on } L,$$

are equivalent. For when (11) holds, $H \supset C_{p+1} \to \Gamma_p + C_p$, $C_p \subset L$. As a boundary is an absolute cycle, $F(\Gamma_p) = -F(C_p) = \gamma_{p-1}$, which implies (12). Conversely when (12) holds, $L \supset C_p \to -\gamma_{p-1}$, $\Gamma_p + C_p$ is an absolute cycle on H, and since its dimension $\neq 1, n$, it is ~ 0 on H, which implies (11).

Since any linear combination of Γ's corresponds to the same linear combination of γ's and conversely, with association of homologies (11) for one type with homologies (12) for the other, we have

$$\text{(13)} \qquad \left.\begin{cases} R_p(H_n; L) = R_{p-1}(L) \\ \theta_i^p(H_n; L) = \theta_i^{p-1}(L) \end{cases}\right\} 1 < p < n.$$

When $p = 1 < n$, the γ_0's must be cycles ~ 0 on H. Since each γ_0 depends upon a single point of L on H, (Ch. I No. 28)

$$\text{(14)} \qquad R_1(H_n; L) = R_0(L) - 1,$$

and there are no zero-divisors since L has none of dimension zero. On the other hand, when $p = n > 1$, (12) is to be replaced by $\gamma_{n-1} \equiv 0$, since L is of dimension $\leqq n-1$. This merely implies however $\Gamma_n = tH_n$ and conversely. This time then the correspondence between the γ's and Γ's associates a relation $\equiv 0$ for the first with a relation $\equiv tH_n$ for the second and conversely. Hence

$$\text{(15)} \qquad R_n(H_n; L) = R_{n-1}(L) + 1,$$

and there are no zero-divisors to be considered. We can use the Kronecker symbols δ to combine (13), (14), (15) into a single formula

$$(16) \qquad R_p(H_n;\, L) = R_{p-1}(L) - \delta_{p1} + \delta_{pn}.$$

We have then in view of (9), (10) and the properties of the δ's the duality relations

$$(17) \qquad R_p(H_n - L) = R_{n-p-1}(L) + \delta_{p0} - \delta_{p,n-1},$$

$$(17') \qquad \theta_i^p(H_n - L) = \theta_i^{n-p-2}(L).$$

When $n = 1$ the verification of (17) is immediate and there are no torsion coefficients, hence our formulas hold without exception.

Formula (17), mod 2, is precisely Alexander's duality relation [8].

Application. For $p = 0$, $R_0(H - L)$ yields the number of parts into which L disconnects H and we have from (17) for that number

$$(18) \qquad R_0(H_n - L) = R_{n-1}(L) + 1.$$

Remark. Alexander's complete result holds for any L immersed in H and not merely for a subcomplex of H. It will be taken up in Ch. VII §§ 3, 4, together with the extension of (7). It may be pointed out incidentally that *no similar extensions for torsion numbers (i. e. of Formula (17′)) are known.* In the passage from a subcomplex of K to a complex merely immersed in it, or to more general closed subsets, one seems to lose a certain refinement in the results.

35. **Application to spheres.** Let $L = H_{n-1}$, a combinatorial $(n-1)$-sphere. Then $R_{n-1}(L) = R_{n-1}(H_{n-1}) = 1$, hence from (18), $R_0(H_n - H_{n-1}) = 2$. Therefore

Theorem I. *If H_{n-1} is a subcomplex of an H_n which is an M_n, it disconnects it into two parts.*

This is an extension of a theorem due to Veblen [2] on the decomposition of space by a polyhedron. For it will be recalled that H_{n-1} might have been assumed merely to be a polyhedral sphere on H_n. Let E_n, E'_n be the two parts into which H_n is decomposed by H_{n-1}. As we shall see, they are cells. Since H_n is a simple circuit, the n- and $(n-1)$-cells of E or E' have the incidence properties of the simple circuit. From (16) mod 2, $R_n(H_n;\, H_{n-1}, 2) = 2$. Hence E, E' can each contain only one n-circuit. Therefore they are simple circuits (Ch. I, No. 35).

Let this time $p > 0$ and let Γ_p of No. 34 be chosen on E. If $p < n$, $\gamma_{p-1} \sim 0$ on H_{n-1}, so that $H_{n-1} \supset C_p \to \gamma_{p-1}$. Therefore $\Gamma_p - C_p$ is an absolute cycle of H_n. But as the latter is a sphere, $\Gamma_p - C_p \sim 0$, $\Gamma_p \sim 0$ mod H_{n-1} on E. Hence the Betti numbers of E mod H_{n-1} for $0 < p < n$ are all zero and it has no torsion coefficients. Since E is a circuit, it fulfills all conditions for the combinatorial n-cell; similarly of course for E'. Therefore

THEOREM II. *The two parts into which H_{n-1} disconnects H_n are both combinatorial n-cells.*

COROLLARY. *Let L be an absolute M_{n-1} which is a subcomplex of K disconnecting it into two parts $K' - L$, $K'' - L$. If K is an absolute M_n, the two parts are n-manifolds with the common regular boundary L and conversely.*

We can take all complexes simplicial. If E_p is any cell of K we have

$$(19) \qquad Lk(E_p)_K = Lk(E_p)_{K'} + Lk(E_p)_{K''},$$

and the two Lk's at the right intersect in $Lk(E_p)_L$. When K is an absolute M_n and E_p a cell of L, $Lk(E_p)_K$ and $Lk(E_p)_L$ are respectively an H_{n-p-1} which is a simple circuit, and an H_{n-p-2} separating it. From this and Th. II follows that E_p satisfies the condition (c) of No. 5 for semi-regularity. A ready consequence is that E_p is semi-regular. The cells of $K' - L$ for example, being manifestly regular the direct Corollary is proved. The converse follows from (19) and the result of Ch. II No. 49.

36. **A new type of invariants and their duality relations.** Up to the present time we have always compared chains on a definite configuration A as regards homologies on A. In a number of questions, in particular, in studying manifolds with regular boundaries, there arises the problem of com-

paring the cycles of A with those of $\bar{A}-A$. We have then a new type of characters by taking as the modulus $\mathfrak{M}$ of the p-cycles, the p-cycles of A alone, and as the corresponding zero-modulus $\mathfrak{M}^0$, the cycles homologous on A to cycles of $\bar{A}-A$. Thus the absolute cycles Γ_p^i are dependent or not according as we do or do not have λ's not all zero such that

$$\bar{A} \supset C_{p+1} \rightarrow \sum \lambda_i \Gamma_p^i + \Gamma_p, \quad \Gamma_p \subset \bar{A}-A.$$

It will be observed that

$$\Gamma_p = F(C_{p+1}) - \sum \lambda_i \Gamma_p^i$$

and, as a boundary is an absolute cycle, Γ_p is also such a cycle. Hence dependence in the new sense is equivalent to a homology

$$\sum \lambda_i \Gamma_p^i \sim \text{ on } \bar{A} \text{ to an absolute cycle on } \bar{A}-A.$$

The Betti and torsion numbers should be denoted by $R_p(\mathfrak{M})$, etc. Actually it will be more suggestive, if less systematic, to designate the Betti numbers by $r_p(A)$, $r_p(A; m)$. However, for the reasons given at the beginning of § 3, we adhere to our omission of m provided that m is a prime, a necessary restriction here. It is understood that it can be inserted at any time and that when $m = 2$ we may dispense with orientability. Since we have no significant theorems involving the torsion coefficients there is no need to assign special notations for them.

It is important to bear in mind that every r is at most equal to the corresponding R, a fact which follows at once from the definition.

Examples. I. Let γ be a circle on an ordinary two-sphere H, and let $A = H - \gamma$. It consists of two regions A', A'' separated by γ. Any two points of A' (or A'') are homologous to one another on A' (or A''), whereas two points, one on A', the other on A'', are not. For in the first case, the two points can be joined by an arc on A, in the second they cannot. Hence $R_0(A) = 2$. However, any point of A can be joined by an arc to a point of $\gamma = \bar{A} - A = H - A$. Hence every point of A is $\sim 0 \bmod \mathfrak{M}$ on A and therefore $r_0(A) = 0$.

II. Let A be the interior of a torus T in ordinary space, so that $\bar{A} = A + T$. Any 1-circuit on A is deformable into a point or into a multiple of the circle γ, locus of the center of the generating circle of T. Hence every 1-circuit $\sim t\gamma$ on A, and γ itself ~ 0 mod T on $\bar{A}$. Hence $r_1(A) = 0$. On the other hand it is easily verified, for example by writing down the incidence matrices for a suitable defining complex, that $R_1(\bar{A}) = 1$, which implies $\gamma \nsim 0$ on $\bar{A}$.

Invariants of the type here defined were first introduced in our papers [12, 14], the Betti numbers being there designated by S which we have had to change, as we wish to reserve S for spaces.

37. We are particularly interested in the case where K is a complex, L a subcomplex of K with $K - L$ a manifold, and we propose to find the corresponding properties of duality.

Let N be, as before, the K-neighborhood of L, K^* the dual of $K - L$. We choose K simplicial and such that N is normal. Designate by e, E the cells of $N - L$ and $K - N$ respectively and by e^*, E^* their duals. $K - K^* = N^*$ is a neighborhood of L wholly on N, and the boundary of N^* is $\Phi = \sum e^*$, so that E^* is an arbitrary cell of $K^* - \Phi$. (No. 18.)

Calculation of $r_p(K - L)$. The incidence matrix of the p- and $(p+1)$-cells of $K - L$ has the following schematic structure:

	e_p	E_p
e_{p+1}	$\boldsymbol{a}$	$\boldsymbol{b}$
E_{p+1}	0	$\boldsymbol{c}$

Designate by α_p, β_p the number of p-cells of the respective types e, E, and by ϱ_p^a, ϱ_p^b, ϱ_p^c, ϱ_p the ranks of $\boldsymbol{a}$, $\boldsymbol{b}$, $\boldsymbol{c}$ and of the total matrix. From the structure of the matrix we can infer at once that the number of $(p+1)$-subcycles of K, composed of cells E alone, is $\beta_{p+1} - \varrho^c_p$, this being the number of linearly independent combinations $\sum t_i E^i_{p+1}$ without boundary.

For p the corresponding number is $\beta_p - \varrho^c_{p-1}$. We must subtract from it the number of linearly independent subcycles combinations of the E's alone which are $\sim 0 \bmod L$ on K.

All told there are ϱ_p linearly independent relative subcycles $\sim 0 \bmod L$. Among them however we have ϱ^a_p linearly independent combinations which also include e-cells. Hence there are $\varrho_p - \varrho^a_p$ relative subcycles $\sim 0 \bmod L$ and consisting only of E cells. But these *relative* cycles are actually *absolute* cycles. For their boundaries must be on L and likewise on the closures of the E cells, which however have no cells on L. We have then exactly $\varrho_p - \varrho^a_p$ distinct subcycles $\sim 0 \bmod L$. Therefore

$$r_p(K - L) = \beta_p - \varrho^c_{p-1} - \varrho_p + \varrho^a_p.$$

Calculation of $r_{n-p}(K-L)$. Except for a possible change in sign of all terms, the incidence matrix for the $(n-p)$- and $(n-p-1)$-cells of K^* has the structure

	$\overset{*}{e}_{n-p-1}$	E^*_{n-p-1}
$\overset{*}{e}_{n-p}$	a'	$0'$
E^*_{n-p}	b'	c'

.

It shows that K^* possesses $\alpha_p + \beta_p - \varrho_p$ distinct $(n-p)$-subcycles. From this we must subtract the number of those $\sim 0 \bmod \Phi$ on K^*, since it is also the number of the cycles $\sim 0 \bmod L$ on K.

We have, to begin with, $\alpha_p - \varrho^a_p$ distinct cycles composed of e-cells alone, hence $\sim 0 \bmod \Phi$. Therefore there are only $\beta_p - \varrho_p + \varrho^a_p$ cycles no linear combination of which is without E^* cells or, what is the same, is on Φ. From this we have to subtract the number of distinct cycles of that type $\sim 0 \bmod \Phi$. For the dimension $n-p-1$ our matrix shows that the number is ϱ_p^c. Therefore for $n-p$ it is ϱ_{p-1}^c. We conclude then

$$r_{n-p}(K-L) = \beta_p - \varrho_p + \varrho_p^a - \varrho_{p-1}^c.$$

Hence

$$(20) \qquad r_p(K-L) = r_{n-p}(K-L).$$

This proves that *the numbers r satisfy the same duality relation for $K-L$ as the Betti numbers.*

Due to a certain lack of symmetry in the parts played by the cells e, e^*, we cannot give any duality relation for the torsion coefficients.

38. **Further simultaneous invariants of K and L and the relations between them.** Dropping for the present the condition that $K-L$ is a manifold, consider the following three sets of p-cycles of $K \bmod L$:

(a) A set composed of cycles on $K-L$ independent mod L. The maximum number of such cycles is $r_p(K-L)$.

(b) A set consisting of true relative cycles, that is actually having boundaries on L, *the boundaries being independent on L.* These boundaries are independent $(p-1)$-cycles on L, but ~ 0 on K. Denote by $s_{p-1}(L)$ the maximum number of such cycles, or the maximum number of independent $(p-1)$-cycles of $L \sim 0$ on K.

(c) A set consisting of $t_p(L)$ relative cycles whose boundaries on L are ~ 0 on it, while the cycles themselves are not only independent mod L, but actually independent on $K \bmod L$ of the cycles on $K-L$, that is of the cycles of type (a). Let C_p be any such cycle, C_{p-1} its boundary on L. Then $L \supset C'_p \to C_{p-1}$. Hence $C_p - C'_p = \Gamma_p$ is an absolute p-cycle on K. The condition that we have imposed, however, implies that the cycles in question are independent of the first type or: *No linear combination of them is homologous to a cycle on $K-L$.* Or again *every cycle depending upon the Γ's meets L.* The integer t is the maximum number of such Γ's.

39. Let now $\{\Gamma_p^{1i}\}$, $\{\Gamma_p^{2i}\}$, $\{\Gamma_p^{3i}\}$ be maximal sets for the three types, the last consisting of absolute cycles (see end

of No. 38). If Γ_p is any cycle mod L not a cycle of $K-L$, we have as a consequence of the definition of the cycles of type (b):

$$xF(\Gamma_p)-\sum x_i F(\Gamma_p^{2i}) \sim 0 \text{ on } L; \quad x \neq 0.$$

Hence the left hand side bounds a chain C_p on L. Therefore $x\Gamma_p - \sum x_i \Gamma_p^{2i} - C_p$ is an absolute cycle of type (c). From the fact that the set of the Γ_p^3's is maximal follows that there exist integers u, u_i with $u \neq 0$, such that $ux\Gamma_p - u\sum x_i \Gamma_p^{2i} - \sum u_i \Gamma_p^{3i}$ depends upon the (absolute) cycles of $K-L$ mod L. And since the Γ_p^1's constitute a maximal set, it depends upon them mod L. Therefore every cycle mod L depends upon those of our three maximal sets. Since the latter are independent, their total number is precisely $R_p(K; L)$. Hence

$$(21) \qquad R_p(K; L) = r_p(K-L) + s_{p-1}(L) + t_p(L).$$

If Γ_p is an absolute cycle it depends on the cycles of type (a), (c) mod L, i. e. the corresponding coefficients x_i are zero. It means that it is homologous to a combination of cycles of the first and third sets plus a cycle on L. From this follows

$$(22) \qquad R_p(K) = R_p(K; L) - s_{p-1}(L) + R_p(L) - s_p(L).$$

Formulas (21) and (22) hold even for $p = 0$ with the convention $s_{-1} = 0$.

40. **Application to a separation problem.** Let us assume that K defines a connected absolute manifold, and consider, for the present only boundary relations mod 2. Since K is an n-circuit

$$R_n(K; 2) = 1.$$

Let us assume in addition that every $(n-1)$-cycle mod 2 bounds on K, so that

$$R_{n-1}(K; 2) = 0.$$

Owing to (3) we have then

$$R_1(K; 2) = 0,$$

hence every one-cycle mod 2 bounds, or every one-cycle mod 2 can be shrunk into a sum of lines taken twice.

From (9) and (21) follows

(23) $$R_0(K-L;\,2) = R_n(K;\,L,\,2);$$

(24) $$R_n(K;\,L,\,2) = r_n(K-L;\,2)+s_{n-1}(L;\,2)+t_n(L;\,2).$$

Since K is an n-circuit every point of K is on every n-subcycle mod 2 of it, hence $r_n = 0$. From our assumption follows that every $(n-1)$-cycle mod 2 of L is ~ 0 on K, hence

$$s_{n-1}(L;\,2) = R_{n-1}(L;\,2).$$

Every n-cycle mod 2 meets L, hence

$$t_n(L;\,2) = R_n(K;\,2) = 1.$$

Adding (23) and (24) we have

(25) $$R_0(K-L;\,2) = R_{n-1}(L;\,2)+1.$$

But the left hand side is merely the number of separate parts into which K is disconnected by L. *Under our assumptions the number of parts into which L disconnects K is equal to* $1+R_{n-1}(L;2)$. *In particular when L is of dimension* $<n-1$, *it does not disconnect K.* This generalizes the separation property described in No. 34 in connection with Alexander's duality relation.

The last part of the statement follows from the fact that when the dimension of L is $<n-1$, its $(n-1)$-st Betti number is zero.

41. **Remarks on the numbers t_p. Application to manifolds with a boundary.** The simplest and in a sense the most interesting formulas are those which involve the fewest invariants. We are going to consider a sufficient condition on L in order that $t_p(L) = 0$. We assume that K is simplicial and that the K-neighborhood of L is normal.

Let σ_p be any cell of K and let $K_{\pi-p-1}$, with π as in No. 3, be the sum of the faces of the cells of the star of σ_p which are opposite to σ_p itself. If C_{q-p-1} is any subchain of $K_{\pi-p-1}$ and C_q its join with σ_p we have by Ch. II, No. 47,

(26) $$C_q \to (-1)^{p+1} (\sigma_p, F(C_{q-p-1})) + \cdots$$

the cells omitted being cells not in the star of σ_p.

Now if D_q is any subchain of K, we may write $D_q = C_q + C'_q$, where C_q consists of cells of the star of σ_p and C'_q has no cell in the star. Therefore $C_q = (\sigma_p, C_{q-p-1})$, where C_{q-p-1} is as above.

The chain C_{q-p-1} is the *trace* of D_q on $K_{\pi-p-1}$. We shall find traces of that type very useful in an important connection in Ch. IV, § 6. We have then

(27) $$D_q = (\sigma_p, \textit{trace } D_q) + \cdots$$

and therefore

(28) $$D_q \to (-1)^{p+1} (\sigma_p, F(\textit{trace } D)) + \cdots.$$

The part of the boundary actually written includes all boundary cells that belong to the star of σ_p. *Therefore, in order that $F(D_q)$ possess no cell in the star of σ_p, it is necessary and sufficient that trace D_q be a cycle.*

42. Let us now introduce the assumption

(29) $$R_{q-p-1}(Lk(\sigma_p)) = 0.$$

Then also

$$R_{q-p-1}(K_{\pi-p-1}) = 0$$

and therefore $K_{\pi-p-1}$ possesses a subchain $C_{q-p} \to s \textit{ trace } D_q, s \neq 0$. Hence

(30) $$\pm (\sigma_p, C_{q-p}) \to s(\sigma_p, \textit{trace } D_q) + \cdots = sD_q - D'_q$$

where D'_q is a chain without cells in the star of σ_p. It follows that $sD_q \sim D'_q$. The chain D' has no cell in common with the star of σ_p. Therefore σ_p is not a cell of it. Thus as a consequence of (29) *every subchain of K to whose boundary σ_p does not belong is related by a homology to a subchain of which σ_p is not a cell.*

43. THEOREM. *Let every cell of L be on the boundary of some cell of dimension* $>q$. *Then if* $R_{q-1}(Lk(\sigma_0)-L)=0$ *for every vertex of L the invariant* $t_q(L)=0$.

Let Γ_q be any subcycle of K. If it has the cell σ_q in common with L, σ_q is a face of a σ_{q+1} of which it is the only q-face belonging to L. We may assume that σ_q and σ_{q+1} are positively related to one another.

Let σ_q enter s times in the expression of Γ_q. Then it does not enter at all in that of

$$\Gamma'_q = \Gamma_q - sF(\sigma_{q+1}) \sim \Gamma_q \text{ on } K.$$

This means that we can suppress step by step the q-cells common to L and Γ_q. From No. 42 and the conditions imposed follows then that from successive subcycles of K, homologous to various multiples of Γ_q, we can remove the vertices on L. In other words, Γ_q is dependent upon the subcycles of K which do not meet L. Since every cycle on K is homologous to a subcycle of it, the result in question holds for every q-cycle, and our proposition is proved.

44. Let the cell σ_p be semi-regular. Then the cycles on K_{n-p-1} are all ~ 0, (29) holds for $q<n$ with $Lk(\sigma_p)$ replaced by $Lk(\sigma_p)-L$ and $s=1$ throughout. Observing that when the cells of L are semi-regular we can replace K by a subdivision without losing that property, and hence take it so that the K-neighborhood of L is normal. Hence

THEOREM I. *When L is of dimension* $<n$ *and its cells are all semi-regular, all cycles of K whose dimension* $<n$ *are homologous to cycles on* $K-L$. *In particular all cycles of L are homologous to cycles of* $K-L$.

We can actually go farther. For let C_p be any chain on K whose boundary Γ_{p-1} does not meet L. We can find a subdivision K^0 of K such that Γ_{p-1} has no points on the K^0-neighborhood of L. Reducing C and Γ to subchains C^0, Γ^0 of K^0 by the fundamental deformation theorem, the same situation prevails. The deformation of Γ being on K^0-L, we have $\Gamma\sim\Gamma^0$ on K^0-L. Since C has no boundary cell on L the same reasoning as above enables us to find a subchain $C_p'^0$ of $K^0 - L$ homologous to C_p on K^0. Since $C_p^0 - C_p$ is a cycle, $F(C_p^0) = F(C_p'^0) = \Gamma_p^0$, and therefore $\Gamma_p^0 \sim 0$ on $K - L$. From this follows that $\Gamma_p \sim 0$ on $K - L$. This shows that if Γ of $K - L$ is ~ 0 on K, it is also ~ 0 on $K - L$. The converse is evident. Therefore we have

THEOREM II. *Under the same circumstances as in Th.* I, *the homology relations between the absolute cycles of* K *and* $K - L$ *whose dimensions are* $< n$ *are the same. This implies that they have the same absolute homology characters.*

Suppose that not all cells of L are regular or semi-regular but only those of $L - L'$, where L' is a subcomplex of L. Observe in passing that the minimal set L' such that all cells of $L - L'$ are regular or semi-regular, is a closed complex: this follows at once from the definition of regularity or semi-regularity. In any case our whole discussion holds with L replaced by $L - L'$. Therefore

THEOREM III. *Let* L' *be a subcomplex of* L *such that all cells of* $L - L'$ *are regular or semi-regular and of dimension* $< n$. *The homology relations between the cycles of* $K - L$ *and of* $K - L'$ *whose dimensions are* $< n$, *are the same and their absolute homology characters are equal.*

In short, in evaluating the homology characters of $K - L$, we can suppress the regular or semi-regular cells of L whose dimension $< n$.

45. **Application to manifolds with a regular boundary.** When K defines an M_n with the regular boundary B, the preceding result holds with $L = B$ or even merely a subcomplex of B. We have therefore in view of the duality relations (9), (10):

THEOREM I. *The homology characters of* $\overline{M}$ *and* $\overline{M} - B$ *are equal. Furthermore*

$$R_p(\overline{M}) = R_{n-p}(\overline{M}; B), \tag{31}$$

$$\theta_i^p(\overline{M}) = \theta_i^{n-p-1}(\overline{M}; B). \tag{32}$$

From (31) follows:

$$\sum (-1)^p R_p(\overline{M}) = (-1)^n \sum (-1)^p R_p(\overline{M}; B). \tag{33}$$

Therefore

THEOREM II. *The Euler-Poincaré characteristics of* $\overline{M}$ *and* $\overline{M}-B$ *are in the ratio* $(-1)^n$.

When n is even the Euler-Poincaré characteristic result can be obtained directly as follows:

(a) Since B is an absolute M_{2n+1}, its characteristic is zero. (No. 28).

(b) In any case remembering the expression of the characteristic in terms of the numbers of cells:

$$\mathit{Char}\ \overline{M} = \mathit{Char}\ B + \mathit{Char}\ \overline{M}-B.$$

Combining the two when n is even, the required result follows.

In a previous paper [11] we had obtained several duality theorems for a manifold with a regular boundary. We can now show that they are consequences of the various relations obtained so far. We take $L = B$, the boundary of M_n defined by K. The first two of the theorems in question written in terms of our invariants correspond to the following relations:

$$\text{I.} \qquad R_p(\overline{M}-B)-r_p(\overline{M}-B) = s_{n-p-1}(B),$$

$$\text{II.} \qquad r_p(\overline{M}-B) = r_{n-p}(\overline{M}-B).$$

The third theorem reduces again to II, owing to the vanishing of the numbers t_p. Of course II is merely (20) written in different form. As for I it follows from (9), (20), and (21) with the t's replaced by zero.

§ 4. INVARIANCE OF MANIFOLDS

46. Since the n-cell is topologically invariant, the invariance of manifolds will be proved if we can establish the

THEOREM. *In order that* $K-L$ *be an* M_n *it is necessary and sufficient that every point* Q *of* $K-L$ *possesses a neighborhood as small as we please which is an n-cell.* (*Van Kampen* [3]).

This theorem is important in itself for it implies a local invariantive characterization of manifolds.

Let us first replace K by some simplicial subdivision K^0 (a derived of some order) with Q for vertex, and whose mesh $< \varepsilon$.

This is permissible since the manifold property is preserved under subdivision.

Since Q is a regular cell of $K-L$, its linked complex is an H_{n-1} and the join $(Q, H)-H$ is the star of Q and therefore a neighborhood of Q as well as an n-cell of diameter $< 2\varepsilon$, (No. 8, I). The condition of the theorem is therefore necessary.

47. The crux of the proof is really in the second part — the sufficiency of the condition. It will be convenient to designate the star of E_p by $St(E_p)$. We assume then that $K-L$ has the required behavior and show that its cells are regular. From No. 15, Th. III, follows that K can be chosen simplicial.

There is nothing to prove for the n-cells, since all n-cells of a simplicial K_n are regular. Let $p<n$ and let us assume that the required result has been established for all cells of more than p dimensions. Then every cell of $St(E_p)$ is regular so that we merely have to show that $Lk(E_p)$ is an H_{n-p-1}. Since K is simplicial Lk may be replaced by its homeomorph, the complex sum of all faces of the cells of $St(E_p)$ opposite to E_p. In order not to complicate matters we also call that new complex Lk for the present.

Our problem is equivalent to the following: *To prove that the homology and circuit characters of Lk are the same as those of an* H_{n-p-1}.

48. The required result will follow from a study of the relationship between the subchains of $\overline{St(E_p)}$ and their traces on $Lk(E_p)$ (No. 41).

Let then C_r be any subchain of $\overline{St}$ and C_q $(q=r-p-1)$ its trace on Lk so that $C_r=(E_p, C_q)$ the join of E_p and C_q.

Let $\Phi=(F(E), Lk)=$ the boundary of St. The boundary relations for joins (Ch. II, No. 47), yield here

$$(34)\quad (-1)^{p+1}F(C_r) \equiv \begin{cases}(E_p, F(\text{trace } C_r)), & r>p+1\\ t\,E_p, & r=p+1\end{cases} \mod \Phi.$$

From this and the linearity of the trace as a chain function follows at once that *the homologies between the chains* C_r *mod* Φ *and trace* C_r *(absolute) are the same for* $r>p+1$.

49. Under the assumption of our theorem a point Q of E_p will possess a neighborhood N which is an n-cell, and whose closure $\bar{N} \subset St$. Since $\bar{N}$ is an n-complex, St includes cells of dimension $\geqq n$ (Ch. II No. 32), and as n is the maximum dimension of its cells, it includes n-cells. Therefore E_p is the face of an n-cell of $\overline{K-L}$ and $\overline{K-L}$ *is the sum of its closed n-cells.* Then, if $\overline{St-N} = \Psi$, a subcycle of $\overline{St}$ mod Φ, such as C_r above, is also a cycle on $\Psi + N$ mod Ψ. Since $r \geqq p+1$, necessarily $r \geqq 1$. Since $\bar{N}$ is a complex $F(N)$ is locally connected, hence by reasoning as in Ch. II, No. 26, we show that $C_r \sim$ mod Ψ to a cycle of $\bar{N}$ mod $F(N)$ on $\overline{St}$. Hence if $r < n$ and since N is an n-cell, $C_r \sim 0$ mod Ψ and therefore $\Psi \supset D_r \sim C_r$. It follows that the boundaries of C and D coincide. Hence $F(D) \subset \Phi$. Since $D \not\supset Q$ there is a unique segment QP from Q to every point P of D. Let R be its intersection with Φ. We can slide P along QR until it comes to R. This deforms D into $D' \subset \Phi$ with the same boundary. Hence $D' \sim D \sim C_r$, and finally $C_r \sim 0$ mod Φ. In other words *the r-subcycles of* $\overline{St}$ *mod* Φ, $p+1 < r < n$, *are all* ~ 0 *mod* Φ.

50. Passing to $r = n$, let first $p = n-1$. Since $F(N)$ is locally connected and $\not\supset Q$, it follows from Ch. II, No. 26, and the deformation theorem for simplicial complexes, that the subcycles of $\overline{St} (= \Psi + N)$ mod Ψ are deformable into simplicial subchains of some subdivision K^0 of K, their boundaries being at a distance from Ψ that tends to zero with the mesh ξ of K^0. Since N is a circuit, K^0 will then carry, for ξ small enough, an n-subchain C_n, whose boundary mod. 2 $\not\supset Q$. This implies that the closed $(n-1)$-cells of K^0 which carry Q are incident with an even number of n-cells. In particular those on E_{n-1} have that property. This can only be if E_{n-1} is itself incident with an even number of n-cells of K. We conclude then: E_{n-1} must be incident with at least two different n-cells, E_n, E'_n of $K-L$.

Let now C_n be a subchain of $\overline{St}(E_{n-1})$. From (34) follows

$$(-1)^n C_n \to t E_{n-1} \text{ mod } \Phi.$$

Let in addition

$$E_n \to \varepsilon E_{n-1}, \; E'_n \to \varepsilon' E_{n-1} \text{ mod } \Phi; \quad \varepsilon, \varepsilon' = \pm 1.$$

Then $(-1)^n C_n - t\varepsilon E_n$ and $\varepsilon E_n - \varepsilon' E'_n$ are subcycles of $\overline{St}(E_{n-1})$ mod Φ and hence they are cycles of $\Psi + N$ mod Ψ. Since N is an n-cell, we have, as in a similar case No. 49, that the two cycles must be dependent. Therefore we have

$$\alpha C_n + \beta E_n + \gamma E'_n \sim 0 \text{ mod } \Phi.$$

Since n is the maximum dimension of all cells and we are dealing only with subchains of $\overline{St}$ this homology reduces to

$$\alpha C_n + \beta E_n + \gamma E'_n \equiv 0.$$

As the cells E, E' are distinct, we cannot have $\alpha = 0$. Therefore C_n is necessarily a linear combination of E, E'. In other words $\overline{St}(E_{n-1})$ consists of two closed n-cells, i. e. *it is a simple circuit.* It follows from this and No. 49:

(a) E_{n-1} *is regular.*

(b) $K-L$ *itself is a sum of a finite number of simple circuits.*

51. We are now ready to draw conclusions regarding $Lk(E_p)$. Since $p = n-1$ has already been treated we can take $p < n-1$.

Now the incidence relations between the $(i+1)$- and i-cells of $St(E_p)$ are the same as for the $(i-p)$- and $(i-p-1)$-cells of Lk. Therefore from Nos. 49, 50 follows that Lk is the sum of its closed $(n-p-1)$-cells and that every $(n-p-2)$-cell of Lk is incident with two $(n-p-1)$-cells. Therefore Lk is a sum of simple circuits. To prove that it is a sphere it is only necessary to show that *the homology relations between its cycles are as for the sphere.* For the cycles of dimension $q \neq 0, n-p-1$, that follows from Nos. 48, 49.

In view of (34), $R_{n-p-1}(Lk) = R_n(\overline{St}; \Phi)$. Now if C_n, C'_n are any two cycles of $\overline{St}$ mod Φ, passing as in No. 49 through their homologies mod Ψ we shall find this time a homology

$$tC_n + t'C'_n \sim 0 \text{ mod } \Phi.$$

Hence there is at most one independent n-cycle and $R_n(\overline{St}; \Phi) \leqq 1$. On the other hand, let D_n be the fundamental n-cycle of the cell N. We will assume that it is already reduced to a simplicial chain after the manner of No. 50. The sum of the closed n-cells of D_n which meet Q

is a chain $D_n^0 \neq 0$. Since $F(D_n) \not\supset Q$, it can be deformed as in No. 49 onto Φ. If we add its deformation chain to D_n we have a new chain D_n' which is a cycle of $\overline{St}$ mod Φ, that we may assume to be simplicial. Now if the initial mesh of D_n has been chosen small enough, and since the various operations involved in passing from D_n to D_n' are on cells sufficiently far from Q, the sum of the closed cells of D_n' meeting Q will still be $D_n^0 \neq 0$. Therefore D_n' is a proper cycle of $\overline{St}$ mod Φ, and $R_n\,(\overline{St};\ \Phi) \geqq 1$. Hence $R_n(\overline{St};\ \Phi) = 1 = R_{n-p-1}\,(Lk)$, Lk is an orientable simple circuit and as it is connected, $R_0\,(Lk) = 1$. *Therefore* Lk *is a sphere.* This completes the proof of van Kampen's theorem.

52. Remarks. I. At the end of No. 49 we had recourse to the invariance of dimensionality in order to show that $\overline{K-L}$ is the sum of its closed n-cells. Actually this follows independently from the fact just proved that $R_n(\overline{St};\ \Phi) = 1$.

II. The same mode of proof establishes the invariance of manifolds mod m.

III. What has actually been established is the invariance of what might be called the *manifold of grade* p, i. e. of a $K-L$ whose cells of dimension $\geqq p$ are regular. For $p = n-1$ it is a sum of simple circuits. In particular our result implies the invariance of the simple circuit, already proved in another manner, Ch. II, No. 37.

IV. Van Kampen's proof published in his thesis [3] is substantially along the following lines: He defines as a *p-cycle at a point* Q on K, any C_p whose boundary $\not\supset Q$. The cycle ~ 0 at Q when $C_p \sim C_p' \not\supset Q$. This enables him to define homology characters at Q which are simultaneous topological invariants of K and Q. It is then shown that when these characters are the same as for an H_{n-1} we have a manifold. The similarity between the two proofs is evident. Our C_r's are the cycles at Q of van Kampen, and to his homologies at Q correspond the relative homologies on N that we had. I am advised by Alexander that an unpublished proof which he obtained several years ago, was practically the same as van Kampen's.

53. We will now prove the *topological invariance of the manifold with regular boundary.* Let K be a simplicial n-complex and A the set of all points of K having the property of possessing an arbitrarily small n-cell for neighborhood. An inner point of A will be called a *regular* point of K.

According to No. 46 every point of a regular E_p is a point of A. But with E_p every cell of its star is regular, hence when E_p is regular $St(E_p) \subset A$, and as St is a neighborhood for every point of E_p, every point of E_p is regular.

Conversely let Q be a regular point of E_p. The reasoning of the preceding numbers shows that $Lk(E_p)$ is then a sphere. But when Q is regular every cell of St also carries regular points. Hence the linked complexes of all cells of St are spheres and since K is simplicial, E_p is regular. Therefore the set B of all regular points coincides with the sum of all regular cells.

Now if E_p is irregular so is every face of E_p. Hence the sum of all irregular cells is a closed subcomplex L of K and $B = K - L$. Since B is characterized by a topological property, the same holds for L.

Suppose that $K - L$ is an M_n with the regular boundary L. Let K^0 be a copy of K and bring into coincidence the points of L with their images on K^0, thus obtaining a new complex $K + K^0$. We now have the following properties:

(a) L is the set of all irregular cells of K.

(b) L is an absolute M_{n-1} which disconnects $K + K^0$ into two parts $K - L$, $K^0 - L$.

(c) $K + K^0$ is an absolute M_n.

The first two are true by assumption or construction, while (c) is a consequence of (b), (No. 35 Corollary). These three properties, which are topological, jointly imply that $K - L$ is an M_n whose boundary L is regular. For (a) implies that L is a closed subcomplex of both K and K^0 and (b), (c) that $K - L$ behaves as asserted (No. 35 Corollary). Since the property of being an M_n with regular boundary is a consequence of three topological properties, it is itself topological.

CHAPTER IV

INTERSECTIONS OF CHAINS ON A MANIFOLD

Given chains C_p, C_q on an orientable n-manifold (absolute or relative) that do not meet each other's boundaries it is possible to associate with them a definite $(p+q-n)$-cycle, their intersection. When $p+q=n$ there is an associated number, the Kronecker index. Both cycle and index have a large degree of invariance under certain modifications of the chains, for example when they undergo sufficiently small deformations. In a recent paper [8] we have given a complete treatment of these questions. We propose to take them up again here with the double object of simplifying the theory and extending its scope.

As in our paper, we first study specialized polyhedral intersections, then treat the general case by means of approximations of the special type. The result, as it is shown with greater precision than in our paper, is wholly independent of the mode of approximation.

Instead of the convex polyhedral intersections first chosen as the special type, we have given the preference to pairs of chains of a manifold $K-L$ and of its dual K^*, a scheme followed with much success by Veblen [6], Weyl [3], also by Poincaré [3] for $n=3$. For the theory the new type has decidedly the best of it but in the applications the reverse is true, since given polyhedral chains in practise rarely have subdivisions related as in the new scheme. Furthermore it seems desirable to have a unique basic type of polyhedral intersections including the two. This we have obtained by means of Brouwer's looping coefficients, which have enabled us to define an intersection chain for two subchains whenever their common cells are of the right dimensions.

It is hardly necessary to point out the importance and practical value of a carefully built up topological theory of intersections of chains. This will be clearly shown by the applications to be found in the following chapters. It is also apparent from the numerous explicit or implicit occurrences of the Kronecker index, the looping coefficients, etc. in the literature. We may mention only the noted double integral of Gauss, and the characteristics and integrals considered by Kronecker and Poincaré. There are also numerous applications to algebraic geometry as will be shown in Ch. VIII.

§ 1. Polyhedral Intersections

1. As in the preceding chapter the fundamental complex K is a normal n-complex with L as its subcomplex and K' as its first derived complex. We assume that $K-L$ is an orientable and oriented n-manifold, although orientability may be dispensed with whenever we deal with intersections mod 2 to be defined later. We denote again by K^* the dual of $K-L$ built up by means of K', and use the same notations for the cells of K' as in Ch. III.

2. Let us first examine the intersection of E_p of $K-L$ with E_q^* of the dual K^*. As we shall see, when it is not vacuous it is an s-cell, $s=p+q-n$, which we shall orient in a definite manner.

The cell E_p is the sum of all cells of K' in whose symbol A^p is the last vertex. On the other hand E_q^* is the sum of all cells of K' in whose symbol A^{n-q} is the first vertex. Since E_p, E_q^* are both exact sums of cells of K', their intersection is also an exact sum of cells of K'. These cells must then all be of type $A^{n-q}\dots A^{ir}\dots A^p$ where the vertices A^{ir} are common to two cells of K', one on E_q^* the other on E_p. It implies also that E_{n-q}, E_r^i, E_p are mutually incident, with $n-q\leqq r\leqq p$; therefore, in particular, that $s=p+q-n\geqq 0$.

The required intersection $\mathfrak{E}_s$ is the sum of all cells of K' having a symbol $\sigma_s = A^{n-q} \dots A^r \dots A^p$. They are all on E_p and they all have A^{n-q} and A^p as their extreme vertices.

We now recall that E_p is an M_p with the regular boundary $F(E_p)$, (Ch. III, No. 8). Since the cells of K' on $\bar{E}_p$ decompose that complex into the cells of a first derived, we find by reference to Ch. III, No. 13, that $\mathfrak{E}_s$ is merely the part of $Tv\,(E_{n-q})_{\bar{E}_p}$, which is on E_p itself and not on its boundary. That part consists precisely of a cell of dimension $p-(n-q) = s$.

The component σ_s of the cell $\mathfrak{E}_s$ is uniquely determined by a mutually incident set. $E_{n-q}, \cdots, E_r, \cdots E_p$ and conversely. Any component σ_{s-1} of $\mathfrak{E}_s$ is then of the form $A^{n-q} \dots A^{r-1} A^{r+1} \dots A^p$, $n-q<r<p$. The σ_s's of which σ_{s-1} is a face are obtained by inserting all vertices A^r that are on a cell E_r which is on $F(E_{r+1})$ and has E_{r-1} for face. Since $F(E_p)$ is an absolute M_p, the same property belongs to the cells of its boundary. Therefore $F(E_{r+1})$ is an absolute M_{r+1} and hence a simple circuit. Consequently there are just two cells such as E_r, and two components σ_s incident with σ_{s-1}, so that $\mathfrak{E}_s$ is a simple circuit.

The boundary of $\mathfrak{E}_s$ is the sum of the $(s-1)$-faces of the cells σ_s in which there enters only one of the vertices A^{n-q}, A^p. Referring to their symbols we find that it is the join of $H_0 = A^{n-q} + A^p$ with $H_{s-2} = Lk(E_{n-q})_{F(E_p)}$ and hence it is an H_{s-1}. Summing up, we have then proved:

THEOREM. *The geometric intersection of E_p of $K-L$ and E^*_q of K^* is either vacuous or else a cell of dimension $s=p+q-n$ which is a simple circuit whose boundary is an $(s-1)$-sphere.*

3. We now consider the orientation of the intersection cell $\mathfrak{E}_s$. Since it is a simple circuit the orientation of all the s-cells, components of K' on $\mathfrak{E}_s$, is completely determined when we have that of one of them, then orient all the others by the condition that any two s-cells of K' on $\mathfrak{E}_s$ with a common $(s-1)$-face are oppositely related to it.

Let $\beta_p A^0 A^1 \cdots A^{n-q-1} \sigma_s = \beta_p\, \sigma_{n-q-1} \sigma_s$ be a cell of K' on E_p and indicatrix of the latter. β_p and the other β's to be used presently are all ± 1. As in Ch. III, § 3 they are introduced in order to take care of the exceptional cases where some of the σ's are of dimension zero. Let similarly E_q^* carry the cell $\beta_q \sigma_s A^{p+1} \ldots A^n = \beta_q \sigma_s \sigma_{n-p-1}$. In the sequence $A^0 \ldots A^{n-q} \ldots A^r \ldots A^p \ldots A^n$ each vertex is on one cell of a mutually incident set of cells of K. In particular A^n is on an n-cell to whose closure all the other vertices belong. Therefore, for proper β_n, $\beta_n \sigma_{n-q-1} \sigma_s \sigma_{n-p-1}$ is a certain n-cell of K', indicatrix of K' or of K itself. As for $\mathfrak{E}_s$ let $\beta_s \sigma_s$ be its indicatrix.

The cell σ_s of K' is a face that may belong to a number of components of E_p or E_q^*. Following our customary conventions the p-cells components of E_p, for example, are to be oriented concordantly with E_p, and similarly for E_q^*. Let us examine what happens when we vary the components having σ_s for face. E_p is decomposed by K' into the cells of an orientable circuit mod $F(E_p)$ with two p-cells incident with, and oppositely related to, each $(p-1)$-cell. Hence if A^t is one of the vertices of σ_{n-q-1}, one of the components of E_p is a cell

$$-\beta_p A^0 A^1 \ldots A^{t-1} A'^t A^{t+1} \ldots A^{n-q-1} \sigma_s = -\beta_p \sigma'_{n-q-1} \sigma_s .$$

But then the corresponding n-cell $\sigma'_{n-q-1} \sigma_s \sigma_{n-p-1}$ has a common face with $\sigma_{n-q-1} \sigma_s \sigma_{n-p-1}$, and as positive cells of K' they must be oppositely related to that face. Hence as indicatrix of K' the new n-cell must be represented by $-\beta_n \sigma'_{n-q-1} \sigma_s \sigma_{n-p-1}$. Therefore on passing from one component of E_p to an adjacent one the product $\beta_p \beta_q \beta_n$ is left unchanged. Since we can thus pass from one component to every other the product is independent of the components chosen on E_p, and likewise of those on E_q^*, whose common face σ_s is. We choose this product as the value of β_s or more symmetrically *we determine* β_s *by the relation.*

$$\beta_p \beta_q \beta_s \beta_n = 1. \tag{1}$$

This gives us a unique orientation for the component σ_s of $\mathfrak{E}_s$ depending solely upon the orientations of $K-L$ and of the intersecting cells.

It remains to be shown that the components of $\mathfrak{E}_s$ are thus concordantly oriented. The reasoning is about the same. It is again sufficient to consider a cell having a common $(s-1)$-face with σ_s, say $\sigma'_s = A^{n-q} \dots A^{r-1} A'^r A^{r+1} \dots A^p$, obtained by replacing A^r by A'^r among the vertices of σ_s. The proper components to be considered may be chosen as $\sigma_{n-q-1} \sigma'_s$, $\sigma'_s \sigma_{n-p-1}$, $\sigma_{n-q-1} \sigma'_s \sigma_{n-p-1}$. The first has a common $(p-1)$-face with $\sigma_{n-q-1} \sigma_s$ and, since as a concordant component of E_p it must be oppositely related to it, it is described properly oriented by $-\beta_p \sigma_{n-q-1} \sigma'_s$. Thus β_p has been changed in sign and similarly for β_q and β_n. From (1) follows then that β_s goes into $-\beta_s$. The orientation that our scheme attributes to the component σ'_s of $\mathfrak{E}_s$ is described by $-\beta_s \sigma'_s$, i. e. it is such that σ_s and σ'_s are oppositely related to their common $(s-1)$-face. In other words both are concordantly oriented relatively to $\mathfrak{E}_s$.

REMARK. It follows in particular from our discussion that the special choice of any order for naming the vertices of the σ's is immaterial — any change in it is always compensated by suitable changes of signs in the β's, but the indicatrix $\beta_s \sigma_s$ remains fixed.

4. We have then determined a unique and well defined orientation for $\mathfrak{E}_s$, the indicatrix being $\beta_s \sigma_s$. The cell thus oriented is by definition the *intersection chain* of the two chains E_p, E^*_q and it shall be designated by $E_p \cdot E^*_q$. There will be no danger of confusion with the "intersection" symbol of point-sets since we will never consider such intersections for chains.

When $p+q = n$, $s = 0$, the intersection consists of a single point; the intersection chain is the point affected with the sign of β_0 determined by

$$(2) \qquad \beta_0 \beta_p \beta_{n-p} \beta_n = 1.$$

The number β_0 is called the *Kronecker index* of the two cells, and designated by $(E_p \cdot E^*_{n-p})$. One may think of it as describing the multiplicity (positive or negative) of intersection of the two cells. It was already considered in Ch. III, No. 24, and the orientation imposed upon E^*_{n-p} there was equivalent to the condition $(E_p \cdot E^*_{n-p}) = 1$. However for the present we wish to make our treatment independent of any particular orientation chosen for the cells.

Here as in Ch. III the two dual cells are symmetrically related with respect to K'. If we permute their parts in the above scheme we find that E^*_q, E_p and K have for respective indicatrices $(-1)^{(s+1)(n-p)} \beta_q \sigma_{n-p-1} \sigma_s$; $(-1)^{(s+1)(n-q)} \beta_p \sigma_s \sigma_{n-q-1}$ and $(-1)^{(s+1)(n-p)+(q+1)(n-q)} \beta_n \sigma_{n-p-1} \sigma_s \sigma_{n-q-1}$. Hence by a very simple calculation, if $\beta'_s \sigma_s$ is the new indicatrix to be assigned to $\mathfrak{E}_s$,

$$\beta'_s = (-1)^{(n-p)(n-q)} \beta_s \tag{3}$$

and therefore

$$E^*_q \cdot E_p = (-1)^{(n-p)(n-q)} E_p \cdot E^*_q. \tag{4}$$

In particular for $s = 0$ from (3) follows

$$(E^*_{n-p} \cdot E_p) = (-1)^{(n+1)p} (E_p \cdot E^*_{n-p}). \tag{5}$$

From (1) and (2) follows also

$$\begin{cases} (-E_p) \cdot E^*_q = -E_p \cdot E^*_q = E_p \cdot (-E^*_q), \\ ((-E_p) \cdot E^*_{n-p}) = -(E_p \cdot E^*_{n-p}) = (E_p \cdot (-E^*_{n-p})). \end{cases} \tag{6}$$

When E_p and E^*_q do not meet we have a null chain or a zero index as the case may be, that is

$$E_p \cdot E^*_q = 0, \qquad (E_p \cdot E^*_{n-p}) = 0 \tag{7}$$

for non-intersecting cells.

In order that there be intersections it is necessary and sufficient that the dual E_{n-q} of E^*_q, be E_p or a boundary cell of E_p. *Hence* $E_p \cdot E^*_q$ *or* $(E_p \cdot E^*_{n-p})$ *are* $\neq 0$ *when and only when* E^*_q *or* E^*_{n-p} *are duals of* E_p *or of cells of the boundary of* E_p.

5. The boundary of the intersection cell. We have seen (No. 2) that the $(s-1)$-faces of the cells σ_s in whose symbols there appear both A^{n-q} and A^p are cells of $\mathfrak{E}_s$ itself and hence they do not belong to its boundary. The only boundary cells of σ_s that belong to that of $\mathfrak{E}_s$ are therefore of the type σ_{s-1} such that $A^{n-q}\sigma_{s-1} = \sigma_s$ or $\sigma_{s-1}A^p = \sigma_s$.

Consider those of the first type. σ_{s-1} has then a symbol $A^{i,n-q+1}\ldots A^{h,p-1}A^p$. It is therefore a component of $E_p \cdot E^{*i}_{q-1}$. Since E^i_{n-q+1} is incident with E_{n-q}, E^{*i}_{q-1} is on the boundary of E^*_q. Furthermore every component of the intersection in question is a cell such as σ_{s-1}, for the only condition imposed is merely that $A^{n-q}A^{i,n-q+1}\ldots A^p$ be a cell of K', hence that $E_{n-q}, E^i_{n-q+1}, \cdots, E_p$ be mutually incident.

As for the orientation of σ_{s-1} relatively to $\mathfrak{E}_s$, let

$$E^*_q \rightarrow \sum \eta_i E^{*i}_{q-1}. \tag{8}$$

We have $E^{*i}_{q-1} = \pm\sigma_{s-1}\sigma_{n-p-1}$. Hence in view of (8) and of the fact that $\beta_q A^{n-q}\sigma_{s-1}\sigma_{n-p-1}$ is an indicatrix of E^*_q, $\eta_i\beta_q\sigma_{s-1}\sigma_{n-p-1}$ is an indicatrix of E^{*i}_{q-1}. From (1) applied to $E_p \cdot E^{*i}_{q-1}$ follows that it has for indicatrix $\eta_i\beta_s\sigma_{s-1}$. Since $\beta_s A^{n-q}\sigma_{s-1}$ is an indicatrix of $\mathfrak{E}_s$ it follows that $E_p \cdot \eta_i E^*_{q-1}$ is positively related to $E_p \cdot E^*_q$. Therefore

$$E_p \cdot E^*_q \rightarrow \sum \eta_i E_p \cdot E^{*i}_{q-1} + \cdots, \tag{9}$$

the unwritten cells being cells σ_{s-1} such that $\sigma_{s-1}A^p = \sigma_s$.

But in view of the symmetry between $E_p \cdot E^*_q$ as regards the intersection and K', if the boundary relation for E_p is

$$E_p \rightarrow \sum \zeta_j E^j_{p-1}$$

then

$$E^*_q \cdot E_p \rightarrow \sum \zeta_j E^*_q \cdot E^j_{p-1} + \cdots, \tag{10}$$

the cells in the sum being those left out in (9) and the cells omitted those in the sum in (9). From (10) and (4) follows

$$E_p \cdot E^*_q \rightarrow (-1)^{n-q} \sum \zeta_j E^j_{p-1} \cdot E^*_q + \cdots. \tag{11}$$

Since the sum in either one of the formulas (9), (11) corresponds to the omitted part in the other, we have finally

(12) $E_p \cdot E_q^* \to \sum \eta_i \, E_p \cdot E_{q-1}^{*i} + (-1)^{n-q} \sum \zeta_j \, E_{p-1}^j \cdot E_q^*$,

which is the desired boundary relation. We shall give it a simpler form in a moment.

6. **Intersection of arbitrary chains. The fundamental boundary relation.** Having numbered and oriented the cells we consider in all cases the symbols $E_p^i \cdot E_q^{*j}$, $(E_p^i \cdot E_{n-p}^{*j})$, although they are $\neq 0$ when and only when E_q^{*j} (or E_{n-p}^{*j} in the case of the index) is the dual of E_p or of a boundary cell of E_p (No. 4). We return to the convention of Ch. III, No. 24, and orient again the starred cells so that

(13) $$(E_p^i \cdot E_{n-p}^{*j}) = \delta_{ij}.$$

Let now C_p, C_q^* be two subchains of $K-L$ and K^*. We define as their intersection a certain s-chain, $s = p+q-n$, to be designated by $C_p \cdot C_q^*$, by the following conditions: (a) It is a *linear* function of C_p or C_q^*, i. e.

(14) $$(tC_p + t' C_p') \cdot C_q^* = tC_p \cdot C_q^* + t' C_p' \cdot C_q^*,$$

and similarly for C_q^*. (b) When the chains are cells the intersections are the chains already defined. From these conditions follows immediately that if

(15) $$C_p = \sum x_i \, E_p^i, \qquad C_q^* = \sum y_j \, E_q^{*j},$$

then

(16) $$C_p \cdot C_q^* = \sum x_i \, y_j \, E_p^i \cdot E_q^{*j}.$$

When $p+q=n$, $s=0$ and the intersection consists of a finite number of points. The Kronecker index, again defined by conditions (a), (b), is designated by $(C_p \cdot C_{n-p}^*)$ and its value is

(17) $$(C_p \cdot C_{n-p}^*) = \sum x_i \, y_j \, (E_p^i \cdot E_{n-p}^{*j}) = \sum x_i \, y_i.$$

Formulas (4), (5), (6) hold with chains in places of cells, i. e. with E replaced by C, and we need not rewrite them.

7. The boundary relation (12) is now seen to be equivalent to

$$F(E_p \cdot E_q^*) = E_p \cdot F(E_q^*) + (-1)^{n-q} \, F(E_p) \cdot E_q^*.$$

When combined with (16) it yields at once the *fundamental boundary relation*

$$(18)\quad F(C_p \cdot C_q^*) = C_p \cdot F(C_q^*) + (-1)^{n-q} F(C_p) \cdot C_q^* .$$

8. **Remark on the Kronecker index.** The index may be approached from a somewhat more general and advantageous angle. We first define it for any C_0 whatever as the number of vertices of C_0 each counted with its coefficient in the chain. In other words designating it by (C_0), if $C_0 = \sum t_i E_0^i$, then $(C_0) = \sum t_i$. From Ch. I, No. 28, Remark, follows then the

THEOREM. *When* $C_0 \sim 0$ *on any* K *that carries it,* $(C_0) = 0$.

Regarding the application to intersections, $(C_p \cdot C_{n-p})$ is merely the index of the zero-cycle $C_p \cdot C_{n-p}$.

9. Let us return now to our general intersections. From (18) follows

$$(19)\quad F(C_{p+1} \cdot C_{n-p}^*) = C_{p+1} \cdot F(C_{n-p}^*) + (-1)^p F(C_{p+1}) \cdot C_{n-p}^* .$$

Since the right hand side is a bounding C_0 we have by the above,

$$(20)\quad (C_{p+1} \cdot F(C_{n-p}^*)) + (-1)^p (F(C_{p+1}) \cdot C_{n-p}^*) = 0 .$$

This noteworthy formula enables us to pass from a Kronecker index of a pair of chains of dimensions $p, n-p$ to the index of another pair of dimensions $p-1, n-p+1$. We will find this most useful in the proof of the topological invariance of the index (§ 4).

10. A series of simple properties of the intersection chains and indices, following at once from our formulas, will now be listed:

I. *If* C_p, $[C_q^*]$, *is a cycle mod* L, [*is an absolute cycle*], *the boundary of the intersection* $C_p \cdot C_q^*$ *is* $\pm$ *the intersection of* C_p, [*of* C_q^*], *with* $F(C_q^*)$, [*with* $F(C_p)$].

II. *If both are cycles, mod L and absolute respectively, their intersection is an absolute cycle.*

III. *When the chains do not meet one another's boundaries their intersection is a cycle.*

IV. *The intersection of two cycles of which one is a zero-divisor, is a zero-divisor.*

For if say, $C_p \to 0, C^*_{q+1} \to t C^*_q, t \neq 0$ then $F(C_p \cdot C^*_{q+1}) = t C_p \cdot C^*_q$.

The intersection cycle or zero-divisor in III or IV are both absolute, the bounding relations referring to $K - L$. An equivalent statement to IV is: *From* $C_p \to 0$, $C^*_q \approx 0$ *follows* $C_p \cdot C^*_q \approx 0$. Here as usual $\approx$ is the symbol of homology with division allowed or with zero-divisors dropped.

V. *When a multiple of* C_p *bounds a* C_{p+1} *which does not intersect* $F(C^*_{n-p})$, $(C_p \cdot C^*_{n-p}) = 0$. *In particular this is the case when* C^*_{n-p} *is an absolute cycle.*

This follows at once from (20), for the first term at the left is then zero.

The following simple special case of V in which $p = n - 1$, is practically equivalent to this well known fact: When an arc AB of S_n has its end points exterior to a region $\mathfrak{R}$, the number of times that a point describing AB in a given sense, enters $\mathfrak{R}$ is equal to the number of times that it leaves the region. Historically Theorem V is also of considerable interest, in that it is through a partial inversion of it that Poincaré attempted first to prove the duality theorem for absolute manifolds in his first paper on Analysis Situs [2].

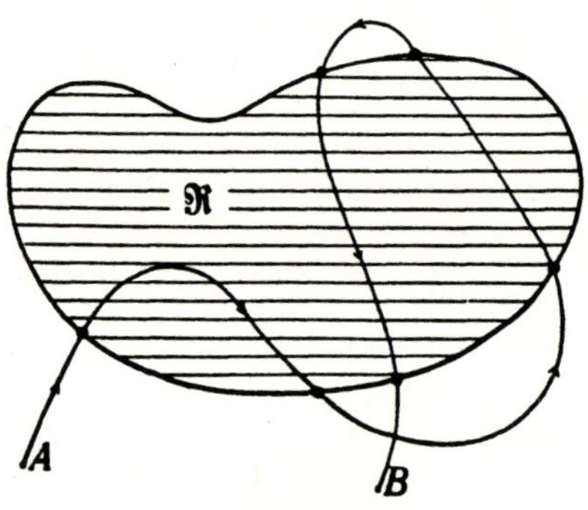

Fig. 14.

Definition. When the chains have their coefficients reduced mod m, the chain obtained from $C_p \cdot C^*_q$ by reducing coefficients mod m is the *intersection mod m*. When $q = n - p$, the algebraic number of points in the chain $C_p \cdot C^*_{n-p}$, reduced mod m, is the *Kronecker index mod m*. *All the results of this section hold for intersections or Kronecker indices mod m, with the usual privilege of neglecting orientability when* $m = 2$.

In the last case the formulas become especially simple. For example (18) becomes

$$(18') \qquad F(C_p \cdot C_q^*) = C_p \cdot F(C_q^*) + F(C_p) \cdot C_q^*.$$

Also $C_p \cdot C_q^* = C_q^* \cdot C_p$, etc.

11. **Extension to an arbitrary number of intersecting chains.** Let K^i be the ith derived complex of K, K^{*i+1} its dual as constructed by means of K^{i+1}, with $K^{*1} = K^*$, so that K may be considered as K^0. We call K^{*j} the jth *dual* of K.

We consider intersections of the two types $C_p \cdot C_q^{*i} \cdots C_r^{*h}$ and $C_q^{*i} \cdots C_r^{*h}$, where C_p is a subchain of K and C^{*j} a subchain of K^{*j}, and propose to define them as certain s-chains, where if ν is the number of intersecting chains, $s = p + q + \cdots \cdots + r - n(\nu - 1)$, or $= q + \cdots + r - n(\nu - 1)$ in the two respective cases.

We will first show that when each chain consists of a single cell the intersection is either vacuous or else an s-cell which is a simple circuit with an $(s-1)$-sphere for boundary. This is the extension of the theorem proved in No. 2.

Consider first the intersection $E_p \cdot E_q^{*i} \cdots E_r^{*h}$. Suppose that the boundary L of K is regular, so that its cells or all those of any simplicial subdivision of it are semi-regular. If E_t is a semi-regular cell of K^j on L, we recall that the part of its dual which is on $K - L$ is also an $(n-t)$-cell which is a simple circuit with a sphere for boundary. For purposes of the proof it is necessary to consider also intersections of the type now taken up, when to the cells of K^{*j+1} are added cells such as the $(n-t)$-cell considered. In other words it is advisable to consider K^{*j+1} as the sum of the parts on $K - L$ of the duals of all cells of K^j. Thus K^{*j+1} is an open complex which as a point set coincides with $K - L$.

Let E_q^{*j+1} denote any one of the starred intersecting cells. It is the dual of a cell E_{n-q}^j of K^j and therefore it is on the star of that cell with respect to K^j. Hence E_q^{*j+1} can only meet E_p if that star does, that is if E_{n-q}^j is a cell of the subdivision induced on $\bar{E}_p$ by K^j. Hence, as in No. 2, their intersection is the part of the dual of E^j_{n-q} relatively to the manifold E_p which is on the cell E_p itself. Since E_p is an M_p with regular boundary that dual is a t-cell $\mathfrak{E}_t$, $t = p + q - n$, of the $(j+1)$st dual of E_p; moreover it is a simple circuit whose boundary is a sphere.

We now return to our group of intersecting cells. If we denote the t associated with $E^{*i}_q, \cdots$ by $t_1, t_2, \cdots$ we find that the intersection of the given cells coincides with that of the cells $\mathfrak{E}_{t_j}$ on E_p, that is with cells of the successive duals of orders $i, \cdots, h$, of E_p. There are now two possibilities: (a) $p < n$. The required result being trivial for $n = 0$, may be assumed for lower values of n, then follows for p and hence for the case here considered of intersections on E_p. (b) $p = n$. Then E_p may be suppressed and the intersection is actually of the second type. For that second type however we use exactly the same reasoning with E_p replaced by the succeeding cell E^{*i}_q and so on. Our proof is thus complete for both types.

REMARK. One may properly ask why in the earlier parts of the book we have not included, among the dual cells, the parts of the duals of semi-regular cells that are on $K - L$. When L is a regular boundary this would be perfectly feasible, but to demand that the boundary be regular is to impose an unnecessary restriction on our treatment.

The orientation is determined by recurrence. Writing for example $E^{*i}_p E^{*j}_q \cdots E^{*h}_r$ for the intersection cell of the second type we define it by

$$E^{*i}_p \cdots E^{*h}_r = E^{*i}_p \cdot [E^{*j}_q \cdots E^{*h}_r]$$

where $[\cdots]$ designates the oriented intersection of the bracketed cells. As between the first cell and the bracketed intersection the situation is essentially as in No. 3 and need not be considered again here.

THEOREM. *The intersection symbols follow the associative law.*

Suppose that we have three cells E_p, E_q, E_r, — we omit the asterisks in their designations—intersecting in an

$s\,(=p+q+r-2n)$-cell $\mathfrak{E}_s$. We can find an indicatrix of K (cell of a suitable derived), $\sigma_n = \sigma_s \sigma_\varkappa \sigma_\varrho \sigma_\pi$, $\pi = q+r-n-s-1$ etc., such that $\sigma_s \sigma_\pi$, $\sigma_s \sigma_\varkappa$, $\sigma_s \sigma_\varrho$ are component cells of $E_q \cdot E_r$, $E_r \cdot E_p$, $E_p \cdot E_q$. Then $\sigma_\varkappa \sigma_\varrho \sigma_s$, $\sigma_\varrho \sigma_s \sigma_\pi$, $\sigma_s \sigma_\pi \sigma_\varkappa$ are components of E_p, E_q, E_r. Let us assume that σ_n and these three new simplexes are respective indicatrices of the corresponding carrier cells—the contrary cases being taken care of as in No. 3 by introducing coefficients β everywhere. Under the circumstances let us orient $\mathfrak{E}_s$ by taking σ_s for indicatrix.

Now from the orientation rule for the intersection of two cells we find that the indicatrix of $E_p \cdot E_q$ is $\sigma_\varrho \sigma_s$ and hence that of $[E_p \cdot E_q] \cdot E_r$ is $(-1)^\alpha \sigma_s$, where $\alpha = n(\varkappa+1)$. By a similar calculation we find the same indicatrix for $E_p \cdot [E_q \cdot E_r]$. Therefore

$$E_p \cdot E_q \cdot E_r = E_p \cdot [E_q \cdot E_r] = [E_p \cdot E_q] \cdot E_r .$$

Suppose now that we have r cells E^ι with $\mathfrak{E}_s$ as the oriented intersection. By definition

$$\mathfrak{E}_s = E^\iota \cdot [E^j \cdots E^h].$$

Assuming that the associative law has been proved for less than r cells we have

$$E^j \cdots E^h = [E^j \cdots E^k] \cdot [E^l \cdots E^h]$$
$$E^i \cdot [E^j \cdots E^k] = E^i \cdot E^j \cdots E^k.$$

Hence

$$\begin{aligned}\mathfrak{E}_s &= E^i \cdot [E^j \cdots E^k] \cdot [E^l \cdots E^h] \\ &= [E^i \cdot \cdots E^k] \cdot [E^l \cdots E^h],\end{aligned}$$

which proves the associative law for r intersecting cells.

The behavior under commutation follows by successive application of the formulas given in No. 4.

From the cells to the chains is but a step. Let C_p, $C_{q_i}^{*h_i}$ be subchains of K and K^{*h_i}. We define the intersections $C_p \prod C_{q_i}^{*h_i}$, $\prod C_{q_i}^{*h_i}$ and when $s=0$, the corresponding Kronecker indices by the condition that the distributive law hold. The associative law follows then automatically. Regarding the fundamental boundary relation (18) we prove first as before, that it holds for two intersecting chains that are respective intersections of several chains. By means of the associative law we can then derive extended boundary relations analogous to (18) for any number of intersecting chains. From these extended boundary relations we can also derive theorems analogous to those of No. 10. Thus when all chains are absolute cycles and one is ≈ 0 on $K-L$ so is their intersection, etc. We return to this later.

§ 2. Special Combinatorial Properties of the Kronecker Index

12. We have already observed that the rule for orienting the dual E^*_{n-p}, of E_p, is equivalent to the condition $(E \cdot E^*) = 1$. By reference to Ch. III, § 3, it will be seen that this implies intimate relations between the combinatorial properties of the index and the duality relations. We now propose to exhibit this relationship, by developing a partial theory of the index that might fit directly into Ch. III, say at the beginning of § 3, and without regard to the general intersection theory of the present chapter.

We take the situation as in Ch. III, Nos. 29, 30 when we proved the fundamental duality relations for the homology characters. The pairs of dual cells of $\mathfrak{K}-\mathfrak{F}$ and $K^*-\Phi$ are called E^i_p, E^{*i}_{n-p}, and the subchains of the two open complexes are called C_p, C^*_{n-p}. We recollect that if C_p is a cycle of $\mathfrak{K}$ mod $\mathfrak{F}$ it defines a unique family of cycles of $K-L^1$ mod L^2 and the homologies mod $\mathfrak{F}$ of the cycles C_p correspond to those mod L^2 for the associated cycles on $K-L^1$. Similarly for C^*_{n-p}, $K^*-\Phi$, except that L^1 and L^2 must be interchanged. To simplify matters we shall understand by "cycle C_p", not only that cycle but the associated family of cycles on $K-L^1$ mod L^2, and when we write $C_p \sim 0$ it is to be understood in the double sense: (a) The subchain C_p of $\mathfrak{K}$ is ~ 0 mod $\mathfrak{F}$ on $\mathfrak{K}$. (b) Any cycle of the associated family is ~ 0 on $K - L^1 \bmod L^2$. A similar remark applies to the starred chains with the appropriate modifications.

13. Our starting point is then the index of

$$(21) \qquad C_p = \sum x_i E_p^i, \qquad C_{n-p}^* = \sum y_i E_{n-p}^{*i}$$

which we define a priori by (17), however we may have happened to obtain that expression:

$$(17) \qquad (C_p \cdot C_{n-p}^*) = \sum x_i y_i.$$

Suppose now that we have two sets of chains:

$$C_p^i, \; i = 1, 2, \cdots, r; \qquad C_{n-p}^{*j}, \; j = 1, 2, \cdots, s.$$

They determine a matrix of indices

$$(22) \qquad \| (C_p^i \cdot C_{n-p}^{*j}) \|$$

whose behavior under linear transformation is noteworthy. Let

$$(23) \qquad \begin{cases} C_p'^i = \sum a_{ih} C_p^h, & i = 1, 2, \cdots, r, \\ C_{n-p}'^{*j} = \sum b_{jk} C_{n-p}^{*k}, & j = 1, 2, \cdots, s. \end{cases}$$

Then, by a simple calculation,

$$(24) \qquad \| (C_p'^i \cdot C_{n-p}'^{*j}) \| = \boldsymbol{a} \cdot \| (C_p^i \cdot C_{n-p}^{*j}) \| \cdot \boldsymbol{b}',$$

which describes the exact behavior of (22) under the transformation (23).

14. The result just derived combined with the considerations of Ch. I, No. 27 yields rapidly results of considerable significance. Consider indeed the boundary relations

$$E_{p+1}^i \to \sum \eta_{ij}^p E_p^j$$

and apply again simultaneously the transformations, loc. cit.,

$$C_p^i = \sum a_{ij}^p E_p^j, \quad \boldsymbol{a}^p \text{ unimodular},$$

giving for the new fundamental sets $\{C_p^i\}$ for p-chains here considered, boundary relations with quasi-canonical matrices of coefficients:

$$C_{p+1}^i \to \zeta_i^p\, C_p^i; \quad \zeta^p = \boldsymbol{a}^{p+1} \cdot \eta^p \cdot (\boldsymbol{a}^p)^{-1}.$$

If η^{*q} are the boundary matrices for the starred chains we have (Ch. III, No. 27)

$$(25) \qquad \eta^{*n-p-1} = (-1)^{p+1} \cdot (\eta^p)'.$$

Hence the matrices

$$(-1)^{p+1}\, (\boldsymbol{a}^p)^{-1\prime}\, \eta^{*n-p-1}\, (\boldsymbol{a}^{p+1})'$$

are likewise quasi-canonical. Therefore the transformations

$$C_{n-p}^{*i} = \sum b_{ij}^{n-p}\, E_{n-p}^{*j}, \quad \boldsymbol{b}^{n-p} = ((\boldsymbol{a}^p)^{-1})',$$

will reduce simultaneously the starred matrices to matrices ζ^{*p}, likewise quasi-canonical.

Now we have by definition

$$\|(E_p^i \cdot E_{n-p}^{*j})\| = \mathbf{1}.$$

Hence, by the result of No. 13:

$$\|(C_p^i \cdot C_{n-p}^{*i})\| = \boldsymbol{a}^p \cdot (\boldsymbol{b}^{n-p})' = \mathbf{1}.$$

15. Let us suppose in particular that the matrices $\boldsymbol{a}^p$, have been so chosen that for the dimensions $p+1$, p and p, $p-1$ the incidence matrices assume the simplified form of Ch. I, No. 27. To make the situation clearer we will designate the non-bounding cycles by Γ, the zero-divisors or bounding cycles by Δ, the remaining chains by D. Then we shall have

I. *Boundary relations for the non-starred fundamental sets*:

$$\begin{cases} D_{p+1}^i \to \zeta_i^p\, \Delta_p^i \sim 0; & i = 1, 2, \cdots, \varrho_p, \\ \Gamma_{p+1}^i \text{ or } \Delta_{p+1}^i \to 0. \end{cases}$$

$$\begin{cases} \Delta_p^i \to 0, \\ D_p^i \to \zeta_i^{p-1}\, \Delta_{p-1}^i, & i = 1, 2, \cdots, \varrho_{p-1}; \\ \Gamma_p^i \to 0, & i = 1, 2, \cdots, R_p. \end{cases}$$

II. *Boundary relations for the starred fundamental sets.* If we substitute the expressions of the ζ's in (25) we verify

$$(25') \qquad \zeta^{*n-p-1} = (-1)^{p+1} \cdot (\zeta^p)',$$

or, that the relation (25) holds with the η matrices replaced by the ζ's. Taking into account the fact that a diagonal matrix and its transverse have the same diagonal terms we see that in passing from non-starred to starred chains the Δ_p's go over into the D^*_{n-p}'s and similarly for the Δ's and D's interchanged. For consider as an example the first boundary relation in the above set. Designating for the moment by C's the associated starred chains we shall have corresponding to the relation in question

$$C^{*i}_{n-p} \rightarrow (-1)^{p+1} \zeta^p_i \, C^{*i}_{n-p-1},$$

which shows that the chain at the right is a Δ^*, the chain at the left a D^*. Thus if the ith p-chain of the non-starred set is a Δ, the ith $(n-p)$-chain of the starred set is a D and similarly with Δ and D, or starred and non-starred chains, interchanged. It follows that the sets Γ, Γ^* can only be permuted. The boundary relations for the starred set assume then the general form

$$\begin{cases} \Gamma^{*i}_{n-p+1} \quad \text{or} \quad \Delta^{*i}_{n-p+1} \rightarrow 0; \\ D^{*i}_{n-p+1} \rightarrow (-1)^p \, \zeta^{p-1}_i \, \Delta^{*i}_{n-p}, \qquad i = 1, 2, \cdots, \varrho_{p-1}; \\ \Gamma^{*i}_{n-p+1} \quad \text{or} \quad \Delta^{*i}_{n-p+1} \rightarrow 0; \end{cases}$$

$$\begin{cases} D^{*i}_{n-p} \rightarrow (-1)^{p+1} \, \zeta^p_i \, \Delta^{*i}_{n-p-1}, & i = 1, 2, \cdots, \varrho_p; \\ \Delta^{*i}_{n-p} \rightarrow 0, & i = 1, 2, \cdots, \varrho_{p-1}; \\ \Gamma^{*i}_{n-p} \rightarrow 0,. & i = 1, 2, \cdots, R_p. \end{cases}$$

16. According to No. 14, the matrix of the indices of the two sets Δ^i_p, D^i_p, Γ^i_p and D^{*i}_{n-p}, Δ^{*i}_{n-p}, Γ^{*i}_{n-p} in the orders named is the identity. Schematically the situation is then represented by the following table, where each term stands for the matrix of indices of the chains whose type is given by the left column and upper row:

	D_{n-p}^*	Δ_{n-p}^*	Γ_{n-p}^*
Δ_p	1	0	0
D_p	0	1	0
Γ_p	0	0	1

In particular

(26) $$\|(\Gamma_p^i \cdot \Gamma_{n-p}^{*j})\| = 1, \quad \text{or} \quad (\Gamma_p^i \cdot \Gamma_{n-p}^{*j}) = \delta_{ij};$$

(27) $$\begin{aligned} &\|(\Gamma_p^i \cdot \Delta_{n-p}^{*j})\| = \|(\Delta_p^i \cdot \Gamma_{n-p}^{*j})\| = \|(\Delta_p^i \cdot \Delta_{n-p}^{*j})\| = 0, \\ &\text{or} \quad (\Gamma_p^i \cdot \Delta_{n-p}^{*j}) = (\Delta_p^i \cdot \Gamma_{n-p}^{*j}) = (\Delta_p^i \cdot \Delta_{n-p}^{*j}) = 0. \end{aligned}$$

Of course (27) is a direct consequence of V of No. 10.

17. From this follow two noteworthy propositions, the first of which goes far towards explaining the fundamental importance of the Kronecker index. They were obtained for an absolute M_3 by Poincaré [3], for an absolute M_n by Veblen [6] and Weyl [3]:

THEOREM I. *In order that a cycle* $\Gamma_p \approx 0$, $[\Gamma_{n-p}^* \approx 0]$, *(i. e. that they be zero-divisors), it is necessary and sufficient that*

$$(\Gamma_p \cdot \Gamma_{n-p}^*) = 0$$

for every Γ_{n-p}^* [*for every* Γ_p] *which is a cycle.*

THEOREM II. *It is possible to find two associated minimum bases of cycles* $\Gamma_p^i, \Gamma_{n-p}^{*j}$, *for the homologies with division allowed,* $\approx$, *such that the corresponding Kronecker index matrix is the identity.*

THEOREM III. *The cycles* Γ_p, Γ_{n-p}^* *are represented in terms of the bases by homologies*

$$\Gamma_p \approx \sum (\Gamma_p \cdot \Gamma_{n-p}^{*i}) \Gamma_p^i,$$
$$\Gamma_{n-p}^* \approx \sum (\Gamma_p^i \cdot \Gamma_{n-p}^*) \Gamma_{n-p}^{*i}.$$

There exists a Γ_p *for which the indices* $(\Gamma_p \cdot \Gamma_{n-p}^{*i})$ *have arbitrarily assigned integral values, and similarly for the starred cycles.*

The restriction to subchains will be removed later (No. 39).

The second theorem merely states in words the same thing as the first matrix equation (26). As to the first, that the condition is necessary constitutes Theorem V of No. 10. Independently also, when $\Gamma_p \approx 0$ we have (Ch. I, No. 27),

$$\Gamma_p = \sum t_i \Delta_p^i.$$

From this and (27) follows

$$(\Gamma_p \cdot \Gamma_{n-p}^{*j}) = \sum t_i (\Delta^i \cdot \Gamma^{*j}) = 0,$$
$$(\Gamma_p \cdot \Delta_{n-p}^{*j}) = \sum t_i (\Delta^i \cdot \Delta^{*j}) = 0.$$

But if Γ_{n-p}^* is an arbitrary cycle we have (loc. cit.)

$$\Gamma^* = \sum x_i \Gamma^{*i} + \sum y_i \Delta^{*i}.$$

Hence

$$(\Gamma \cdot \Gamma^*) = \sum x_i (\Gamma \cdot \Gamma^{*i}) + \sum y_i (\Gamma \cdot \Delta^{*i}) = 0.$$

Conversely let the condition in question be fulfilled for a cycle Γ_p. We have

$$\Gamma_p = \sum x_i \Gamma^i + \sum y_i \Delta^i.$$

Hence

$$(\Gamma_p \cdot \Gamma_{n-p}^{*j}) = 0 = \sum x_i (\Gamma^i \cdot \Gamma^{*j}) = x_j,$$

so that Γ_p is a linear combination of the Δ's alone, i. e. $\Gamma_p \approx 0$. Therefore Th. I is completely proved.

Since every $\Delta \approx 0$, we have

$$\Gamma_p \approx \sum x_i \Gamma_p^i,$$

hence

$$(\Gamma_p \cdot \Gamma_{n-p}^{*i}) = \sum x_j (\Gamma_p^j \cdot \Gamma_{n-p}^{*i}) = x_i,$$

which proves the first part of Th. III for the Γ's. Its second part for the same cycles is obvious. The treatment of the Γ^*'s is the same.

18. Theorem I implies, what we already know, that the number of independent cycles Γ_p and Γ_{n-p}^* is the same, or *the duality relation for the Betti numbers* of Ch. III, No. 33.

Historically this is of interest, for Poincaré's first attempt at a proof of the duality theorem for the Betti numbers of an absolute M_n, rested precisely on a wrong proof of the second part of Th. I, as was pointed out later by Heegard [1]. See Poincaré's papers [2], [3], notably the introduction to the latter. It is in connection with his second and correct proof that Poincaré introduced the boundary relations and matrices.

As the situation presents itself now the duality relations are obtained before Th. I. On the other hand, as just stated and first observed by Poincaré [2], the duality relations are an immediate corollary of that theorem. For let there be given s independent cycles γ_p^i, and let t be the maximum number of independent cycles Γ^*_{n-p}, with $\{\gamma^{*j}_{n-p}\}$ as a maximal set. The rank of

$$\|(\gamma_p^i \cdot \gamma^{*j}_{n-p})\|$$

does not exceed s or t. Assume $t < s$. There exists then a cycle

$$\Gamma_p = \sum x_i \cdot \Gamma_p^i \not\approx 0,$$

yet such that every $(\Gamma_p \cdot \gamma^{*j}_{n-p}) = 0$. Since $\{\gamma^{*j}_{n-p}\}$ is a maximal set for cycles of its type and homologies $\approx$, any such cycle, say

$$\Gamma^*_{n-p} \approx \sum y_j \, \gamma^{*j}_{n-p}, \quad y_j \text{ rational.}$$

Hence by V of No. 10,

$$(\Gamma_p \cdot \Gamma^*_{n-p}) = \sum y_j (\gamma_p \cdot \gamma^{*j}_{n-p}) = 0,$$

and by Th. I, $\Gamma_p \approx 0$, a contradiction. Therefore $t \geqq s$. Similarly if s is maximal, $s \geqq t$, hence $s = t$, which is the duality relation for the Betti numbers.

19. The duality relations for the numbers r_p of Ch. III, No. 37 can be proved in similar fashion. In the notations of Ch. III, Nos. 29, 30, let the subcycles of K^* be subdivided into two types: (a) The subcycles of Φ. (b) The subcycles independent of those of Φ. Due to the sliding properties of our neighborhoods and to the fact that $\Phi \subset N$, the cycles of type (a)

are homologous to cycles on L. Conversely if $\Gamma^*_{n-p} \sim 0 \mod L$, it is homologous to a subcycle on Φ (loc. cit.). Therefore r_{n-p} is the number of independent $(n-p)$-cycles of type (b). As for r_p it is the number of independent absolute subcycles of $K-L$.

Now let Γ_p be a subcycle of $K-L$, γ^*_{n-p} a subcycle of Φ. Since they do not meet, $(\Gamma \cdot \gamma^*) = 0$. Hence in order that $\Gamma \approx 0 \mod L$, it is necessary and sufficient (No. 17, Th. I) that $(\Gamma_p \cdot \Gamma^*_{n-p}) = 0$ for every Γ^* of type (b). On the strength of this we prove as in the preceding number that the number of independent Γ's cannot exceed that of independent Γ^*'s, so that $r_p \leqq r_{n-p}$. But since these numbers are symmetrically defined relatively to L, we have also $r_{n-p} \leqq r_p$, hence $r_p = r_{n-p}$, which is the required duality relation.

20. Returning to the earlier situation we have the following theorem relating independent sets for the starred and non-starred cycles.

THEOREM. *Given a set of s independent cycles γ^i_p, there exists an associated set of as many independent cycles γ^{*i}_{n-p} such that the determinant of order s*

$$D = |(\gamma^i_p \cdot \gamma^{*j}_{n-p})| \neq 0.$$

Conversely, when $D \neq 0$, the starred cycles are independent. When the first set is maximal, i. e. when $s = R_p$, we can take as the second an arbitrary maximal set. In that case in order that both be bases for $\approx$ it is necessary and sufficient that $D = \pm 1$.

For in terms of the minimum base let

$$\gamma^i \approx \sum x_{ij} \Gamma^j, \qquad i = 1, 2, \cdots, s.$$

The γ^*'s, if they can be found, are expressible in terms of their own minimum base in the form

$$\gamma^{*i} \approx \sum y_{ij} \Gamma^{*j}$$

and the question is to determine the matrix $\boldsymbol{y}$.

By (24) of No. 13

$$D = |(\gamma^i \cdot \gamma^{*j})| = \left|\boldsymbol{x} \cdot \|(\Gamma^i \cdot \Gamma^{*j})\| \cdot \boldsymbol{y}'\right| = |\boldsymbol{x} \cdot \boldsymbol{y}'|.$$

Since $\boldsymbol{x}$ is of rank s we can always choose $\boldsymbol{y}$ of rank s such that $D = |\boldsymbol{x} \cdot \boldsymbol{y}'|$. When $s = R_p$, $D \neq 0$ for every $\boldsymbol{y}$ of rank R_p, hence for every maximal independent set $\{\gamma^{*j}\}$. In order that $\{\gamma^i\}$, $\{\gamma^{*j}\}$ be minimum bases it is necessary and sufficient that $\boldsymbol{x}$, $\boldsymbol{y}$ be unimodular. When this condition is fulfilled $D = |\boldsymbol{x}| \cdot |\boldsymbol{y}| = \pm 1$. Conversely when $D = \pm 1$, $|\boldsymbol{x}|$ and $|\boldsymbol{y}|$, being integers that divide 1, are both ± 1, so that $\boldsymbol{x}$ and $\boldsymbol{y}$ are unimodular and the sets are both minimal bases. The theorem is thus completely proved.

REMARK. When we allow rational chains we can always choose $\{\gamma^{*j}\}$ for given $\{\gamma^i\}$ such that $D = 1$, for we can always find a $\boldsymbol{y}$ of rank s with rational elements such that $|\boldsymbol{x} \cdot \boldsymbol{y}'| = 1$.

§ 3. INTERSECTIONS OF ARBITRARY CHAINS AND THEIR COMBINATORIAL INVARIANCE

21. Intersections of arbitrary chains present a problem which belongs essentially to continuous topology, to the same order of ideas as the Jordan curve theory, the deformation theorems of Ch. II, etc. We shall treat here solely the case of a manifold $K - L$, L a subcomplex of K. The case of a more general L is reserved for Ch. VII.

The actual point-set common to C_p, C_q may be very involved. In any case however the only part of it that interests us is the part on the closed manifold $\overline{K - L}$. We shall call that part the *geometric intersection* of the two chains and designate it by G.

If we wish to have an intersection which comes within the scope of our homology theory, it must be a cycle of some sort and it must be obtained by means of some approximation process. In other words the natural procedure would be to substitute for C_p, C_q, polyhedral approximating chains, take their intersection, and attach a cycle to the family of all such intersections. Of course one will expect polyhedral approximations such as those of § 1, by pairs of subchains of $K - L$ and K^*. By reference to the fundamental boundary formula (18) we find that the intersection of such chains can only be a cycle when they do not meet each other's boundaries on L. This in turn demands that C_p, C_q have that property. In other words we need the

FUNDAMENTAL ASSUMPTION: *The chains C_p, C_q do not meet each other's boundaries on $K-L$.*

All this is clearly illustrated by the simple example of two intersecting arcs γ, δ on a plane π. We assume that they are in "general position", and hence only have a finite number of crossings. Let A be one of them. We may modify slightly γ or δ so as to add two new crossings in the vicinity of A but if we take two inscribed polygonal lines whose sides are small enough the cycle $\gamma \cdot \delta$ will remain within a family of homologous cycles and $(\gamma \cdot \delta)$ will not change. Both properties are lost however if either γ or δ has an end point on the other. Therefore our assumption is quite essential.

22. The accurate description of the intersection cycle rests upon certain notions that we now introduce.

DEFINITIONS. A sequence $\{\Gamma_s^i\}$ of s-cycles mod L on K is said to be *homologous mod L about the closed set $A \subset K$* whenever: (a) On every neighborhood $\mathfrak{N}^A$ of A on K there is at least one Γ. (b) For every $\mathfrak{N}^A$ there is an $\mathfrak{N}'^A \subset \mathfrak{N}^A$, such that all Γ's $\subset \mathfrak{N}'$ are homologous mod L on $\mathfrak{N}$.

Two sequences $\{\Gamma^i\}$, $\{\Gamma'^i\}$ of the above type are *equivalent* whenever $\Gamma^1, \Gamma'^1, \Gamma^2, \Gamma'^2, \cdots$ is itself such a sequence. The sum $\mathfrak{S}$ of any number (denumerable or not) of equivalent sequences is a *homologous set mod L about A*. The total set $\mathfrak{S}$ may conceivably fail to possess the property characteristic of its individual sequences. For it is quite possible that for any given $\mathfrak{N}$, $\Pi \mathfrak{N}' = A$, where the $\mathfrak{N}'$'s are as above for the individual subsequences $\{\Gamma^i\}$ of $\mathfrak{S}$.

One may ask why we introduce the sequences at all. The answer is simply that our construction will yield us a category of cycles which possesses the property just introduced but not—so far as we have been able to prove—the stronger property that we get when the total set $\mathfrak{S}$ behaves like the sequences ($\Pi \mathfrak{N}'$ is a neighborhood of A). It would be interesting to know if the two properties are identical. Such is certainly the case when $\mathfrak{S}$ is the sum of a finite number of sequences, but beyond that the identity is doubtful.

23. We are going to establish that the intersections of certain polyhedral chains approximating C_p and C_q constitute a family of s-cycles mod L, $s = p+q-n$, homologous mod L about G, the geometric intersection of the two given chains. When $s = 0$ there will be a fixed Kronecker index, characteristic of the family. We shall show that there is a maximal family $\{\Gamma_s\}$, having the following properties:

(a) It is independent of the polyhedral approximations for a fixed K. This is in a sense combinatorial invariance.

(b) When K is replaced by any other complex with the new $K-L$ oriented concordantly with the old, the family is again unchanged. This is invariance with change of straightness or topological invariance proper.

The family $\{\Gamma_s\}$ is then essentially our "intersection", and it is a true topological attribute of the ordered pair C_p, C_q. Similarly the Kronecker index is a topological invariant of the ordered pair C_p, C_{n-p}.

24. In our paper [8], where these questions were treated for the first time, L did not occur, and in place of homology about G we had merely ordinary homology. The new elements thus introduced do not greatly increase the difficulty of exposition while they yield in truth much sharper information. Regarding the second an example will show this clearly. Let Γ_2, Γ_2' be cycles on H_3. Their intersection in our sense is any cycle of a certain class $\{\Gamma_1\}$. Without homology about a set all that we know concerning the class is that its cycles are homologous on H—a statement of little value since all one-cycles of H are ~ 0. Bring in homology about a set and let Δ be a linear cycle not passing through any point of intersection of Γ_2 and Γ_2'. Then we can show that all cycles of $\{\Gamma_1\}$ sufficiently near to the set of points G common to Γ_2 and Γ_2' will either loop or not loop Δ (i. e. meet it, or not meet it when shrunk to a point). Such examples can of course be multiplied indefinitely.

25. It is clear that if K^0 is a subdivision whose mesh is sufficiently small, and $\mathfrak{N}^0$ the K^0-neighborhood of the geometric intersection G, we can replace at any time an assigned neighborhood $\mathfrak{N}$ by $\mathfrak{N}^0$.

Since we are free to replace K by any subdivision we can assume that it is simplicial and that N, the K-neighborhood of L, is normal. In this case the two neighborhoods N, N' of Ch. II, No. 25 can be taken coincident with the present N. It follows then, as shown there, that we have a correspondence between cycles Γ_p mod L and Δ_p mod N, with the following properties:

(a) Given Γ to obtain Δ we subdivide Γ in such manner that the Γ-neighborhood of L is on N, then remove that Γ-neighborhood from Γ.

(b) Given Δ to obtain Γ we deform $F(\Delta)$ onto L along the projecting segments of N through its points, then add the deformation chain to Δ.

(c) If $\Gamma \sim 0 \bmod L$, evidently $\Delta \sim 0 \bmod N$. Conversely let $C_{p+1} \to \Delta \bmod N$. Then $C_{p+1} \to \Delta + \Delta'$, $\Delta' \subset N$. Deform Δ' onto L in the same manner as above and let C'_{p+1} be its deformation chain and $-\Delta''_p$ that of $F(\Delta')$. Since $\Delta + \Delta'$ is a cycle, $F(\Delta) = -F(\Delta')$, hence Δ'' is the deformation chain of $F(\Delta)$ and therefore $\Delta + \Delta'' = \Gamma$. Now

$$C'_{p+1} \to -\Delta' + \Delta'' \bmod L.$$

Hence

$$C_{p+1} + C'_{p+1} \to \Delta + \Delta'' = \Gamma \bmod L.$$

Thus if Δ bounds C_{p+1}, Γ bounds C_{p+1} plus a chain C'_{p+1} which is on the set of cells of N that meet C_{p+1}. This result is more precise than that of Ch. II, No. 25 as regards the comparison of the Γ' and Δ's, and that additional precision is needed in the sequel.

26. We shall now hold fixed the two neighborhoods N of L, and $\mathfrak{N}$ of G. In view of the fundamental assumption as to C_p, C_q and for evident reasons of continuity we have the following two properties:

PROPERTY I. *On $K-N$ the distances from G to $F(C_p)$, $F(C_q)$ and $K-\mathfrak{N}$ have a lower bound $\varrho>0$.*

PROPERTY II. *For every positive ε there is a positive $\eta(\varepsilon)$, tending to zero with ε, such that if A, B are sets on $\overline{K-N}$ within a distance η from C_p, C_q respectively, their geometric intersection is within a distance ε from G.*

These two properties will form the basis of the approximations by which we define our homologous family about G (= the intersection cycle).

27. Let then K^α designate a simplicial subdivision of K whose mesh is α. Denote by K'^α, $K^{*\alpha}$, N^α, N'^α, a first derived complex of K^α, the corresponding dual, and the K^α and K'^α-neighborhoods of L. We recall that $K-N'^\alpha=K^{*\alpha}\supset\Phi^\alpha$, the boundary of N'^α (Ch. III, No. 18).

Applying the deformation theorem of Ch. II, No. 45, we reduce C_p to a subchain C_p^α of K^α and C_q to a subchain C_q^α of $K^{*\alpha}+N'^\alpha$. This involves respectively α and 2α deformations. For each is less than the mesh of the corresponding complex and since every cell of $K^{*\alpha}$ is a sum of cells of K'^α with a common vertex, (mesh of $K^{*\alpha}+N'^\alpha$) $\leqq 2$ (mesh K'^α) $\leqq 2$ (mesh K^α) $= 2\alpha$.

We have to assign to α some fixed upper bound ξ whose actual value is immaterial. To have ample latitude let us take for ξ the least of the two numbers $\frac{1}{10}\varrho$ and $\frac{1}{10}\eta(\frac{1}{10}\varrho)$.

Under the circumstances, the displacements of $F(C_p)$, $F(C_q)$ and the distance on $K-N$ from G to the geometric intersection of the displaced chains C_p^α, C_q^α, being all $<\frac{1}{10}\varrho$, the latter continue to possess Properties I, II, with a new geometric intersection G^α, perhaps a new ϱ, and the same $\mathfrak{N}$.

Let us remove from C_q^α the cells that belong to N'^α, which turns it into a subchain $C_q^{*\alpha}$ of $K^{*\alpha}$ itself. The chains C_p^α, $C_q^{*\alpha}$ are related to one another like the intersecting chains of § 1. Therefore they have a well defined intersection $C_p^\alpha \cdot C_q^{*\alpha}$ in the sense of § 1. We will now examine that intersection.

If we compare C_q^α and $C_q^{*\alpha}$ we find that $F(C_q^\alpha-C_q^{*\alpha})\subset\overline{N}'^\alpha$. Therefore we have

$$F(C_q^{*\alpha}) = F'+F''; \qquad F'\subset F(C_q^\alpha), \quad F''\subset \Phi^\alpha.$$

Since Property I holds for the chains C_p^α, C_q^α, their boundaries do not meet their geometric intersection G^α on $K-N$. Hence F' does not meet C_p^α on $K-N$, else it would meet G^α there. For a similar reason, $F(C_p^\alpha)$ does not meet there $C_q^{*\alpha}$, which is a subchain of C_q^α. Therefore, by the fundamental boundary formula (18) and mod N,

$$C_p^\alpha \cdot C_q^{*\alpha} \to (-1)^{n-q}\, C_p^\alpha \cdot F'' \subset \Phi^\alpha \subset \bar{N}'^\alpha \subset N,$$

which means that $C_p^\alpha \cdot C_q^{*\alpha} \to 0 \mod N$, so that the intersection at the left is an s-cycle mod N, $s = p+q-n$.

Since Property I holds for C_p^α and C_q^α with the same $\mathfrak{N}$, $G^\alpha \subset \mathfrak{N}$. Hence $C_p^\alpha \cdot C_q^{*\alpha} \subset G^\alpha \subset \mathfrak{N}$. Thus $C_p^\alpha \cdot C_q^{*\alpha}$ *is an s-cycle mod N on $\mathfrak{N}$*. It is essentially the class of all such cycles that will enable us to construct the homologous family about G, by which we shall define $C_p \cdot C_q$.

28. Let us suppose that we have obtained by any method whatsoever a subchain D_p of K^α, such that $C_p^\alpha - D_p = F(C_{p+1})$ where C_{p+1} is a subchain of K^α whose points are not farther than 10ξ from C_p. Again by (18) we have

$$\begin{aligned}(28)\qquad C_{p+1} \cdot C_q^{*\alpha} &\to C_{p+1} \cdot F(C_q^{*\alpha}) + (-1)^{n-q}\,(C_p^\alpha - D_p) \cdot C_q^{*\alpha} \\ &= C_{p+1} \cdot (F' + F'') + (-1)^{n-q}\,(C^\alpha - D) \cdot C_q^{*\alpha}.\end{aligned}$$

I say that C_{p+1} and F' do not meet on $K-N$. For let there exist a point of intersection P of C_{p+1} and F' on $K-N$. Since C_{p+1} and F' are not farther than $\eta\,(\frac{1}{10}\varrho)$ from C_p and C_q respectively on $K-N$, by Property II, P is not farther than $\frac{1}{10}\varrho$ from G. On the other hand P being on F', is not farther than $2\xi < \frac{1}{5}\varrho$ from $F(C_q)$ on $K-N$. Hence $F(C_q)$ would have a point on $K-N$ not farther than $\frac{1}{5}\varrho + \frac{1}{10}\varrho = \frac{3}{10}\varrho$ from G, which contradicts Property I.

Since C_{p+1} and F' do not meet on $K-N$, and $F'' \subset \Phi^\alpha$,

$$(29)\qquad C_{p+1} \cdot C_q^{*\alpha} \to (-1)^{n-q}\,(C_p^\alpha - D_p) \cdot C_q^{*\alpha} \mod N.$$

Let now $\bar{E}_{p+1}$ be a cell of C_{p+1}, on $K-N$ and with points on $K-\mathfrak{N}$. We have

$$E \cdot C_q^{*\alpha} \rightarrow E \cdot (F' + F'') + (-1)^{n-q} F(E) \cdot C_q^{*\alpha}.$$

I say that every term at the right is zero. (a) Since $\bar{E} \subset K - N$, $\Phi^\alpha \subset N$, E does not meet Φ^α, and hence not F''. Therefore $E \cdot F'' = 0$. (b) Since C_{p+1} and F' do not meet, and E is a cell of C_{p+1}, E and F' do not meet. Therefore $E \cdot F' = 0$. (c) Suppose now that $F(E)$ meets $C_q^{*\alpha}$. $F(E)$ belongs to the part of C_{p+1} on $K - N$, and hence it is not farther than $10\xi < \eta(\frac{1}{10}\varrho)$ from C_p. Also $C_q^{*\alpha}$ is not farther than $2\xi < \frac{1}{5}\eta(\frac{1}{10}\varrho)$ from C_q. Therefore the geometric intersection of $\bar{E}$ and $C^{*\alpha}$, if it exists, is not farther than $\frac{1}{10}\varrho$ from G. But the diameter of $\bar{E} < \xi < \frac{1}{10}\varrho$ and by assumption it has points on $K - \mathfrak{N}$ and hence at least as far as ϱ from G. Hence the distance from $\bar{E}$ to G is at least $\frac{9}{10}\varrho$. This contradiction shows that $\bar{E}$, and hence $F(E)$ cannot meet $C_q^{*\alpha}$ and therefore, also $F(E) \cdot C_q^{*\alpha} = 0$.

We conclude then: $E \cdot C_q^{*\alpha} \rightarrow 0$. This shows that we can suppress E from C_{p+1} in (29) without disturbing that boundary relation.

We may therefore reduce C_{p+1} to its cells on $\mathfrak{N} + N$ without disturbing (29). Hence that boundary relation implies

$$(30) \qquad C_p^\alpha \cdot C_q^{*\alpha} \sim D_p \cdot C_q^{*\alpha} \quad \text{mod } N \text{ on } \mathfrak{N}.$$

29. From the preceding result we shall derive an important consequence. Suppose that C_p is 8ξ-deformed in any manner whatever (after having perhaps been subdivided) into a subchain D_p of $\overline{K^\alpha - L}$. Let C'_{p+1} and $-D'_p$, be the differences between the deformation chains of C_p and $F(C_p)$ first into D_p and $F(D_p)$, then into C_p^α and $F(C_p^\alpha)$. We have

$$C'_{p+1} \rightarrow D_p + D'_p - C_p^\alpha.$$

Applying the deformation theorem for simplicial complexes we now reduce C'_{p+1} to a subchain of $\overline{K^\alpha - L}$. This can be done without displacing $D_p - C_p^\alpha$ since it is already a subchain of $\overline{K^\alpha - L}$. Let C'_{p+1}, D'_p be reduced to C''_{p+1}, D''_p. We have

$$C''_{p+1} \rightarrow D_p + D''_p - C_p^\alpha.$$

Now the deformations of C_p whose deformation chains have for difference C'_{p+1} are 8ξ and α, hence both $<8\xi$. The deformation from C'_{p+1} to C''_{p+1} is $\alpha<\xi$. Hence the points of C''_{p+1} are not farther than 9ξ from C_p. Therefore, from (30):

$$(D_p+D''_p)\cdot C_q^{*\alpha} \sim C_p^\alpha \cdot C_q^{*\alpha} \bmod N \text{ on } \mathfrak{R}.$$

But D''_p is not farther than $9\xi<\eta(\frac{1}{10}\varrho)$ from $F(C_p)$, and $C_q^{*\alpha}$ is still nearer to C_q. Therefore the points of $D''_p\cdot C_q^{*\alpha}$ on $K-N$, if there any, are not farther than $\frac{1}{10}\varrho$ from G. These points being also on D''_p would not be farther than 9ξ from $F(C_p)$ and hence the latter would have points on $K-N$ not farther than $10\xi<\varrho$ from G, which contradicts Property I.

We have then $D''_p\cdot C_q^{*\alpha}\subset N$, hence

$$D_p\cdot C_q^* \sim C_p^\alpha\cdot C_q^{*\alpha} \bmod N \text{ on } \mathfrak{R}.$$

The same reasoning holds throughout with C_p^α and $C_q^{*\alpha}$ interchanged. Therefore *all the possible cycles* $C_p^\alpha\cdot C_q^{*\alpha}$, *where* C_p^α, $C_q^{*\alpha}$ *are derived from* C_p, C_q *by as much as* 8ξ *deformations, are homologous to one another mod* N *on* $\mathfrak{R}$.

REMARK. The result just obtained holds whether K^α is simplicial or not since we have not utilized that property in deriving it.

30. THEOREM. *The cycles* $C_p^\alpha\cdot C_q^{*\alpha}$ *obtained by suitably small deformations from* C_p, C_q, *are all homologous mod* N *on* $\mathfrak{R}$.

We will show that the deformations may be taken as much as 4ξ. Let K^β be another complex analogous to K^α, and let $K'^\beta\cdots$ be the analogues of $K'^\alpha\cdots$ associated with it. We prove that for α, $\beta<\xi$, and deformations $<4\xi$,

$$C_p^\alpha\cdot C_q^{*\alpha} \sim C_p^\beta\cdot C_q^{*\beta} \bmod N \text{ on } \mathfrak{R}. \tag{31}$$

As shown in Ch. I, No. 61, the two complexes K^α, K^β have a common simplicial subdivision obtained from one or the other complex by a succession of sections of single cells. Therefore everything comes down to proving (31) for a convex K^α and a K^β obtained from it by section of a single cell E_r.

As we are only interested in the cells of $K^\alpha - N$ we are free to assume that E_r and its faces are all regular.

31. Let E'_r, E''_r be the two r-cells resulting from the section of E_r, the section cell being E_{r-1}. Let A'^r, A''^r, A^{r-1} be the vertices of K'^β on E'_r, E''_r, E_{r-1}, and A^r the vertex of K'^α on E_r. The vertices of K'^β not on E_r are to be the same as those of K'^α. The duals $K^{*\alpha}$, $K^{*\beta}$ differ only in the cells that are duals of cells on $\bar{E}_r$, and in fact only in the parts of those duals that are on the star of E_r. For if we write down the sum of cells of K'^β which gives rise to any cell of $K^{*\beta}$ we find that it also represents a cell of $K^{*\alpha}$, unless the sum includes cells of K'^β having A'^r, A''^r, A^{r-1} among their vertices.

We will now transform K'^β into K'^α by a transformation T defined as follows: (a) The vertices A'^r, A''^r, A^{r-1} go into A^r and the other vertices remain fixed. (b) The vertices of any simplex of K'^β go into those of a simplex of K'^α of the same or lower dimension. (c) The transformation T is extended to the whole of K'^β by making it barycentric between any simplex of K'^β and the simplex of K'^α into whose vertices its own are transformed. Every point not on the star of E_r is its own T transform, so that outside that star T reduces to the identity.

To show that (a), (b), (c) are consistent we merely have to prove that (b) follows from (a). Now the only cells of K'^β affected by T are of type $A^s \cdots (\,) \cdots A^t$ where the parenthesis stands for one of the five groups of vertices: A'^r, A''^r, A^{r-1}, $A^{r-1}A'^r$, $A^{r-1}A''^r$. They correspond to mutually incident cells $E_s, \cdots, (\,), \cdots, E_t$ of K'^β those in parenthesis being respectively E'_r; E''_r; E_{r-1}; E_{r-1}, E'_r; E_{r-1}, E''_r. But in each case we have a corresponding mutually incident group of cells of K'^α by merely replacing the parenthesis by E_r. This shows that the cell of K'^β written above is transformed by T into the cell $A^s \cdots A^r \cdots A^t$ of K'^α. It is a degenerate simplex when and only when $(\,) = A^{r-1}A'^r$ or $A^{r-1}A''^r$.

Remembering now that the dual of any cell of $K^\beta - L$ is the sum of all cells of K'^β which have just one vertex on that cell, we verify the following property: The only cells of $K^{*\beta}$ that undergo any modification are the duals of cells on $\bar{E}_r$, and the modifications occur only in the star of E_r, but not outside of it. Furthermore, each cell of $K^{*\beta}$ is turned into a definite cell of $K^{*\alpha}$ of the same dimension, exception being made for the dual of E_{r-1}, which is turned into the dual of E_r.

If P is any point of K, its transform $Q = T \cdot P$ is on the same closed cell of K^α, hence the two points determine a segment PQ on K. Therefore T is a deformation with the segments PQ as the paths of the points of K.

32. We will now take two pairs of chains C_p^α, $C_q^{*\alpha}$, C_p^β, $C_q^{*\beta}$, making use of the freedom that is allowed by the result of No. 29. Since E_r is regular, its star ($= E_r$ itself when $r = n$) is an n-cell. Let us replace in K^α that whole star by the unique n-cell and let K^0 be the new complex obtained. Suppose that $p < n$. We can reduce C_p to a subchain C_p^0 of K^0. This subchain will not include the new n-cell and hence it will be a subchain of both K^α and K^β. Since the new n-cell is of diameter $< 2\alpha < 2\xi$, the deformation from C_p to C_p^0 is at most 2ξ. Therefore by No. 29, C_p^0 can be taken as C_p^α or C_p^β. In other words we may assume $C_p^\alpha = C_p^\beta$.

On the other hand, T will transform $C_q^{*\beta}$ into a subchain $C_q^{*\tau}$ of $K^{*\alpha}$ which coincides with it outside of the star of E_r, and hence intersects $C_p^\beta = C_p^\alpha$ in the same chain. But the deformation from $C^{*\beta}$ to $C^{*\tau}$ is again at most 2ξ, and hence from C_q to $C_q^{*\tau}$ at most 6ξ. Hence $C_q^{*\tau}$ can be taken as $C_q^{*\alpha}$ (No. 29). Therefore

$$(32) \qquad C_p^\beta \cdot C_q^{*\beta} = C_p^\alpha \cdot C_q^{*\beta} = C_p^\alpha \cdot C_q^{*\tau} = C_p^\alpha \cdot C_q^{*\alpha}.$$

In other words, we can choose the α and β chains such that the intersections are actually identical and therefore *a fortiori* homologous. This proves (31) for $p < n$.

33. The disposition of the case $p = n$ rests on the following

LEMMA. *The notations being as in* § 1, *let* Γ_n *be the cycle mod* L *sum of the n-cells of* $K - L$. *Then*

$$(33) \qquad \Gamma_n \cdot C_q^* = C_q^*.$$

For let σ_q be any cell of K' component of E_q^* and let it be on E_n of $K-L$. Then if the β's of No. 3 refer to the intersection of E_n with E_q^* we have $\beta_p = \beta_n$, hence $\beta_s = \beta_q$ by (1). Therefore if $\beta_q \sigma_q$ is a positive component of E_q^* it is also a positive component of $E_n \cdot E_q^*$. Therefore

$$\Gamma_n \cdot E_q^* = E_q^*,$$

which implies (32).

Let us return now to our problem. As the argument of No. 32 shows, we merely have to prove that

$$(32') \qquad C_n^\alpha \cdot C_q^{*\alpha} \sim C_n^\beta \cdot C_q^{*\beta} \bmod N \text{ on } \mathfrak{N},$$

where, as in No. 32,

$$C_q^{*\alpha} = T \cdot C_q^{*\beta}.$$

We will designate by Γ_n^α, the sum of the n-cells of K^α and similarly for Γ_n^β and K^β. They are merely the subdivisions of Γ_n induced by K^α and K^β. We have, by the Lemma,

$$(33) \qquad \Gamma_n^\alpha \cdot C_q^{*\alpha} = C_q^{*\alpha}, \qquad \Gamma_n^\beta \cdot C_q^{*\beta} = C_q^{*\beta}.$$

We are only interested in the case where C_n^α has cells in the star of E_r intersecting cells of $C_q^{*\beta}$ modified by T. For since the latter are on the star, when that case does not arise the reasoning leading to (32) continues to hold.

Now when the exceptional mode of intersection just mentioned occurs $F(C_n^\alpha)$ cannot have cells in the star of E_r. For such cells have E_r as a face, and therefore meet all the modified closed cells of $C_q^{*\beta}$ since these modified cells all meet E_r^q. Hence $F(C_n^\alpha)$ would meet $C_q^{*\beta}$ on $K-N$. From this would follow as previously that when the deformations from C_p, C_q are $< 4\xi$, $F(C_n^\alpha)$ is nearer to G on $K-N$ than is allowed by Property I. For a similar reason we can assume that in the exceptional case $F(C_q^{*\beta})$ includes no cell modified by T.

As a first consequence of what precedes, taking into account (7) of Ch. II applied to T, and since the deformations only move points over $\mathfrak{N}$, we have

(34) $$C_q^{*\alpha} \sim C_q^{*\beta} \bmod N \text{ on } \mathfrak{N}.$$

Next, since $E_{r'}$ is regular, its star is an n-cell which is a simple circuit. But C_n^α is a cycle of the star relatively to its boundary. Hence the sum of its cells in the star is t times the sum of the oriented cells of the latter. Hence $t\Gamma_n^\alpha - C_n^\alpha = D_n$ has no cells in the star. Since the modifications in $C_q^{*\beta}$ due to T are wholly in the star, we have then

(35) $$D_n \cdot C_q^{*\alpha} = D_n \cdot C_q^{*\beta}.$$

Also by (33′) and (34)

$$t\Gamma_n^\alpha \cdot C_q^{*\alpha} = tC_q^{*\alpha} \sim tC_q^{*\beta} = t\Gamma_n^\beta \cdot C_q^{*\beta} \bmod N \text{ on } \mathfrak{N},$$

from which, together with (35), follows (32′), and the proof of our theorem is complete.

34. From the cycles $C_p^\alpha \cdot C_q^{*\alpha} = \Delta_s^\alpha$ to the homologous family about G by which we are to define $C_p \cdot C_q$, is but a step. We recollect (No. 27) that

$$F(\Delta_s^\alpha) \subset \Phi^\alpha \subset \bar{N}'^\alpha \subset N^\alpha$$

where N^α is the K^α-neighborhood of L. Since K^α is a subdivision of K, $\bar{N}^\alpha$ is a subdivision of $\bar{N}$, and since N is normal, so is N^α. We are therefore in a position to deform $F(\Delta_s^\alpha)$ over N^α onto L, as in No. 25, and derive from Δ_s^α a certain cycle mod L, Γ_s^α. Now we have

THEOREM. *The family* $\{\Gamma_s^\alpha\}$ *is homologous about* G *mod* L.

Given the neighborhood $\mathfrak{N}$ of G let us construct another $\mathfrak{N}'$ such that $\bar{\mathfrak{N}}' \subset \mathfrak{N}$. There exists then a simplicial subdivision of the initial K, which we call K to simplify matters, such that N, the K-neighborhood of L, is normal, and its mesh is so small that every cell of N with points on $\mathfrak{N}'$ is on $\mathfrak{N}$.

According to the theorem just proved, the cycles Δ^α, for ξ and α sufficiently small, will all be homologous to one another mod N on $\mathfrak{N}'$. Therefore there exists a new neighborhood of G, $\mathfrak{N}'' \subset \mathfrak{N}'$, such that when Δ^α and $\Delta^\beta \subset \mathfrak{N}''$, we have $\Delta^\alpha \sim \Delta^\beta \bmod N$ on $\mathfrak{N}'$.

Let now Γ^α, $\Gamma^\beta \subset \mathfrak{N}''$. Since Δ^α and Δ^β are subchains of Γ^α and Γ^β respectively,

$$\mathfrak{N}' \supset C_{s+1} \to \Delta_s^\beta - \Delta_s^\alpha \bmod N.$$

It follows from No. 25, property (c), that there exists a C'_{s+1} such that $C'_{s+1} - C_{s+1}$ is on the sum of the cells of N that meet C_{s+1}, while the boundary of C'_{s+1} mod L is $\Gamma_s^\beta - \Gamma_s^\alpha$. Due to the choice of N, $C'_{s+1} \subset \mathfrak{N}$. This means that Γ_s^α and Γ_s^β are homologous mod L on $\mathfrak{N}$, and proves the theorem.

DEFINITION. *We understand by the intersection cycle $C_p \cdot C_q$ the family of s-cycles $\{\Gamma_s^\alpha\}$ homologous mod L about G, the cycles Γ being the same as the cycles $C_p^\alpha \cdot C_q^{*\alpha}$ plus a chain on the K^α neighborhood of L.*

The extension to the new cycles of the usual operations on cycles is trivial.

35. **An important special case.** It corresponds to $G \subset K - L$, i. e. the geometric intersection G does not meet L, and was alone studied in [8]. *The intersection is then an absolute cycle on $K - L$ and the neighborhoods (closed) may be confined to $K - L$.*

More generally, suppose that G breaks up into any finite number of closed connected subsets G^i without any common points on $K - L$. Then we can select an $\mathfrak{N} = \sum \mathfrak{N}^i$, where $\mathfrak{N}^i$ is a neighborhood of G^i and $\mathfrak{N}^i \cdot \mathfrak{N}^j = 0$ for $i \neq j$. We have in that case

$$\Gamma_s^\alpha = \sum \Gamma_s^{i\alpha}, \qquad \Gamma^{i\alpha} \subset \mathfrak{N}^i,$$

and $\{\Gamma_s^{i\alpha}\}$ is a homologous family about G^i. We thus have one family of that nature for every G^i.

36. The first derived of any complex K may be obtained by successive insertion of vertices on a single cell E_r of K and joining it then to the boundary cells of every cell of the star of E_r.

Let K^α be again as before and K^β obtained from it by the above operation effected on the cell E_r. Then by a reasoning which is essentially parallel to that of Nos. 30—33 —the removed star is now that of E_r— we can again establish (31). From this follows once more that (31) holds when K^α, K^β are of mesh $<\xi$ and have a common derived of some order.

It follows, in particular, that (31) holds for $\alpha<\frac{1}{2}\xi$ and $K^\beta=K^{*\alpha}$. Let us approximate C_p by $C_p^{*\alpha}$ of $K^{*\alpha}$ and C_q by C_q^α of K^α. This time $K^\beta=K^{*\alpha}+\cdots$, $K^{*\beta}=K^\alpha+\cdots$ the neglected parts being on N and therefore indifferent as regards homologies mod N. Therefore from (31) follows

$$C_p^\alpha\cdot C_q^{*\alpha}\sim C_p^{*\alpha}\cdot C_q^\alpha \bmod N \text{ on } \mathfrak{N}.$$

This means, when we follow it up in the light of No. 34, that *we can interchange throughout the roles of C_p and C_q without changing their intersection cycle mod L on $\mathfrak{N}$.*

37. **Case of the Kronecker index.** Everything that we have obtained holds for $p+q=n$, $s=0$. However unless G includes a maximal connected part which does not meet L, it has little significance. For every C_0 on a connected part containing points of L, is ~ 0 mod L on that part, and the part of $C_p\cdot C_{n-p}$ on such connected pieces is ~ 0 mod L on any $\mathfrak{N}$.

We are more fortunate when G does not meet L, for there we deal only with absolute cycles and homologies not open to the same objections. *Thus when the geometric intersection of C_p and C_{n-p} does not meet L, there is a uniquely defined cycle $C_p\cdot C_{n-p}$ attached to this intersection.* All the cycles of the family defining the intersection cycle which conform with the theorem of No. 30, are homologous on $\mathfrak{N}$ and hence their Kronecker index is the same. It is called the Kronecker index of the intersection cycle and designated as usual by $(C_p\cdot C_{n-p})$.

38. REMARK. Suppose in particular that as in No. 35, $G=\sum G^i$ a sum of disjointed closed sets. Then the intersection cycle $\Gamma_0=\sum\Gamma_0^i$, $\Gamma_0^i\subset\mathfrak{N}^i$ ($\mathfrak{N}^i$ as in No. 34). Hence $(\Gamma_0)=(C_p\cdot C_q)=\sum(\Gamma_0^i)$. Of more importance however is the fact that *now instead of one index characteristic of the intersection, we have as many as there are sets G^i.* Each index (Γ_0^i) may be considered as the Kronecker index of the

corresponding set G^i. In particular, if the G's are isolated points, we have thus a definite Kronecker index attached to each point.

EXAMPLE. Let γ, δ be two one-cycles on an H_2 and let them intersect in ν distinct points $P^1, \cdots, P^\nu$. Then

$$\gamma \cdot \delta = \sum_{i=1}^{\nu} \Gamma_0^i \quad \text{and} \quad (\gamma \cdot \delta) = 0.$$

But the numbers (Γ_0^i) may well be $\neq 0$ and each will characterize the multiplicity of intersection of $\gamma \cdot \delta$ at the corresponding point P^i. Each will be the index of the intersection zero-cycle of a pair of approximations C_1^α, $C_1^{*\alpha}$ on the corresponding $\mathfrak{N}^i$.

39. **Properties of the intersection elements.** The effect of permutation of the chains on the new intersection elements is as for cells (No. 4). They also obey the distributive law. Finally Th. V of No. 10 holds in the following form: *If C_p bounds C_{p+1} which does not meet C_{n-p} on L nor $F(C_{n-p})$ anywhere, then $(C_p \cdot C_{n-p}) = 0$.* To prove this take K^α of mesh $< \frac{1}{4} d(C_{p+1}, F(C_{n-p}))$ which is >0. Reduce C_{p+1} to the subchain C_{p+1}^α of K^α with C_p^α corresponding to C_p. Subdivide C_{n-p} into C'_{n-p} of mesh $< \alpha$, remove the cells whose closure is not on $K^{*\alpha}$ from it and reduce the remaining chain C'_{n-p} to the subchain $C_{n-p}^{*\alpha}$ of $K^{*\alpha}$. We manifestly have

$$(C_p \cdot C_{n-p}) = (C_p^\alpha \cdot C_{n-p}^{*\alpha}).$$

But the chains with the superscript α are exactly in the mutual relation of C_{p+1}, C_p, C_{n-p}^* of No. 10. Applying Th. V, No. 10 to them we have $(C_p^\alpha \cdot C_{n-p}^{*\alpha}) = 0$, from which the required result follows.

The theorems of Nos. 17, 20 hold likewise when the intersecting cycles are arbitrary cycles of their type. For on passing from them to the approximations the indices and the homologies between the cycles of each type are preserved, the homologies for the approximations being understood as in No. 12.

40. Extension to more than two intersecting chains. A similar treatment is applicable to the intersection of chains C_{q_i}, $i = 1, 2, \cdots, \nu > 2$, subjected to the condition that for every i, $F(C_{q_i})$ and the chains C_{q_h}, $h \neq i$, have no common points on $K - L$. The approximations are by means of subchains $C^{h_i}_{q_i}$ of the duals K^{*h_i}, $h_i < h_{i+1}$, whose meshes are sufficiently small. We show as before that $\prod C^{h_i}_{q_i}$ is a cycle mod L, Γ_s, $s = \sum q_i - (\nu - 1)n$, homologous to itself when the orders of the h's are indefinitely increased. The cycles thus obtained are homologous mod L. They have all the properties of the polyhedral intersections defined in No. 11. They belong to a set homologous mod L about the geometric intersection G of the given chains on $K - L$. That set is by definition the intersection cycle of the chains C.

If the h's are replaced by any set of ν distinct integers the intersection is unchanged. For with obvious meaning of all symbols, by the associative law and the result of No. 36,

$$\begin{aligned}\Gamma_s &\sim [\cdots]\cdot[C^{h_i}_{q_i} \cdot C^{h_{i+1}}_{q_{i+1}}]\cdot[\cdots]\\ &\sim [\cdots]\cdot[C^{h_{i+1}}_{q_i} \cdot C^{h_i}_{q_i+1}]\cdot[\cdots].\end{aligned}$$

Hence two consecutive h's can be permuted and therefore their order is immaterial. Thus here again although the construction of Γ_s is unsymmetrical the family of cycles $\sim \Gamma_s$ is in a certain symmetrical relation to the C's.

The extension of the theorems of Nos. 34, 39, offers no difficulty. We give explicitly that of the second as we shall use it in an important case later (No. 47). *Let the dimensions of the chains C_p, C_{q_i}, $i = 1, 2, \cdots, \nu$ satisfy $p + \sum q_i - n\nu = 0$ so that the intersection cycle, if any is at hand, is of dimension zero. Let C_p bound C_{p+1} which does not meet some one C_q on L nor intersect any set of chains such as $F(C_{q_1})$, $C_{q_2}, \cdots, C_{q_\nu}$ anywhere. Then we have* $(C_p \cdot C_{q_1} \cdots C_{q_\nu}) = 0$. The proof is as in No. 39. For if we designate the approximation to any chain by accenting its symbol, we find by repeated application of formula (18) that $F(C'_{q_1} \cdots C'_{q_\nu})$ is a sum of chains of type

$\pm F(C'_{q_1}) \cdot C'_{q_2} \cdots C'_{q_\nu}$. When the approximation is sufficiently fine (α small enough), none of these chains will meet C'_{p+1}. From Th. V of No. 10 applied to C'_{p+1} and $C'_{q_1} \cdots C'_{q_\nu}$ we find $(C'_p \cdot C'_{q_1} \cdots C'_{q_\nu}) = 0$ from which the desired result follows as in the previous number.

§ 4. Topological Invariance of the Intersection Elements

41. We have just defined, say for two chains C_p, C_q, an intersection cycle $C_p \cdot C_q$ and an index when $p + q = n$. The intersection cycle was just a name given to a suitable family of polyhedral s-cycles homologous about the geometric intersection G. When $p + q = n$ the Kronecker index of the polyhedral zero-cycles turned out to be fixed and was by definition $(C_p \cdot C_{n-p})$.

Suppose however that we find another pair $\mathfrak{K}$, $\mathfrak{L}$ in point to point coincidence with K and L. We can then determine other intersection elements—let us designate them by $\widehat{C_p \cdot C_q}$ or $(\widehat{C_p \cdot C_{n-p}})$—and the question is whether they are the same as before. The answer is indeed affirmative as we shall now prove, and this fact is expressible as a

Fundamental Theorem. *The intersection elements are topological invariants, provided that all the coincident manifolds K—L considered are concordantly oriented.*

Regarding the intersection index this means that

$$(36) \qquad (\widehat{C_p \cdot C_{n-p}}) = (C_p \cdot C_{n-p}).$$

As for the cycle the theorem asserts that the maximal family of homologous cycles about G which serves to define it is the same for K and $\mathfrak{K}$. In the light of § 3 and in an obvious notation we need merely to prove that any two component sequences $\{C_p^{\alpha_i} \cdot C_q^{*\alpha_i}\}$, $\{\widehat{C_p^{\beta_i} \cdot C_q^{*\beta_i}}\}$ for the intersections mod $\bar{N}$ are equivalent, $\bar{N}$ being fixed, or again that for α, β small enough,

$$(37) \qquad C_p^{\alpha} \cdot C_q^{*\alpha} \sim \widehat{C_p^{\beta} \cdot C_q^{*\beta}} \mod \bar{N} \text{ on } \mathfrak{N}.$$

Instead of N as in § 3 we take here $\bar{N}$, which being a closed complex is somewhat more convenient. There is of course neither gain nor loss of generality thereby, since N can always be replaced by a similar N' such that $\bar{N}' \subset N$.

For transparent reasons of continuity, if we have taken α sufficiently small, C_p^α, $C_q^{*\alpha}$ may replace throughout C_p and C_q. This means that we may assume that the situation is as follows:

(a) The two intersecting chains are subchains C_p, C_q^* of K and K^* which do not meet one another's boundaries on $K - N$.

(b) They are approximated by subchains C_p^β, $C_q^{*\beta}$ of $\mathfrak{K}^\beta$, $\mathfrak{K}^{*\beta}$ (a subdivision of $\mathfrak{K}$ and its dual).

To prove (37) it is sufficient to establish that when β is small enough

$$(37') \qquad C_p^\beta \cdot C_q^{*\beta} \sim C_p \cdot C_q^* \text{ mod } \bar{N} \text{ on } \mathfrak{M}.$$

To simplify matters we shall prove (37′) taking for N the K-neighborhood of L, which is permissible provided that we take care to replace, if need be, the initial complex by another whose mesh is small enough. For a similar reason we may assume that N is normal.

In the case of the index the matter can be simplified. Indeed, as we are then only interested in the points away from L, we can immediately remove from suitable subdivisions of the initial chains the cells that meet L, and therefore assume that the chains C_p, C_{n-p} do not meet L. It becomes then obvious that K, C_p, C_{n-p} may be replaced by K^α, C_p^α, $C_{n-p}^{*\alpha}$ in computing $(\widehat{C_p \cdot C_{n-p}})$, or without loss of generality, that we deal at the very outset with subchains C_p, C_{n-p}^* of K and K^*. What we have to prove then in place of (36) is the relation

$$(36') \qquad (\widehat{C_p \cdot C_{n-p}^*}) = (C_p \cdot C_{n-p}^*).$$

Similar remarks hold for the intersections of any number of chains.

42. **Invariance of the index.** In view of the distributive law, valid for both old and new indices, we need only to

prove (36′) when the chains are a pair of dual cells E_p, E^*_{n-p}, i. e. it is merely necessary to show that

$$(38) \qquad (E_p \cdot E^*_{n-p}) = +1 = (\widehat{E_p \cdot E^*_{n-p}}).$$

I say that if $p<n$ we can replace it by $p+1$ in our proof. For let E_{p+1} be a cell of the star of E_p positively related to ηE_p. Then its dual E^*_{n-p-1} is positively related to $(-1)^{p+1}\eta E^*_{n-p}$, (Ch. III, No. 26). The only cell of $F(E^*_{n-p})$ that meets E_{p+1} is its dual E^*_{n-p-1} (No. 2). Similarly, the only cell of $F(E_{p+1})$ that meets E^*_{n-p} is E_p. Therefore formula (20) yields here

$$(39) \quad (E_{p+1} \cdot (-1)^{p+1}\eta E^*_{n-p-1}) + (-1)^p (\eta E_p \cdot E^*_{n-p}) = 0,$$

and this relation is identically satisfied in view of

$$(E_p \cdot E^*_{n-p}) = (E_{p+1} \cdot E^*_{n-p-1}) = 1,$$

which merely describes our convention for orienting the starred cells.

Let now $\mathfrak{K}^\beta$ be a subdivision of $\mathfrak{K}$ of mesh β and $\mathfrak{K}^{*\beta}$ its dual. Apply the proper deformation theorems to E_{p+1} and E^*_{n-p} to rêduce them respectively to subchains of $\mathfrak{K}^\beta$ and $\mathfrak{K}^{*\beta}$. We denote the deformed elements by replacing wherever necessary E by $\mathfrak{E}$. When β is taken sufficiently small, $\mathfrak{E}_{p+1}$ may meet the chain $\mathfrak{E}^*_{n-p-1}$ but not $F(\mathfrak{E}^*_{n-p}) - \mathfrak{E}^*_{n-p-1}$, since this holds for the chains E which they approximate. Similarly $\mathfrak{E}^*_{n-p}$ may meet $\mathfrak{E}_p$ but not $F(\mathfrak{E}_{p+1}) - \mathfrak{E}_p$. Applying (20) to the intersection of $\mathfrak{E}_{p+1}$ with $\mathfrak{E}^*_{n-p}$ we find

$$(\mathfrak{E}_{p+1} \cdot (-1)^{p+1}\eta \mathfrak{E}^*_{n-p-1}) + (-1)^p (\eta \mathfrak{E}_p \cdot \mathfrak{E}^*_{n-p}) = 0.$$

This, however, is equivalent to

$$(\widehat{E_{p+1} \cdot E^*_{n-p-1}}) = (\widehat{E_p \cdot E^*_{n-p}}).$$

By comparing with (39) we see then at once that (38) will follow if we can prove that

$$(E_{p+1} \cdot E^*_{n-p-1}) = (\widehat{E_{p+1} \cdot E^*_{n-p-1}}),$$

or more generally if we can prove (36) with p replaced by $p+1$, as asserted.

43. In the last analysis, then, in establishing (36) we may take $p = n$. In that case $C^*_{n-p} = C^*_0$, consists merely of a finite number of oriented points. By the distributive law for the index it is sufficient to assume that it consists of a single point A taken positively, while $C_n = E_n$, the cell of $K-L$ that carries A. We have therefore to prove

$$(E_n \cdot A) = (\widehat{E_n \cdot A}). \tag{40}$$

Let G be the maximal connected subset of $K-L$ to which A belongs. Since $K-L$ is a manifold $\bar{G}$ is a simple orientable n-circuit (Ch. III, No. 12). Therefore the sum of its n-cells oriented with E_n as indicatrix is an n-cycle that we still denote by G to simplify matters. Since A does not meet any cell of G other than E_n we have

$$(E \cdot A) = (G \cdot A); \quad (\widehat{E \cdot A}) = (\widehat{G \cdot A}).$$

Now if B is any point of (the open set) G, $G \supset E_1 \to A - B$, hence $A \sim B$ on $K-L$ and

$$(G \cdot A) = (G \cdot B), \quad (\widehat{G \cdot A}) = (\widehat{G \cdot B}).$$

It follows that the value of $(G \cdot A)$ depends solely upon the orientation of G but not upon A. It is $+1$ when G is concordantly oriented with $K-L$, i. e. if E_n is likewise an indicatrix of $K-L$, and -1 otherwise. Similarly, of course, for $(\widehat{G \cdot A})$ and $\mathfrak{K}-\mathfrak{L}$.

Let $K-L$ break up into q circuits G^i, $i = 1, 2, \cdots, q$, $G^1 = G$. We designate also by G^i one of the fundamental cycles corresponding to its circuit. We agree to orient throughout the corresponding set of n-cells of $K-L$ or $\mathfrak{K}-\mathfrak{L}$ in such manner that their sums $\sim G^i \bmod L$. This fixes the orientation of all the n-cells of $K-L$ or $\mathfrak{K}-\mathfrak{L}$ and we shall have

$$(G^i \cdot A) = +1, \quad (\widehat{G^i \cdot A}) = +1$$

when $A \subset G_i$, for every i. Then (40) will hold, and hence (36). *We have therefore proved that the Kronecker index does not depend upon the particular choice of complex K turning the point set $K-L$ into a manifold and having L for subcomplex, provided that the orientations chosen for the circuits G^i are concordant for all choices of K.*

The preceding result implies *the invariance of the index with respect to all homeomorphisms of $K-L$ into a similar manifold $K'-L'$ provided that the transforms of the cycles G^i on $K-L$ are the corresponding cycles for $K'-L'$.*

44. The extension to the index of any number of intersecting chains offers no difficulty. Suppose that we have, for example, C_p, C_q, C_r; $p+q+r=2n$. The crucial point is to replace $(C_p \cdot C_q \cdot C_r)$ by a similar index with a third term of dimension $r+1$ instead of r. This is done by extending (18) as pointed out at the end of No. 11. From the new boundary relation we deduce the proper extension of (20), which will be a formula in three terms, in two of which the last chain is an $(r+1)$-chain while in the third it is an r-chain. It is this formula which will enable us to replace r by $r+1$ in the treatment, and hence ultimately to have a sum of intersections, any one of which has the form $(C_p \cdot C_{n-p} \cdot C_n)$ with $C_p \cdot C_{n-p}$ and $C_p \cdot C_{n-p} \cdot C_n$ consisting of a single point. Now applying the theorem just proved for the index of two cycles, first to $C_p \cdot C_{n-p}$, then to this intersection taken as intersecting C_n, we obtain the invariance of $(C_p \cdot C_{n-p} \cdot C_n)$. This completes the proof.

45. **Invariance of the intersection cycles.** In No. 41 our task has already been narrowed down to the proof of formula (37′).

Let $\mathfrak{N}'$ be the K'-neighborhood of the geometric intersection of C_p, C_q^*, where K' is the first derived complex of K. Since $C_p \cdot C_q^*$ is an exact sum of cells of K', $\overline{\mathfrak{N}}'$ is a subcomplex of K'. Furthermore, since K may be replaced by a subdivision of arbitrarily small mesh, we may assume $\overline{\mathfrak{N}}' \subset \mathfrak{N}$.

Referring to the operations by means of which we pass from C_p and C_q^* to the corresponding β chains (subdivision, deformation at most 2β, removal of cells whose closure meets L) we see that we are free to choose β such that there exist C_{p+1} and C_{q+1} whose points are not farther than 2β from C_p or C_p^β, and C_q^* or $C_q^{*\beta}$ respectively, and that

$$C_{p+1} \to C_p - C_p^\beta \bmod \bar{N}; \qquad C_{q+1} \to C_q^* - C_q^{*\beta} \bmod \bar{N};$$
$$C_p^\beta \cdot C_q^{*\beta} \subset \mathfrak{N}' \subset \mathfrak{N}.$$

The upper limits imposed upon β here and later do not restrict us in any way since we need merely to prove (37′) for any β as small as we please.

By passing the usual projecting segments in $\mathfrak{N}'$ through all the points of $C_p^\beta \cdot C_q^{*\beta}$ and sliding them along these segments we can deform the cycle onto $C_p \cdot C_q^*$. Applying then the deformation theorem of Ch. II, No. 45 to the reduced cycle relatively to the complex K_s whose cells are those of $C_p \cdot C_q^*$ (each the intersection of an E_p of K with an E_q^* of K^*) we reduce finally that s-cycle, over K_s, to a subchain Γ_s of K_s and we have therefore

$$C_p^\beta \cdot C_q^{*\beta} \sim \Gamma_s \quad \bmod \bar{N} \text{ on } \mathfrak{N}'.$$

It will be remembered that we are taking for N the K-neighborhood of L, (No. 41).

46. Any s-cell of $C_p \cdot C_q^*$ is of type $E_p^i \cdot E_q^{*j}$ where the first cell belongs to $K - L$ and the second to K^*. Since $\bar{N}$ is an exact sum of cells of K, E_p^i, hence also the s-cell, belongs wholly to $\bar{N}$ or else has no point on it at all. Let us designate by $\mathfrak{E}_s^i$, $i = 1, 2, \cdots$ the s-cells in question $\not\subset \bar{N}$. Then

$$C_p \cdot C_q^* \;=\; \sum x_i \, \mathfrak{E}_s^i \;\bmod \bar{N},$$
$$C_p^\beta \cdot C_q^{*\beta} \sim \Gamma_s = \sum y_i \, \mathfrak{E}_s^i \;\bmod \bar{N} \text{ on } \mathfrak{N}'.$$

In order to prove (37′) it is sufficient to show that $x_i = y_i$, for in that case

$$C_p \cdot C_q^* \sim C_p^\beta \cdot C_q^{*\beta} \bmod \bar{N} \text{ on } \mathfrak{N}',$$

hence also on $\mathfrak{N}$.

The $(n-s)$-cells, E_{n-s}^{**h}, of the second dual K^{**} of K, intersect $E_p^i \cdot E_q^{*j}$ in a single point or not at all. Let the double-starred cells be so numbered that the k-th is the one which meets the particular cell $E_p^i \cdot E_q^{*j}$ that we call $\mathfrak{E}_s^k$. We have then by No. 11 plus the definition of the index of more than two chains:

$$(\mathfrak{E}_s^k \cdot E_{n-s}^{**k}) = (E_p^i \cdot E_q^{*j} \cdot E_{n-s}^{**k}) = +1,$$

the sign $+$ before 1, being merely a matter of properly orienting the double-starred cells.

We now impose upon β the further restriction that it be less than half the least distance from the point of intersection of any pair $\mathfrak{E}_s^h$, E_{n-s}^{**h} to the boundary of E_{n-s}^{**h}. This is in no sense restrictive since β can be taken as small as we please. Under the circumstances, however, throughout the deformations applied to $C_p^\beta \cdot C_q^{*\beta}$, it will never meet the boundaries in question. Therefore there exists on $\mathfrak{M}'$ a C_{s+1} that does not meet the boundary of the E^{**} in question, and such that

$$C_{s+1} \rightarrow C_p^\beta \cdot C_q^{*\beta} - \sum y_i \mathfrak{E}_s^i \mod \bar{N}.$$

Since the double-starred cells are cells of a dual of a manifold mod $\bar{N}$ they do not meet $\bar{N}$. We can therefore apply the theorem of No. 39 and find:

$$((C_p^\beta \cdot C_q^{*\beta} - \sum y_i \mathfrak{E}_s^i) \cdot E_{n-s}^{**h}) = 0.$$

Since $K-L$ and $\mathfrak{K}-\mathfrak{L}$ are concordantly oriented, we have then in view of the invariance of the index and of the associative law:

$$(C_p^\beta \cdot C_q^{*\beta} \cdot E_{n-s}^{**h}) = y_h.$$

On the other hand,

$$(C_p \cdot C_q^* \cdot E_{n-s}^{**h}) = (\sum x_i \mathfrak{E}^i \cdot E_{n-s}^{**h}) = x_h.$$

47. Consider now again the complexes C_{p+1}, C_{q+1} of No. 45. The boundary of the first is $C_p - C_p^\beta \mod \bar{N}$ and its points are within a distance of 2β from C_p or C_p^β. Since C_p does

not meet $F(C_q^*)$ on $K-\bar{N}$, for β small enough C_{p+1} will also fail to meet it on $K-\bar{N}$, hence it will not meet $F(C_q^*)\cdot E^{**h}$. Similarly, since $C_p \cdot C_q^*$ does not meet $F(E^{**h})$, C_{p+1} will not meet $C_q^* \cdot F(E^{**h})$. Hence C_{p+1} will not meet $F(C_q^* \cdot E^{**h})$ $= C_q^* \cdot F(E^{**h}) + (-1)^s F(C_q^*) \cdot E^{**h}$. From the theorem of No. 39 follows once more

$$(C_p - C_p^\beta \cdot (C_q^* \cdot E_{n-s}^{**h})) = 0,$$

and since the associative and distributive laws hold

$$(C_p \cdot C_q^* \cdot E_{n-s}^{**h}) = (C_p^\beta \cdot C_q^* \cdot E_{n-s}^{**h}).$$

Similarly, C_{q+1} will not meet $F(C_p^\beta)$ and E_{n-s}^{**h}, nor C_p^β and $F(E_{n-s}^{**h})$ on $K-\bar{N}$. Hence (No. 39),

$$(C_p^\beta \cdot C_q^* \cdot E_{n-s}^{**h}) = (C_p^\beta \cdot C_q^{*\beta} \cdot E_{n-s}^{**h}).$$

so that finally,

$$(C_p \cdot C_q^* \cdot E_{n-s}^{**h}) = (C_p^\beta \cdot C_q^{*\beta} \cdot E_{n-s}^{**h}).$$

From this and No. 46 follows $x_h = y_h$. *The topological invariance of the intersection cycle is therefore proved.*

The extension to more than two intersecting chains is automatic.

In the case of the cycles as in that of the Kronecker index the proof actually shows that *the intersection cycles are invariant relatively to homeomorphisms of the open manifold $K-L$, provided that the orientations of the manifold and of its transforms are concordant.*

§ 5. Looping Coefficients

48. Let Γ_p, Γ_{n-p-1} $(p<n)$ be two non-intersecting cycles on an n-sphere H_n which is an M_n, and suppose that $p \neq 0$, $n-1$ or else for $p = 0$, $n-1$ that $\Gamma_0 \sim 0$.

Then there are on the sphere two chains $C_{p+1} \to \Gamma_p$, $C_{n-p} \to \Gamma_{n-p-1}$. If C'_{n-p} is any other chain $\to \Gamma_{n-p-1}$, $C_{n-p} - C'_{n-p} \to 0$ and since $p \neq 0$ or n, the cycle at the left ~ 0. Therefore by No. 39, $(\Gamma_p \cdot C_{n-p}) = (\Gamma_p \cdot C'_{n-p})$. Thus the number $(\Gamma_p \cdot C_{n-p})$ depends solely upon the two given cycles. It is called the *looping coefficient* of the two cycles (Brouwer) and will be designated by $Lc(\Gamma_p, \Gamma_{n-p-1})$. We have similarly

$$Lc(\Gamma_{n-p-1}, \Gamma_p) = (\Gamma_{n-p-1} \cdot C_{p+1}).$$

Let us now subdivide H_n into a complex of suitably small mesh, reduce C_{p+1} to a subchain of it, C_{n-p} to a subchain of its dual. Keeping the same designations for the reduced chains, and also, since their indices are by definition those of the initial chains, we have by (20) and (5),

$$\begin{aligned}(C_{p+1} \cdot \Gamma_{n-p-1}) &= (-1)^{p+1}(\Gamma_p \cdot C_{n-p}) \\ &= (-1)^{(n+1)(p+1)}(\Gamma_{n-p-1} \cdot C_{p+1}),\end{aligned}$$

from which follows

$$Lc(\Gamma_p, \Gamma_{n-p-1}) = (-1)^{n(p+1)} Lc(\Gamma_{n-p-1}, \Gamma_p). \tag{41}$$

Thus the looping coefficient is a symmetric or skew-symmetric function of the two cycles.

Geometrically the looping coefficient of Γ_p, Γ_{n-p-1} represents the algebraic number of times that Γ_p must cross Γ_{n-p-1} when it is shrunk to a point over H_n.

49. For $n = 3$, $p = 1$, $n - p - 1 = 1$, we have as a special case the looping coefficient of two circuits on a Euclidean three-sphere. By the usual transformation of H_3 minus a point into an S_3, we can pass to the looping coefficients of two circuits on the latter. The looping coefficient is then nothing more nor less than the classic integral of Gauss attached to the two circuits. The general case of Euclidean H_n or S_n was investigated by Brouwer [15].

He proved that these numbers are functions of the two Γ's, but not that they are independent of the mode of straightness adopted for H. Complete topological invariance (independence from straightness) follows from the fact that the looping coefficient is defined as a Kronecker index and that the latter is a topological invariant (§ 4).

50. The notion behind the looping coefficients is susceptible of immediate extension. For the sequel we may confine ourselves to polyhedral cycles throughout.

Suppose then that we have Γ_p, Γ_q, on our former H_n, polyhedral and such that their common points coincide with a polyhedral K_s, $s = p + q - n$. We can define their looping cycle mod K_s as the intersection mod K_s, $\Gamma_p \cdot C_{q+1}$ of Γ_p with $C_{q+1} \rightarrow \Gamma_q$. This intersection is well defined in the sense of § 3 and with a variation in C may change only by an absolute cycle. If we consider then the looping relative cycle only modulo absolute cycles (that is, as unchanged when an absolute cycle is added), we have a uniquely defined attribute of Γ_p, Γ_q which we designate by $L\gamma(\Gamma_p, \Gamma_q)$. The extension to several cycles is immediate and obvious, and when the looping cycle is of dimension zero there is a uniquely defined looping coefficient. The uniqueness of the latter follows from the fact that the addition of an absolute cycle to the intersection of the C's involved does not alter the Kronecker index whose value is the looping coefficient.

51. The looping coefficients may be introduced in another manner which is somewhat more symmetrical as regards the two chains. Let A be any point not on H_n and consider the join (A, H_n). That join is an $\bar{E}_{n+1}$ whose boundary is H_n. Introduce now the subchains of the cells

$$\Delta_{p+1} = (A, \Gamma_p) \rightarrow \Gamma_p, \qquad \Delta_{n-p} = (A, \Gamma_{n-p-1}) \rightarrow \Gamma_{n-p-1}.$$

Since H_n is an M_n, E_{n+1} is an open manifold with H_n for regular boundary. For we can assume that H_n is a simplicial complex and $\bar{E}_{n+1}$ that complex plus the unique cell E_{n+1}. Then all that is necessary is to show that if E_p

is any cell of H_n, $Lk(E_p)_{\bar{E}_{n+1}}$ is a closed cell of the type required by Ch. III, No. 5, condition (c). But that follows immediately from Ch. II No. 49 plus the fact that it is the join of A and the sphere $Lk(E_p)_H$, which is a simple circuit since H is an M_n. The two chains $\varDelta$ meet on E and not on H, hence their index is uniquely defined and I say that

$$(42)\qquad Lc(\varGamma_p, \varGamma_{n-p-1}) = (\varDelta_{p+1} \cdot \varDelta_{n-p}).$$

This relation will readily follow from another. Let us take another copy $\bar{E}'_{n+1}$ of $\bar{E}_{n+1}$, unite the corresponding points of the boundary spheres, then orient E'_{n+1} in such manner that $E'_{n+1} \to -H_n$. The result is a normally oriented H_{n+1} (Ch. II, No. 49), and this H_{n+1} is an absolute manifold (Ch. III, No. 35, Corollary).

Let now T be the homeomorphism between $\bar{E}_{n+1}$ and $\bar{E}'_{n+1}$ which associates each point of $\bar{E}_{n+1}$ with its image on its copy $\bar{E}'_{n+1}$. Let also $A' = T \cdot A$. We have

$$\varDelta'_{p+1} = T \cdot \varDelta_{p+1} \to \varGamma_p.$$

Therefore,

$$G_{p+1} = \varDelta - \varDelta' \to 0,$$

that is, G is a cycle.

Let us now replace H_n by a subdivision of as small a mesh as we please, and continue to call H_n the new subdivision. Let H^*_n be a dual of it. The joins of $A + A'$ with the cells of H^*_n give rise to a dual H^*_{n+1} of H_{n+1} according to our systematic construction. Finally, referring to our definition of Kronecker indices and hence looping coefficients by suitable approximations, we see that we may assume that G_{p+1}, $\varGamma_p$ are respectively subchains of H^*_{n+1}, H^*_n, and C_{n-p}, $\varGamma_{n-p-1}$ subchains of H_n.

Under the circumstances

$$(43)\qquad (\varGamma_p \cdot C_{n-p})_{H_n} = (G_{p+1} \cdot C_{n-p})_{H_{n+1}},$$

where the subscripts after the Kronecker indices designate the containing manifolds relatively to which they are computed. To prove (43) we observe that Γ and G meet C at the same points, finite in number, as the chains are subchains of two dual complexes. The relation follows then automatically on comparing indicatrices.

Now $C_{n-p} \to \Gamma_{n-p-1} \leftarrow \Delta_{n-p}$. Hence $C_{n-p} - \Delta_{n-p} \to 0$, hence ~ 0 on H_{n+1}, since $0 \leqq p \leqq n-1$, and therefore $n-p \neq 0$ or $n+1$. It follows, since G is a cycle, that

$$(G_{p+1} \cdot C_{n-p}) = (G_{p+1} \cdot \Delta_{n-p}).$$

But Δ_{n-p} and Δ'_{p+1} do not meet. Hence

$$(G_{p+1} \cdot \Delta_{n-p}) = (\Delta_{p+1} \cdot \Delta_{n-p}), \tag{44}$$

from which and (43) the required relation (42) follows.

Observe that (41) is an immediate corollary of (42) and the commutation law for the Kronecker index applied on H_{n+1} (formula (5)).

The extension of what precedes to the looping cycles and coefficients of any number of cycles (No. 50) is more or less automatic. By reference to H_{n+1} we have this important new result: *the looping cycles and coefficients obey the associative law.*

52. The new approach to the looping coefficients suggests a useful interpretation for the looping coefficients in the exceptional cases. We consider only two cycles, Γ_{-1}, Γ_n, but what we shall say is readily transcribed for the looping cycles and coefficients of any number of them.

Let Γ^0_{-1} be a symbol that we write down for the (vacuous) boundary of any point taken with the sign $+$. Then if C_0 is any zero-chain and (C_0) its Kronecker index, $(C_0)\,\Gamma^0_{-1}$ will be considered as its boundary and designated generically by Γ_{-1}, the symbol of a cycle of dimensionality -1. We can also write as usual

$$C_0 \to \Gamma_{-1}, \qquad F(C_0) = \Gamma_{-1}.$$

These (-1)-dimensional cycles are thus pure dummy symbols which, however, will enable us to simplify many statements later. This is entirely in line with the conventions of Ch. III, No. 5 regarding the dimensionality -1.

Let there be given *a priori* $\Gamma_{-1} = x\Gamma^0_{-1}$ considered as a cycle on H_n. We take $\Delta_0 = xA$ where A is a point of E_{n+1}. Then $\Gamma_n = yH_n$, $\Delta_{n+1} = yE_{n+1}$ and (42) yields as definition for Lc:

$$(42') \qquad Lc(\Gamma_{-1}, \Gamma_n) = (\Delta_0 \cdot \Delta_{n+1}) = xy.$$

All this assumes $n \geqq 0$. On the same principle, however, we can consider a symbolical H_{-1} ((-1)-dimensional sphere, boundary of a zero-cell) and cycles Γ_{-1}, Γ'_{-1} on it. If $\Gamma_{-1} = x\Gamma^0_{-1}$, $\Gamma'_{-1} = y\Gamma^0_{-1}$, we shall have $\Delta_0 = xE_0$, $\Delta'_0 = yE_0$, and we define again in accordance with (42),

$$(42'') \qquad Lc(\Gamma_{-1}, \Gamma'_{-1}) = (\Delta_0 \cdot \Delta'_0) = xy.$$

§ 6. A New Definition of Intersections of Chains

53. The procedure which we have followed in dealing with intersections is well adapted to the use principally made of it so far — the treatment of questions of invariance. The polyhedral chains considered are however of too restricted nature for most practical questions. Consider for example two oriented chains which are made up of cells of subdivisions of two (Euclidean) hemispheres and let their geometric intersection be of the same type. Our method does not enable us to obtain the intersection chain directly in terms of the cells of the geometric intersection, even when the latter is of the right dimensionality ($s = p + q - n$). For we have no means of subdividing the containing manifold, say even a region of ordinary space, in such manner that one of the two intersecting chains is a subchain of a certain $K - L$ while the other is one of its dual.

In that respect the method of intersection of *convex* chains of our earlier work [8] had decidedly the best of it. For whenever the geometric intersection proved to be of the right dimension the intersection chain was expressed in terms of cells of a subdivision of it.

We will say that C_p, C_q have a *regular intersection* whenever their intersection on $K-L$ is a sum of cells of dimension $\leqq s = p+q-n$.

The difficulty that we have indicated will be met by a definition to be introduced presently which fulfills the following requirements:

(a) *It yields a unique intersection* $C_p \cdot C_q$ *when* C_p, C_q *are subchains of* $K-L$ *such that*

$$\left.\begin{matrix} C_p \\ F(C_p) \end{matrix}\right\} \quad \textit{intersect both} \quad \left\{\begin{matrix} C_q \\ F(C_q) \end{matrix}\right\} \textit{regularly.}$$

The intersection is then an s-subchain of $K-L$ *whose cells are common to the two intersecting chains.*

(b) *The new intersection preserves the chief properties of the old. In particular formula (18) remains true.* Essentially this means that the invariance theory goes through as before.

(c) *When the chains are specialized as in* § 1 *or as in our paper* [8] *(convex chains) the new definition reduces to those already given.*

54. For the sequel we need an extension of the notion of *trace* of a chain on the linked complex of a cell. This notion, as introduced in Ch. III, No. 41, was valid for a simplicial K, and the customary substitute for the linked complexes of such a K. It did not require that the cells be regular. We need now the trace of a subchain C_p of $K-L$ on $Lk(E_s)$ where E_s is a cell of the manifold $K-L$. While the situation is somewhat more restricted than before, we do not restrict the cells to the simplicial type. This is important here, in view of the applications to intersections, since we wish to introduce intersections of subchains of a given *unmodified* $K-L$, without having to subdivide it.

Let then E^*_{n-s} be the dual of E_s. The boundary of E^*_{n-s} coincides with $Lk(E_s)$, and in fact the latter is merely a first derived of $F(E^*_{n-s})$. We will designate the chain $F(E^*_{n-s})$ by $Lk^*(E_s)$, thus keeping in mind: (a) that it coincides with Lk; (b) that it is a subchain of K^*.

Since C_p is a subchain of $K-L$ and Lk^* one of its dual, they have a well defined intersection in the sense of § 1. We now define the intersection $C_p \cdot Lk^*(E_s)$ as the *trace* of C_p on $Lk(E_s)$. When the cells are all simplexes, it is a simple matter to show that the present trace has the structure of a first derived of the former trace. Furthermore it can be deformed into that former trace barycentrically over the star of E_s, each simplex going into the corresponding one of a first derived of the former trace.

Since Lk^* is an absolute cycle the fundamental boundary relation (18) gives here

$$F(C_p \cdot Lk^*) = (-1)^{s-1} F(C_p) \cdot Lk^*,$$

which is equivalent to

$$(45) \qquad F(\textit{trace } C_p) = (-1)^{s-1} \textit{ trace } F(C_p).$$

When E_s is not a boundary cell of C_p the right hand side $\equiv 0$ and the trace is a cycle.

It is convenient to complete the definition of the trace as follows:

(a) When $p = s$ and E_s occurs in C_s with the coefficient x_i we set (see No. 52)

$$\textit{trace } C_s = \Gamma_{-1} = x_i \Gamma^0_{-1}.$$

As a consequence (45) still holds when $p = s+1$.

(b) When E_s is not a cell of C_p we set

$$\textit{trace } C_p \equiv 0$$

which manifestly does not alter (45).

55. **New definition of intersections.** In terms of traces and looping coefficients we define the intersection of two chains positionally restricted as in No. 53 as the s-chain, $s = p+q-n$:

$$(46) \quad \begin{cases} C_p \cdot C_q = \sum x_i E^i_s; \\ \quad x_i = Lc\,(\textit{trace } C_p,\ \textit{trace } C_q) \text{ on } Lk(E^i_s). \end{cases}$$

The definitions of the Kronecker index and the intersections of more than two chains follow automatically.

Under our new definition the commutation formula (4) (for chains) follows from (41). Likewise the distributive law (formula (14)) is a direct consequence of the same law for the trace and Lc. The extension to the intersection of several chains is made by means of the looping coefficients of their traces and it obeys the associative law owing to the associative property of the looping cycles (No. 51). A somewhat more difficult matter is to extend the fundamental boundary relation (18).

Let E_{s-1} be any particular $(s-1)$-cell of $K-L$. The problem is to find its coefficient in the expression of the $(s-1)$-chain $F(C_p \cdot C_q)$. Owing to our positional assumptions, E_{s-1} belongs at most to one of the two chains $F(C_p)$, $F(C_q)$ but not to both. Suppose first that it is a cell of $F(C_q)$. We must show then that its coefficient ξ in $F(C_p \cdot C_q)$ is the same as its coefficient θ in $C_p \cdot F(C_q)$. If E_s^i is any s-cell incident with E_{s-1} it will be convenient to designate the looping coefficients and Kronecker indices relatively to $Lk(E_s^i)$ or $Lk(E_{s-1})$ by adding to their symbols a subscript Lk^i or Lk as the case may be.

Then by (46) and (45) and the definition of Lc,

$$\begin{aligned} \theta &= Lc(\textit{trace } C_p, \textit{ trace } F(C_q))_{Lk} \\ &= (-1)^s (\textit{trace } C_p \cdot \textit{trace } C_q)_{Lk}. \end{aligned} \tag{47}$$

On the other hand, by (46), if $\eta_i E_s^i$ is positively related to E_{s-1},

$$\xi = \sum \eta_i Lc(\textit{trace } C_p, \textit{ trace } C_q)_{Lk^i}. \tag{48}$$

Observe now that by assumption the points common to C_p and C_q constitute an s-subcomplex of K; those common to *trace* C_p, *trace* C_q on Lk are then at the intersection of an s-subcomplex of K with an $(n-s)$-subcomplex of its dual K^*, namely $Lk^*(E_{s-1})$, which coincides with Lk. The geometric intersection consists then of isolated points on the $(n-s)$-cells of Lk^*. But the latter are precisely the cells E_{n-s}^{*i} bounded by the spheres Lk^i.

We find by the use of the new definitions of intersections and consideration of indicatrices that $(-1)^s \eta_i E_{n-s}^{*i}$ is a positive cell of Lk. If we prefer, the sum of cells $(-1)^s \sum \eta_i E^{*i}$ constitutes another normal orientation of the circuit Lk concordant with the primitive orientation. Since *trace* C_p and *trace* C_q meet only on points of the cells E^{*i}_{n-s}, if Δ^i, D^i represent the intersections with, or "traces" on, E^{*i}_{n-s}, we have

$$(trace\ C_p,\ trace\ C_q)_{Lk} = (-1)^s \sum \eta_i (\Delta^i \cdot D^i)_{E^{*i}_{n-s}}.$$

Now by No. 51

$$(\Delta^i \cdot D^i)_{E^{*i}_{n-s}} = Lc\,(trace\ C_p,\ trace\ C_q)_{Lk^i}.$$

From these two relations and (47), (48) follows $\xi = \theta$ as asserted.

If E_{s-1} is a cell of $F(C_p)$ we proceed as in No. 5 by considering the intersection $C_q \cdot C_p$ and applying the result just obtained. Thus formula (18) is completely proved.

REMARK. It is to be observed that owing to the complementary conventions concerning traces and looping coefficients, all possible cases, even $s = p$ or q, $s = p = q = n$, are covered at one and the same time, thus obviating the necessity of taking up two troublesome special cases.

56. To carry through the invariance theory we still need

LEMMA. *Let K^0, C^0_p, C^0_q be a simplicial subdivision of K and the associated subdivisions of its subchains C_p, C_q. If the latter are positionally restricted as in No. 53, so are their subdivisions, and $C^0_p \cdot C^0_q$ is the subdivision of $C_p \cdot C_q$ induced by K^0.*

That the positional restriction is preserved follows from the fact that the cells of K^0 on an s-cell of K are of dimension $\leqq s$ and the fact that boundary relations are unaltered by subdivision. The rest follows from the existence of an orientation-preserving homeomorphism between $Lk(E^0_s)$ and $Lk(E_s)$, where E^0_s is a cell of K^0 on E_s, under which homeomorphism the new traces correspond to the old.

As a matter of fact we could well content ourselves with regular subdivision instead of the more general type. In that case $Lk(E^0_s)$ has the structure of a first derived of $Lk(E_s)$ (Ch. III, No. 4) and then the above mentioned homeomorphism is evident.

The invariance theory will now go through with ease. We approximate C_p, C_q by subchains C_p^i, C_q^i of an i-th derived K^i of K, positionally restricted as in No. 53. To show that such chains exist we merely take subchains of K^{i-1} and its dual K^{*i} respectively and the subdivisions induced on them by K^i. In comparing $C_p^i \cdot C_q^i$ with $C_p^j \cdot C_q^j$, $j > i$, we replace C_p^i, C_q^i by the subdivisions $C_p'^j$, $C_q'^j$ induced on them by K^j. By the lemma this is permissible and we merely have to compare $C_p'^j \cdot C_q'^j$ with $C_p^j \cdot C_q^j$, both intersections being subchains of K^j. The desired homologies are established by means of the fundamental boundary relation (18), as we now show.

57. We can in fact prove an important additional point. The reasoning of No. 42 shows that whenever two modes of defining intersections of chains lead to the same value of indices $(A \cdot K)$ (A any point of $K - L$) they yield the same values for all indices, and the corresponding relative cycles are homologous in the suitable sense pertaining to each case. Now it is a very simple matter to verify that our new intersections give for $(A \cdot K)$ the value $+1$ like the old. *Therefore the new Kronecker indices are equal to the old and the new intersection cycles homologous to the old in the proper sense corresponding to each cycle.*

Two of the three requirements of No. 53 are thus more than fulfilled by the new intersections. It is now easy to show that the third, (c), is likewise satisfied. In the first place, with the situation as in § 1, in the notation of No. 3 we must show that when E_p, E_q^* are considered as subchains of K' their intersection is $\mathfrak{E}_s$ when taken in the new manner. Let E_s' be a cell of K' on $\mathfrak{E}_s$. We have to show that its coefficients t, t', as assigned to it by the old or new types of intersection are equal. To do this consider as a new L the set L' of all cells of K' that do *not* belong to the star of E_s'. The new M is now $K' - L'$. The two intersections are then tE_s', $t'E_s'$ mod L'. Since they are homologous mod L', $t = t'$, as asserted.

The convex intersections of our paper [8] can be identified with the present type in substantially the same way. For when C_p,

C_q are convex polyhedral chains on a convex manifold $K—L$, there is a simplicial subdivision K^0 of K with certain simplicial subdivisions C_p^0, C_q^0 of the given convex chains as subchains. The identification of $C_p^0 \cdot C_q^0$, understood in the new sense, with a subdivision of $C_p \cdot C_q$, understood as in [8], proceeds then exactly as above, since the above remarks on the Kronecker index apply also to convex intersections.

CONCLUSION. We have introduced a definition of intersection of subchains of $K—L$ which not only includes the two known polyhedral types of intersections as special cases but possesses all their properties. We have thus merged the two theories of intersections that we have developed into a single one, more general, yet no less powerful, and better adapted to all applications.

58. **Suggestions for an alternate treatment of intersections.** The generalized definition of the intersection of subchains of $K—L$ rests upon a preliminary treatment of looping coefficients. The latter require a preliminary treatment analogous to that of §§ 1, 2. This suggests a development of the theory *ab initio* along the following lines:

(a) Define the looping coefficients for subchains of H_n by a set of properties insuring that they behave as desired.

(b) Define the intersections as in the present section.

It would seem quite feasible to set up this scheme and the resulting theory should prove particularly attractive on the score of symmetry and simplicity.

§ 7. MISCELLANEOUS QUESTIONS

59. **A new type of topological invariants.** We confine our attention to absolute manifolds and cycles—the extension to the relative field being obvious.

We consider then with Veblen and Alexander [15, 16, 17, 18] the r minimum bases $\{\Gamma_\alpha^i\}$, $\{\Gamma_\beta^j\}$, $\cdots$, $\{\Gamma_\lambda^h\}$, and also the minimum base $\{\Gamma_p^k\}$ for $p = \alpha + \beta + \cdots + \lambda - (r-1)n$, all relative to $\approx$. We assume the dimensions such that $p \geqq 0$. The index i runs from 1 to $R_\alpha(M)$, and similarly for the others. We have also

$$\Gamma_\alpha^i \cdot \Gamma_\beta^j \cdots \Gamma_\lambda^h \approx \sum x_{ij\cdots h}^k \Gamma_p^k,$$

and for $p = 0$ and M_n connected, due to $R_0 = 1$, merely the Kronecker index

$$(\Gamma_\alpha^i \cdot \Gamma_\beta^j \cdots \Gamma_\lambda^h) = x_{ij\cdots h}.$$

Now there is nothing absolute about the choice of bases. If we adopt new bases $\{\Gamma_\alpha'^i\}, \cdots$, they are related to the old by homologies

$$\Gamma_\alpha'^i \approx \sum y_{ia} \Gamma_\alpha^a, \quad \text{etc.}$$

The corresponding x's, say $x_{ij\cdots h}'^k$, can be calculated in terms of the old, and by a simple computation it is found that they are transformed like the elements of what is known as a *mixed tensor* in modern differential geometry. *That tensor is then an attribute of M_n as a whole and evidently a topological invariant.* We have thus a new topological invariant whenever we are assured that for a given set of dimensions $n, \alpha, \beta, \cdots, \lambda$, the tensor is not independent of M_n.

60. The simplest case is where $r = 2$, $\alpha + \beta = n$, $p = 0$. We have then the tensor $(\Gamma_\alpha^i \cdot \Gamma_{n-\alpha}^j)$. But by the theorem of Veblen-Weyl (No. 17, Theorem II) we can always choose the bases such that

$$(\Gamma_\alpha^i \cdot \Gamma_{n-\alpha}^j) = \delta_{ij}.$$

In other words the tensor in this case depends only on $R_\alpha(M_n)$, and no new topological invariant arises.

Weyl [3] restores the possibility of a new invariant by specialising the bases. In particular for n a multiple of four, i. e. for M_{4n} and $\alpha = 2n$, he takes the two bases as coincident. Since

$$(\Gamma_{2n}^i \cdot \Gamma_{2n}^j) = (\Gamma^j \cdot \Gamma^i),$$

the tensor x_{ij} is then symmetrical, i. e. we have the symmetrical unimodular matrix $\boldsymbol{X}$, (unimodular owing to the Veblen-Weyl theorem). The different tensors that may occur *a priori* are the tensors of the coefficients of the various classes of quadratic forms $\sum x_{ij} m_i m_j$ inequivalent in the sense of Gauss, whose determinant is ± 1, and we have the possibility of topological invariants and of distinguishing between manifolds by their values. Thus for $R_{2n} = 2$, we have two possible non-equivalent forms: $x^2 + y^2$, $x^2 - y^2$. More generally, the signature of a form is a topological invariant of the associated M_{4n}.

It may be observed that the $(2n+1)$-cycles of an M_{4n+2} again give rise to a tensor dependent solely on n but not on M, and hence yield no new invariant. However, M_{4n+2} has an interesting property which may be brought out as follows. We can take a unique minimum base $\{\gamma^i_{2n+1}\}$ and we shall have

$$\|(\gamma^i \cdot \gamma^j)\| = \pm 1.$$

But this determinant is skew-symmetric, therefore its order can only be even. Hence *the number* R_{2n+1} *of every absolute orientable* M_{4n+2} *is even.* From the classical theorems on the reduction of alternate matrices we also learn that the set $\{\gamma^i_{2n+1}\}$ can be so chosen that $\|(\gamma^i \cdot \gamma^j)\|$ is a diagonal matrix of order R_{2n+1} whose elements are matrices of order two, those in the main diagonal being $\left\|\begin{matrix} 0 & 1 \\ -1 & 0 \end{matrix}\right\|$. These propositions are well known for Riemann surfaces, $\frac{1}{2} R_{2n+1}$ being then the genus p of the surface. For an arbitrary n they were given by Weyl [3].

The next case was considered by Alexander [15, 16, 17, 18]. We take $n = 3$, $r = 2$, $\alpha = \beta = 2$, $p = 1$. The tensor is then x^k_{ij}, and Alexander shows by actual examples that there are manifolds with the same homology characters and the same group for which the tensors are different.

It is scarcely necessary to point out that absolute cycles can be replaced everywhere by cycles mod m.

61. **On the intersection of subsets of a manifold.** Let A, B be two subsets of M_n. The number ϱ_p of cycles of a given dimension p that are on A and independent on M is $\leqq R_p(M_n)$. Let Γ^i_p, $i = 1, 2, \cdots, \varrho_p$ be a maximal independent set of them and similarly Γ^j_q, $j = 1, 2, \cdots, \varrho_q$ such a set for the q cycles on B. Let finally

$$\Gamma_s^h, \quad h = 1, 2, \cdots, R_s, \quad s = p + q - n,$$

be a similar set for all s-cycles of M_n. We assume $p + q \geqq n$, or $s \geqq 0$. Under the circumstances

$$\Gamma_p^i \cdot \Gamma_q^j \approx \sum x_{ij}^h \Gamma_s^h$$

and as before x_{ij}^h is a mixed tensor of order three, simultaneous topological invariant of M_n, A, B. When $s = 0$, and M_n is connected, i. e. $R_0(M_n) = 1$, we have merely a tensor of order two, x_{ij}, as our invariant. *A necessary condition that A, B do not meet is that all tensors of the type just indicated be* $\equiv 0$. For evidently the intersection cycles are then $\equiv 0$.

The extension to the relative and modular fields is again obvious.

CHAPTER V

PRODUCT COMPLEXES

§ 1. Generalities on Product Sets and Spaces

1. Given two abstract sets of elements $X = \{x\}$, $Y = \{y\}$ we can consider a new set Z whose elements are the ordered pairs (x, y), two elements (x, y) and (x', y') of Z being considered as identical when and only when $x = x'$ and $y = y'$. In particular $(x, y) \neq (y, x)$ unless $x = y$. The set Z is called the *product* of X by Y and designated generically by $X \times Y$; X, Y are the *factors* of Z. We have $X \times Y \neq Y \times X$ unless $X = Y$. These very convenient notations go back to Steinitz.

We may similarly consider the product $X_1 \times X_2 \cdots \times X_n$ of several sets, or even of an infinite number of them constituting a set of sets of any particular cardinal number.

The Steinitz products obey the associative and distributive laws:

$$(X_1 \times X_2) \times X_3 = X_1 \times X_2 \times X_3, \text{ etc.}$$

$$(X_1 + X_2) \times Y = X_1 \times Y + X_2 \times Y.$$

In general, however, the commutative law does not hold.

2. Simple examples of product sets are readily constructed. Thus if X consists of three objects x_1, x_2, x_3 and Y of two objects y_1, y_2, the product set $X \times Y$ consists of the six pairs (x_i, y_j), while $Y \times X$ consists of the pairs (y_j, x_i). To take an interesting example of more geometric nature, let x, y be two arbitrary real numbers. The product set $X \times Y = (x, y)$ is the set of all ordered pairs of real numbers. It is in one-to-

one correspondence with the set of all points of a cartesian plane S_2.

The earliest sharp applications of product spaces were made in algebraic geometry. Algebraic correspondences between two algebraic curves A, B are represented by algebraic curves on $A \times B$ which is an algebraic surface (Segre). The latter configuration has been studied for its own sake by a number of authors, notably by De Franchis, Severi, Picard, Lefschetz. Severi in particular utilized this point of view with telling effect in the theory of algebraic correspondences and involutions. On this question see our paper [10], where detailed references to the work of the preceding authors will be found.

The theory of product complexes to be taken up presently was the essential factor used in our papers [8, 9] to solve a basic problem on continuous transformations of manifolds. See also Ch. VI.

3. **Product spaces.** Suppose that X, Y are spaces with $\{N\}$, $\{N'\}$ as the corresponding sets of neighborhoods. Then the set of all pairs (x, y) associated with the subsets $N \times N'$ as neighborhoods, is likewise a space. It is that space which we now designate by $X \times Y$ and call the *product* space of the spaces X, Y, its *factors*.

The following properties can be verified immediately by reference to the *Introduction*:

I. Let $A \subset X$, $B \subset Y$. Then $A \times B \subset X \times Y$, and if x, y, are inner points of A, B then $x \times y$ is an inner point of their product. When A, B are both open or closed so is their product. Furthermore,

$$(1)\quad F(A \times B) = A \times F(B) + F(A) \times B + F(A) \times F(B).$$

$$(2)\quad F(\bar{A} \times \bar{B}) \subset \bar{A} \times F(B) + F(A) \times \bar{B}.$$

$$(3)\quad \overline{A \times B} = \bar{A} \times \bar{B}.$$

II. When X, Y are respectively homeomorphic to X', Y', then $X \times Y$ is homeomorphic to $X' \times Y'$.

III. The product of two Hausdorff, separable or compact metric spaces is respectively a Hausdorff, separable or compact space.

IV. The product of two metric spaces is a metric space.

For let $\mathfrak{z} = (x, y)$, $\mathfrak{z}' = (x', y')$. Then it is easy to verify that if $d(x, x')$, $d'(y, y')$ are the distance functions for X, Y, the function

$$(4) \qquad D(\mathfrak{z}, \mathfrak{z}') = d(x, x') + d'(y, y')$$

satisfies the two conditions for distance functions given in the *Introduction* relatively to the product space. Therefore the latter is also a metric space.

Since we are dealing throughout exclusively with metric spaces, whenever we speak of the product of two metric spaces X, Y, we shall understand the associated metric space Z, i. e. the set of all pairs (x, y) together with some distance function, for example (4).

§ 2. Product Complexes

4. Let A be a configuration of an S_p referred to coördinates x_i, B a configuration of an S_q referred to coördinates y_j. We can consider the x's and y's together as point coördinates of an S_{p+q}. Then if x_i are the coördinates of a point ξ of A, y_j those of a point η of B, when ξ, η describe A and B the point ζ of S_{p+q} whose coördinates are x_i, y_j has a range C which is the homeomorph of $A \times B$. On the strength of this we prove at once the following propositions in which we assume, as we shall until further notice, that all cells or spheres are ordinary (non-combinatorial) cells or spheres.

I. *The product of a p-cell by a q-cell is a* $(p+q)$*-cell.*

For we can take as E_p and E_q the regions of S_p and S_q represented by

$$0 < x_i < 1, \quad 0 < y_j < 1.$$

Then $E_p \times E_q$ is the region of S_{p+q} represented by the two inequalities taken together, which is manifestly a $(p+q)$-cell.

II. *If the boundaries of the two cells are respectively* $(p-1)$- *and* $(q-1)$- *spheres, that of the* $(p+q)$-*cell, their product, is likewise a* $(p+q-1)$-*sphere.*

The reasoning is the same with $<$ replaced by $\leqq$.

As a consequence, taking into account formula (1) we have

III. *Let* H_{p-1} *and* H_{q-1} *be the boundary spheres of* E_p *and* E_q. *Then* $H_{p-1} \times E_q + E_p \times H_{q-1} + H_{p-1} \times H_{q-1}$ *is a* $(p+q-1)$-*sphere, boundary of the product cell* $E_p \times E_q$.

REMARKS. I. The case $p = 0$, $q = 0$ falls in naturally with what precedes: The product of two points is a point. In particular the product of two points ξ, η of A, B, is the point $\xi \times \eta$ of the product configuration $A \times B$. Every point P of the latter is of type $\xi \times \eta$. The point ξ of A is called the *projection* of P on A. Similarly η is its projection on B. Through every point $\xi \times \eta$ of $A \times B$ there pass homeomorphs of A and B, namely the configurations $A \times \eta$ and $\xi \times B$.

II. The join (A, B) of two configurations A, B, defined in Ch. II, § 8, is intimately related to the product. Assume again as loc. cit. that A, B are immersed in some S_n, separated by some S_{n-1} of S_n, and such that segments with different end points on them never meet. Then the intersection of (A, B) with S_{n-1} is precisely $A \times B$. The latter may also be considered as the image of the set of segments PQ, $P \subset A$, $Q \subset B$. When A and B are complexes (A, B) corresponds cell for cell to $A \times B$ in such manner that to each p-cell of the latter corresponds a $(p+1)$-cell of (A, B). This explains why certain relations developed by Newman for the join are quite similar to those for the product. Regarding the homology relations, when A, B are complexes, those for $K = (A, B) \bmod (A+B)$ and the dimension $p+1$, are the same as those for $A \times B$ and the dimension p.

5. It is important to observe that even though the factors are simplexes, the product, as described at the beginning of No. 4, is not a simplex. Take for instance $p = q = 1$. The product cell is the product of two segments and therefore it has the structure of a rectangle. Thus *simpliciality* is not preserved through the operation of Steinitz

multiplication. By the second theorem *normality* is preserved: the product of two normal (ordinary) cells is still a normal cell. *Convexity* is likewise preserved under suitable definition of rectilinearity: We assume A, B, to be convex regions of S_p, S_q and take for rectilinear segments of $A \times B$ the images of the rectilinear segments of S_{p+q} on the image of $A \times B$ in that space as defined above. It is convenient to identify for the present $A \times B$ with that image. Let $\alpha \times \beta$, $\alpha' \times \beta'$ be two points of $A \times B$. The segments $\alpha\alpha'$ and $\beta\beta'$ of S_p and S_q belong wholly to A and B. Let x_i, x_i' be the coördinates of α, α' and y_j, y_j', those of β, β'. Consider the two points α_t, β_t of $\alpha\alpha'$, $\beta\beta'$ with coördinates $(1-t)x_i + tx_i'$, $(1-t)y_j + ty_j'$, $0 \leqq t \leqq 1$. The point $\alpha_t \times \beta_t$ belongs to $A \times B$ and as t varies it describes a segment whose extremities are precisely $\alpha \times \beta$ and $\alpha' \times \beta'$. Therefore $A \times B$ is convex. Incidentally this implies the results of No. 4.

6. **Products of oriented cells.** We take advantage of the result just obtained in assigning a fixed orientation to the product of two oriented simplexes $\eta\sigma_p$, $\eta'\sigma_q$. The product cell $E_{p+q} = \sigma_p \times \sigma_q$ is convex and under our definition of rectilinearity it carries the simplex $\sigma_{p+q} = A^0 \times B^0\, A^1 \times B^0 \cdots A^p \times B^0\, A^0 \times B^1 \cdots A^0 \times B^q$. Let $|\boldsymbol{x}|$ be the determinant analogous to that of Ch. I, No. 46, for σ_p, $|\boldsymbol{y}|$ its analogue for σ_q. Then by a simple calculation we find that the similar determinant attached to σ_{p+q} is $|\boldsymbol{x}| \cdot |\boldsymbol{y}|$. Since σ_p, σ_q are simplexes, both $|\boldsymbol{x}|$ and $|\boldsymbol{y}|$ are $\neq 0$, hence $|\boldsymbol{x}| \cdot |\boldsymbol{y}| \neq 0$ and the vertices of σ_{p+q} are *independent*, i. e. σ_{p+q} is a true $(p+q)$-simplex, which is not obvious *a priori*.

Suppose next that we replace σ_p, σ_q by other indicatrices σ_p', σ_q' concordant for their cells. Then σ_{p+q} is replaced by σ_{p+q}'. But $|\boldsymbol{x}_1|$, $|\boldsymbol{y}_1|$, the new determinants analogous to the above for the indicatrices, are respectively of the same signs as the old, for this is a necessary and sufficient condition for concordance (Ch. I, No. 46). It follows that $|\boldsymbol{x}| \cdot |\boldsymbol{y}|$ and $|\boldsymbol{x}_1| \cdot |\boldsymbol{y}_1|$ have the same sign and σ_{p+q}' is concordant with σ_{p+q} as indicatrix of $E_p \times E_q$. In other words *the class of indicatrices* $\{\sigma_{p+q}\}$ is unique.

All this applies without change to a product of convex cells $E_p \times E_q$: We take $\eta\,\sigma_p$, $\eta'\,\sigma_q$ indicatrices of the cells and $\eta\,\eta'\,\sigma_{p+q}$ is by definition the indicatrix of the product. The uniqueness of its class is proved as above.

From now on $E_p \times E_q$ will stand for the product cell with its orientation as just defined. By examining the effect of permuting two A's or two B's we verify readily that

$$E_p \times E_q = -(-E_p) \times E_q = -E_p \times (-E_q).$$

7. **Products of complexes or chains.** Let K_p, K'_q be two oriented simplicial, or for that matter convex complexes, E^i_h, E'^j_k their oriented cells. The oriented cells $E^i_h \times E'^j_k$ are all convex and do not meet one another. From the boundary relation (1) follows also that the boundary of any cell of the set consists of a sum of cells of the set. Therefore the set of oriented cells $\{E^i_h \times E'^j_k\}$ is a convex $(p+q)$-complex. It is the product, $K_p \times K'_q$, of the two given complexes.

By the product $C_p \times C'_q$ of the subchains $C_p = \sum t_i\, E^i_p$, $C'_q = \sum u_j\, E'^j_q$ we understand the $(p+q)$-subchain of $K_p \times K'_q$,

$$C_p \times C'_q = \sum t_i\, u_j\, E^i_p \times E'^j_q.$$

The chain product obeys the associative and distributive laws.

Remark. The restriction to subchains of the complexes is in no wise essential. We can consider singular cells E_h, E'_k on K, K', and their product $E_h \times E'_k$ is a singular $(p+q)$-cell on the product complex. If E_h, E'_k are the singular images of two non-singular cells $\mathfrak{E}_h$, $\mathfrak{E}'_k$, then $E_h \times E'_k$ is the singular image of $\mathfrak{E}_h \times \mathfrak{E}'_k$.

8. **Boundary relations.** For the product-cell they offer little difficulty. Let E_{p-1} be a face of E_p so that $E_{p-1} \times E_q$ is a face of $E_p \times E_q$. Choose σ_p with its face $\sigma_{p-1} = A^0 A^1 \cdots A^{p-1}$ on E_{p-1}, and let $\eta\,\sigma_{p-1}$ be an indicatrix of E_{p-1}. Then $(-1)^p\,\eta\,E_{p-1}$ is positively related to E_p. But if we write down the indicatrix of $E_{p-1} \times E_q$ according to our orientation rule, we find it to be $\eta\, A^0 \times B^0 A^1 \times B^0 \cdots A^{p-1} \times B^0 A^0 \times B^1$

$\dots A^0 \times B^q$. Hence $(-1)^p \eta\, E_{p-1} \times E_q$ is positively related to $E_p \times E_q$. In other words, the relation between the two is the same as between E_p and E_{p-1}.

Similarly if E_{q-1} is a face of E_q and related to it with the sign of $\eta' = \pm 1$, we find that $(-1)^p \eta' E_p \times E_{q-1}$ is positively related to $E_p \times E_q$. Since the two types considered include all the $(p+q-1)$-cells of the boundary of the product cell, its boundary relation is

$$(5) \qquad F(E_p \times E_q) = F(E_p) \times E_q + (-1)^p E_p \times F(E_q).$$

From this follows for chain products the general formula:

$$(6) \qquad F(C_p \times C_q) = F(C_p) \times C_q + (-1)^p C_p \times F(C_q).$$

These formulas offer much similarity with (18) of Ch. IV.

Let L, L' be subcomplexes of K, K' and set $\Lambda = L \times K' + K \times L'$, a subcomplex of $K \times K'$. We have evidently

$$K \times K' - \Lambda = (K - L) \times (K' - L'),$$

since both represent the sum of all cells $E \times E'$ such that $E \not\subset L$, $E' \not\subset L'$.

From (6) we derive at once the following noteworthy results:

I. *The product of two subcycles of the factor complexes mod L and L' respectively, is a subcycle of the product complex mod Λ.*

II. *The product of two absolute subcycles of the factors is an absolute subcycle of the product.*

III. *If one of the two subcycles in I or II is ≈ 0 mod L, mod L' or absolutely as the case may be, the same holds for the product subcycle mod Λ or absolute.*

REMARK. Here again as in No. 7 the restriction to subcycles is inessential and we obtain true theorems by replacing everywhere "subcycle" by "cycle". However, the theorems as stated are more precise if more narrow, and that added precision is useful, for example, in calculating the homology invariants of the product in terms of those of the factors.

9. The first two propositions just proved are to some extent invertible. We obtain the inversion by considering the boundary relations for the product complex.

We have shown in Ch. I, No. 27, that there exists a base for the s-subchains of $K-L$ consisting of three groups of elements: (a) A set a_s^i of independent cycles mod L. (b) A set b_s^i of cycles ≈ 0 mod L (zero-divisors or bounding cycles). (c) A set of chains c_s^i no linear combination of which $\rightarrow 0$ mod L (i. e. is a cycle mod L). The corresponding elements for K' and L' will be designated by a', b', c'.

For the dimension k the fundamental chains of $K \times K'$ mod Δ are then $\overset{i}{a_s} \times a_{k-s}'^{j}$, $\overset{i}{a_s} \times b_{k-s}'^{j}$, $\cdots$, $\overset{i}{c_s} \times c_{k-s}'^{j}$, $s=0, 1, \cdots, k$. Omitting the indices i, j, the fundamental boundary relations assume the form

$$(7) \quad \begin{cases} a_s \times c_{k-s+1}' \rightarrow (-1)^s t' \, a_s \times b_{k-s}', \\ c_{s+1} \times a_{k-s}' \rightarrow t \, b_s \times a_{k-s}', \end{cases}$$

$$(8) \quad \begin{cases} b_s \times c_{k-s+1}' \rightarrow (-1)^s t' \, b_s \times b_{k-s}', \\ c_{s+1} \times b_{k-s}' \rightarrow t \, b_s \times b_{k-s}', \end{cases}$$

$$(9) \quad c_{s+1} \times c_{k-s}' \rightarrow t \, b_s \times c_{k-s}' + (-1)^{s+1} t' \, c_{s+1} \times b_{k-s-1}',$$

$$(10) \quad a_s \times a_{k-s}' \rightarrow 0,$$

$$(11) \quad \begin{cases} a_s \times b_{k-s}' \rightarrow 0, \\ b_s \times a_{k-s}' \rightarrow 0, \\ b_s \times b_{k-s}' \rightarrow 0. \end{cases}$$

In these relations t, t' are $+1$ or else torsion coefficients of K mod L or of K' mod L', and in any case $\neq 0$, which is our sole concern here. The boundary relations are all mod Δ.

If we examine these relations we find that the fundamental k-chains of $K \times K'$ mod Δ include the following:

(a) Cycles $a \times a'$. Since they do not occur in the right hand side of the fundamental boundary relations they are independent.

(b) Cycles $a \times b'$, $b \times a'$, $b \times b'$. As shown by (7) or (8), they are all ≈ 0.

(c) Chains $a \times c'$, $b \times c'$, $c \times a'$, $c \times b'$, $c \times c'$.

10. We obtain a base for all cycles by adding to the first two types the terms of a base for all the cycles made up of chains (c). I say that the latter cycles are all ≈ 0.

Let indeed Γ_k be a cycle sum of chains (c). From (9) follows

$$t\, b_s \times c'_{k-s} \sim (-1)^s\, t'\, c_{s+1} \times b'_{k-s-1}.$$

Hence in some multiple of Γ every $b \times c'$ may be replaced by a chain $c \times b'$. Thus we obtain a new cycle $\Gamma^0 \sim r\Gamma$, $r \neq 0$, of same type as Γ but with chains $b \times c'$ absent in its expression. Since it is a cycle its boundary must be $\equiv 0$. If we write it down we find that it consists on the one hand, by the first relation (7), of terms $a \times b'$ on boundaries of the components $a \times c'$, on the other, of terms without a. Therefore the boundaries of the terms $a \times c'$ must cancel one another. This means that if in Γ^0 we have a sum of terms

$$a_s^i \times \sum x_j\, c'^j_{k-s},$$

we must have

$$a_s^i \times \sum x_j\, t'_j\, b'^j_{k-s-1} \equiv 0.$$

But the b''s are all different terms of a base for the $(k-s-1)$-cycles mod L' of the complex K' (the terms ≈ 0). Hence such an identity can only be trivial, hence every $x_j = 0$.

It follows that Γ^0 includes no term in $a \times c'$. The same argument shows that the other terms are also necessarily lacking. Therefore $\Gamma^0 \equiv 0$, $r\Gamma \sim 0$, $\Gamma \approx 0$, which proves our assertion.

11. Thus the cycles, sums of chains (c), are at best ≈ 0. The only cycles of the base that are $\not\approx 0$ are then those of type (a) and they are independent. Hence

THEOREM. *The k-cycles $a \times a'$ form a minimum base for all k-cycles of $K \times K'$ mod Δ and the homologies $\approx$.*

In other words, any k-cycle mod Δ of the product complex, say Γ_k, satisfies a relation with integer coefficients:

$$\Gamma_k \approx \sum t_{ij}^s\, a_s^i \times a'^j_{k-s} \bmod \Delta.$$

From this follows at once

$$(12) \qquad R_k(K \times K'; \varDelta) = \sum R_s(K; L)\, R_{k-s}(\dot{K}'; L').$$

As usual, this relation is true for any modulus m.

12. We now pass to the zero-divisors and torsion coefficients. The fundamental homologies are

$$(13) \qquad t' a_s \times b'_{k-s} \sim 0, \quad t b_s \times a'_{k-s} \sim 0;$$

$$(14) \qquad t' b_s \times b'_{k-s} \sim 0, \quad t b_s \times b'_{k-s} \sim 0;$$

$$(15) \qquad t b_s \times c'_{k-s} + (-1)^{s+1}\, t' c_{s+1} \times b'_{k-s-1} \sim 0.$$

To obtain a fundamental set for these homologies we must first reduce them to the minimum number. A reduction is possible for the type (14) but not for the other two. Any element $b_s \times b'_{k-s}$ will be found in two homologies (14). If θ is the H. C. F. of the corresponding coefficients t, t', both homologies (14) can be replaced by the unique

$$(14') \qquad \theta\, b_s \times b'_{k-s} \sim 0.$$

The homologies (13), (14′), (15) constitute a fundamental set since each contains elements found in no other. That is obvious for the first two and for the other it follows from the fact that a specific $b_s \times c'_{k-s}$ of a homology (15) is on the right hand side of a uniquely determined boundary relation (9)—namely the one whose $c_{s+1} \to t b_s$.

13. The matrix of the coefficients of the new fundamental homologies is not a diagonal matrix. To reduce it to that form we change the base for all k-chains as follows: The two terms in a given homology (15) are found in that homology and in no other. We replace them by two new k-chains A_k, B_k, such that

$$(16) \qquad \begin{aligned} b_s \times c'_{k-s} &= xA + yB, \\ c_{s+1} \times b'_{k-s-1} &= x'A + y'B \end{aligned}$$

where the coefficients are such that

$$xt + (-1)^{s+1} x' t' = \theta,$$

the H. C. F. of t and t',

$$y = (-1)^s \frac{t'}{\theta}, \qquad y' = \frac{t}{\theta},$$

so that

$$xy' - yx' = 1.$$

As a consequence, the sets of chains obtained by replacing the left hand sides by A, B in the base for all k-chains, is still a base for all k-chains. However, (15) will have been replaced by

(15′) $$\theta A \sim 0,$$

and the homologies (13), (14′), (15′) still constitute a minimum base for all homologies between k-chains.

We now have a diagonal matrix whose terms are the numbers t, t' and certain H. C. F.'s of pairs t, t'. Once we have a matrix of that sort it is an elementary problem to calculate its invariant factors. Let its terms be $x_1, \cdots, x_r$. The first, e_1, is the H. C. F. of all numbers x_i; the second that of all pairs $x_i x_j$ divided by e_1 etc. The numbers $e > 1$ are the torsion coefficients of $K \times K'$ mod Λ for the dimension k.

The following useful proposition is a direct consequence of our discussion: *When K, K' have no torsion coefficients mod L, L' for all dimensions $\leqq k$, their product has none mod Λ for the dimension k.*

REMARKS I. As usual everything that we have said holds mod m, with orientation disregarded when $m = 2$.

II. Regarding the homology characters of a product see our paper [8], also Künneth [3, 4].

14. **Product of circuits.** Let $K - L$ and $K' - L'$ be circuits. By referring to Ch. I, No. 35, we have, if E, E' denote generically the cells of $K - L$ and $K' - L'$:

$$\sum E_p^i \to 0 \bmod (L, 2) \text{ on } K,$$
$$\sum E_q^{\prime j} \to 0 \bmod (L', 2) \text{ on } K';$$

the sums being extended to *all* p- and q-cells respectively, with no subset of them satisfying such a relation. Furthermore, the (closed) circuits coincide with the p- and q-chains at the left. From this follows first that

$$\Gamma_{p+q} = \sum E_p^i \times E_q'^j \to 0 \bmod (\Delta, 2) \text{ on } K \times K',$$

next that Γ_{p+q} coincides with the point set $\overline{K \times K' - \Delta}$. Also (loc. cit.),

$$R_p(K; L, 2) = R_q(K'; L', 2) = 1.$$

Therefore by (12)

$$R_{p+q}(K \times K'; \Delta, 2) = 1,$$

which implies that no proper subset of the cells $E_p^i \times E_q'^j$ has for sum a cycle mod $(\Delta, 2)$. Therefore $K \times K' - \Delta$ is a circuit.

If in addition the given circuits are orientable, we have (Ch. I, No. 36)

$$R_p(K; L) = R_q(K'; L') = 1,$$

hence again

$$R_{p+q}(K \times K'; \Delta) = 1.$$

Thus $K \times K'$ is likewise orientable.

Finally suppose that both circuits are simple. Then, say E_{p-1} of $K-L$ is incident with two p-cells E_p^i, E_p^j of it. Therefore the $(p+q-1)$-cell, $E_{p-1} \times E_q'$ of the product circuit $K \times K' - \Delta$ is incident with the two $(p+q)$-cells, $E_p^i \times E_q'$, $E_p^j \times E_q'$ and with no others, as follows from the boundary relation (5). Similarly $E_p \times E_{q-1}'$ is incident with only two $(p+q)$-cells of the product circuit. Since all $(p+q-1)$-cells are of one of the two types just considered, the product circuit is simple. We have therefore proved

THEOREM. *If $K-L$, $K'-L'$ are circuits their product is also a circuit, and when both factors are orientable or simple so is the product.*

15. Products of combinatorial cells and complexes. From the preceding results it is a simple matter to deduce the following propositions.

I. *The product of two combinatorial cells is a combinatorial cell.*

For let $E_p = K-L$, $E'_q = K'-L'$ be two combinatorial cells. Since E, E' are orientable circuits, $E \times E'$ is an orientable circuit. Since the numbers $R_\mu(K; L)$, $R_\nu(K'; L')$, are all zero for $0 < \mu < p$, $0 < \nu < q$, from (12) we deduce that $R_k(K \times K'; \Lambda) = 0$ for $0 < k < p+q$. Similarly there being no torsion coefficients for the factors mod L, L', from No. 13 follows that there are none for the product mod Λ. It follows that the product satisfies all conditions required of a combinatorial cell when p, q are not both zero. For $p = q = 0$ this is obviously the case, hence I is true without restriction.

16. II. *When the factor cells have for boundaries combinatorial* $(p-1)$- *and* $(q-1)$-*spheres, that of the product cell is a* $(p+q-1)$-*sphere.*

For let L, L' be respectively combinatorial $(p-1)$- and $(q-1)$-spheres. For the present E_p, E'_q will denote the oriented cells considered as chains. Their boundaries are cycles H_{p-1}, H'_{q-1}, which are fundamental oriented circuits on L and L'. We have then

$$E_p \rightarrow H_{p-1}, \qquad E'_q \rightarrow H'_{q-1}.$$

Hence by (6),

$$E_p \times E'_q \rightarrow H_{p-1} \times E'_q + (-1)^p E_p \times H'_{q-1} = H_{p+q-1}.$$

For the present H_{p+q-1} is merely the boundary of the cell-product, which is a $(p+q)$-chain, in fact a circuit. As far as the cells of H_{p+q-1} are concerned they are those of $L \times K' + K \times L' = \Lambda$. The problem is then to show that Λ is a $(p+q-1)$-sphere.

Given a subcycle, say Γ_k, of Λ, we have $\Gamma_k = C_k + C'_k$, where C_k is a subchain of $K \times L'$ and C'_k one of $L \times K'$. These two complexes have in common only the subcomplex $L \times L'$ and since $C_k + C'_k \rightarrow 0$, C_k and C'_k can have their boundaries only on $L \times L'$. Therefore they are respective cycles of $K \times L'$ and $L \times K'$ mod $L \times L'$.

Now K has no torsion coefficients mod L, and L' has no absolute torsion coefficients. Therefore the cycles of $K \times L'$

mod $L \times L'$ are all homologous to linear combinations of products of cycles of K mod L and of absolute cycles of L'. Since $K - L$ is a p-cell and L' a $(q-1)$-sphere, the only subcycles $\not\sim 0$ mod $L \times L'$ are of the form $t\, E_p \times H'_{q-1}$.

Similarly, the only subcycles of $L \times K' \not\sim 0$ are of the form $t'\, H_{p-1} \times E_q$. Consequently either $\Gamma_k \sim$ a cycle on $L \times L'$ or else $k = p + q - 1$ and

$$\Gamma_k = t\, E_p \times H'_{q-1} + t'\, H_{p-1} \times E'_q.$$

By (6) however

$$\Gamma_{p+q-1} \rightarrow (t + (-1)^{p-1}\, t')\, H \times H' \equiv 0.$$

Hence

$$t' = (-1)^p\, t, \qquad \Gamma_{p+q-1} = t'\, H_{p+q-1}.$$

Suppose now that Γ_k is a subcycle of $L \times L'$. Again, as above, the only cycles $\not\sim 0$ on $L \times L'$ depend upon $H \times H'$, $H \times A'$, $A \times H'$, $A \times A'$, where A, A' are points of L, L', and we have

$$\left.\begin{array}{l} E \times A' \rightarrow H \times A' \\ A \times E' \rightarrow A \times H' \\ E \times H' \rightarrow H \times H' \end{array}\right\} \sim 0 \text{ on } \Delta.$$

Hence $R_{p+q-1}(\Delta) = R_0(\Delta) = 1$. Since any mod m may be treated in exactly the same way, we have also $R_{p+q-1}(\Delta; 2) = 1$. Regarding the $(p+q-1)$-cycles, there is actually no other subcycle of Δ mod 2 for that dimension than the multiples of H_{p+q-1} as our discussion clearly shows. Since all $(p+q-1)$-cells of Δ appear in it, and since every cell of Δ is a face of one of its $(p+q-1)$-cells, the two circuit conditions (Ch. I, No. 35) are fulfilled. This completes the proof of II.

17. III. *The product of two combinatorial normal cells is a combinatorial normal cell.*

It follows from I, II that when E_p, E'_q are normal, $E_p \times E'_q$ is a $(p+q)$-cell whose boundary is an H_{p+q-1}. Since E_p and E'_q are sums of simplexes with common vertices A, A', $E_p \times E'_q$ is a sum of convex cells with the common vertex

$A \times A'$. From this follows immediately that $E_p \times E'_q$ is the join of $A \times A'$ and H_{p+q-1}, and hence that it is normal.

IV. *If E_{p-1}, E'_{q-1} are combinatorial cells respectively on the boundaries of the combinatorial normal cells E_p, E'_q, the sum $E_p \times E'_{q-1} + E_{p-1} \times E'_q$ is a combinatorial $(p+q-1)$-cell.*

This proposition is proved like II. We give it here in view of an application later (No. 20).

From the first three propositions just proved follows immediately:

V. *The product of two combinatorial complexes is a combinatorial complex, and it is normal if the factors are.*

REMARK. Henceforth, unless otherwise stated, all cells and spheres are understood to be combinatorial.

§ 3. PRODUCTS OF MANIFOLDS

18. Let now our factor complexes K, K' be normal. We consider any cell of a complex as one of the cells of its own star. Hence if E_h, E'_k are cells of K, K' and E_i, E'_j cells of their respective stars, the star of $E_h \times E'_k$ with respect to $K \times K'$ is the sum of all cells $E_i \times E'_j$. This follows at once from (2).

Let A^i, B^j be the respective vertices of first derived complexes K^0, K'^0 of K and K'; here as in Ch. III, i, j denote the *type* of the vertex, that is the dimension of the carrying cell of K or K'. The points $A^i \times B^j$ of $K \times K'$ are so distributed that one and only one is found on each cell of that complex. We can therefore take them as vertices of a first derived complex K^0_{p+q} of the product complex. In particular $A^i \times B^j$ is of type $i+j$ relatively to $K \times K'$. It is to be observed that $K^0_{p+q} \neq K^0 \times K'^0$, since the cells of the latter product are not simplexes. (We assume tacitly that neither p nor q is zero, which is quite justified since these values are without interest).

19. A further study of the relation between K^0_{p+q} and $K^0 \times K'^0$ will yield important results concerning the transverse complexes of the product cells.

Every simplex of K^0_{p+q} is a face of a simplex whose type has a set of consecutive integers. Therefore K^0_{p+q} is the sum of all simplexes $\sigma = A^i \times B^j A^h \times B^k \cdots$ such that:

(a) The differences $h-i+k-j, \cdots$ are all $+1$. Hence in both sequences $i, h, \cdots$ and $j, k \cdots$ the jump from each term to the next is 0 or $+1$.

(b) The cells $E_i \times E'_j, E_h \times E'_k, \cdots$ are mutually incident. According to what has been stated at the beginning of No. 18, it follows that the sets $E_i, E_h, \cdots$ and $E'_j, \cdots E'_k$ are separately mutually incident.

From (b) follows that $A^i A^h \cdots = \sigma^0$, $B^j B^k \cdots = \sigma'^0$ are respectively cells of K^0 and K'^0. Therefore the vertices of the cell σ of K^0_{p+q} are also vertices of the cell $\sigma^0 \times \sigma'^0$ of $K^0 \times K'^0$. Since $\sigma^0 \times \sigma'^0$ is convex and polyhedral, it follows that σ is on that cell and that every one of its faces is either on that cell or on a unique face of it, since σ is a simplex.

We have thus shown that every cell of K^0_{p+q} (either a cell such as σ or a face of such a cell) is on a unique cell of $K^0 \times K'^0$. Every cell of the latter is then an exact sum of cells of K^0_{p+q} and K^0_{p+q} is a simplicial subdivision not only of $K \times K'$ but likewise of $K^0 \times K'^0$.

20. The transverse complex $Tv(E_i \times E'_j)$ is the sum of all cells of K^0_{p+q} such as σ, having $A^i \times B^j$ as their first vertex. A given σ is on a definite pair $\sigma^0 \times \sigma'^0$, where σ^0 is a cell of K^0 such as above not meeting E_i but having its vertex A^i on it and σ'^0 an analogous cell for E'_j with its vertex B^j on the latter.

On the other hand every cell of K^0_{p+q} with $A^i \times B^j$ as its first vertex is a σ or a face of a σ. In particular, every cell of K^0_{p+q} on $\sigma^0 \times \sigma'^0$ with $A^i \times B^j$ as a vertex (then necessarily its first vertex) is of that nature. Therefore the sum of all cells of K^0_{p+q} with $A^i \times B^j$ as their first vertex, or $Tv(E_i \times E'_j)$, is a subdivision of the sum of all cells $\sigma^0 \times \sigma'^0$ of the nature considered.

Now $Tv(E_i)$ is the sum of the cells σ^0, and similarly for E'_j and the cells σ'^0. Hence finally

THEOREM. *The transverse complex of* $E_i \times E'_j$ *is a subdivision of the product of the transverse complexes of* E_i *and* E'_j.

21. Let again L, L' be subcomplexes of $K \times K'$. Since Lk is the boundary of Tv, the preceding theorem, together with those of Nos. 15, 16, 17 and the definitions of manifold and regular and semi-regular cells (Ch. III, No. 5), enables us to prove by simultaneous recurrence:

THEOREM I. *The product of two (absolute or orientable) manifolds is a manifold (absolute or orientable). If the boundaries* L, L' *of the factors are regular so is the boundary* Λ *of the product.*

THEOREM II. *The product of two regular cells is regular; the product of a regular by a semi-regular cell, or of two semi-regular cells, is semi-regular.*

22. **Application to products of spheres.** Let H_p, H'_q be two spheres which are manifolds. The product $H \times H'$ is an M_{p+q}. It has no torsion coefficients since the factors have none. Therefore the products of the cycles of the minimum bases for H and H', whose dimension is k, constitute a minimum base for the k-cycles of $H \times H'$, as regards both $\sim$ and $\approx$. We have two possibilities: (a) $p \neq q$. Then there are cycles $\nsim 0$ for the dimensions 0, p, q and $p+q$. The corresponding bases are $A \times A'$; $H \times A'$; $A \times H'$; $H \times H'$; where A, A' are points of H, H'. The Betti numbers are

$$R_k(H \times H') = \begin{cases} 1 \text{ for } k = 0,\ p,\ q,\ p+q, \\ 0 \text{ otherwise,} \end{cases}$$

and there are no torsion coefficients. (b) $p = q$. Then $H \times A'$, $A \times H'$ are two independent p-cycles and the homology characters are as before except that $R_p(H \times H') = 2$.

The same scheme can be extended to a product of several spheres. The most interesting is

$$M_p = H_1^1 \times H_1^2 \times \cdots \times H_1^p,$$

a product of p circumferences. Here again there are no torsion coefficients. The bases are obtained by inserting vertices in place of the H's in all possible ways in the symbol for M_p. It follows that

$$R_k(M_p) = \binom{p}{k}.$$

M_p is called a *p-ring*. The two-ring is the well known anchor ring. The $2p$-ring plays a fundamental role in investigations on multiply periodic functions and their applications to algebraic geometry. See in this regard the last chapter of our Borel series monograph [5].

23. Returning to our general product of two manifolds, let E_i, E'_j be two cells of the same manifolds $K - L$, $K' - L'$ that we have considered so far. Their transverse complexes are E^*_{p-i}, E'^*_{q-j}, where the starred cells are the duals of the given cells. Their product $E^*_{p-i} \times E'^*_{q-j}$ is the transverse complex of $E_i \times E'_j$, i. e. its dual cell. This proves the

THEOREM. *The product of the duals of two regular cells is the dual of their product. The dual complex of $K \times K' - \Delta$ is the product of those of $K - L$ and $K' - L'$.*

Regarding the orientation of the dual cell and with notations as in No. 14 let E_i, E^*_{p-i}, E'_j, E'^*_{q-j} have for indicatrices $\eta A^0 \cdots A^i$, $\eta' A^i \cdots A^p$, $\xi B^0 \cdots B^j$, $\xi' B^j \cdots B^q$. In view of our conventions for orienting dual cells (Ch. III, No. 24), the indicatrices of $K - L$ and $K' - L'$ at the points A^i, B^j are $\eta\eta' A^0 \cdots A^p$, $\xi\xi' B^0 \cdots B^q$. Hence that of $K \times K' - \Delta$ at $A^i \times B^j$ is

$$(-1)^{i+j} \xi\xi' \eta\eta' A^i \times B^j A^0 \times B^j \cdots A^{i-1} \times B^j A^{i+1} \times B^j$$
$$\cdots A^p \times B^j A^i \times B^0 \cdots A^i \times B^{j-1} A^i \times B^{j+1} \cdots A^i \times B^q.$$

For the product cells, on the other hand, they are

$$E_i \times E'_j:$$
$$(-1)^{i+j} \xi\eta A^i \times B^j A^0 \times B^j \cdots A^{i-1} \times B^j A^i \times B^0 \cdots A^i \times B^{j-1},$$
$$E^*_{p-i} \times E'^*_{q-j}:$$
$$\xi'\eta' A^i \times B^j A^{i+1} \times B^j \cdots A^p \times B^j A^i \times B^{j+1} \cdots A^i \times B^q.$$

Comparing the two with that of $K \times K' - \Delta$ at $A^i \times B^j$ we find, in view of the convention for orienting dual cells, that the *oriented dual* of $E_i \times E'_j$ is $(-1)^{(p-i)j} E^*_{p-i} \times E'^*_{q-j}$.

24. The preceding result is susceptible of another interpretation. For the Kronecker index of a cell and its transverse being $+1$, we have

$$(17)\quad \begin{aligned} (E_i \times E'_j \cdot (-1)^{(p-i)j} E^*_{p-i} \times E'^*_{q-j}) &= +1, \\ (E_i \times E'_j \cdot E^*_{p-i} \times E'^*_{q-j}) &= (-1)^{(p-i)j}. \end{aligned}$$

If we denote the cells by E^m_i, E'^{μ}_j and their duals by E^{*m}_{p-i}, $E'^{*\mu}_{q-j}$, we find that unless both $m = n$ and $\mu = \nu$, $E^m_i \times E'^{\mu}_j$ and $E^{*n}_{p-i} \times E'^{*\nu}_{q-j}$ do not meet, hence their index is zero. Therefore

$$(E^m_i \times E'_j \cdot E^{*n}_{p-i} \times E'^{*\nu}_{q-j}) = (-1)^{(p-i)j} \delta_{mn} \delta_{\mu\nu}.$$

We have then

$$(18)\ (E^m_i \times E'^{\mu}_j \cdot E^{*n}_{p-i} \times E'^{*\nu}_{q-j}) = (-1)^{(p-i)j} (E^m_i \cdot E^{*n}_{p-i})(E'^{\mu}_j \cdot E'^{*\nu}_{q-j}).$$

From this follows that if C_i, C'_j, C^*_{p-i}, C'^*_{q-j} are subchains of $K - L$, $K' - L'$ and of the associated duals, we have the important relation

$$(19)\quad (C_i \times C'_j \cdot C^*_{p-i} \times C'^*_{q-j}) = (-1)^{(p-i)j} (C_i \cdot C^*_{p-i})(C'_j \cdot C'^*_{q-j}).$$

25. The extension to intersections of arbitrary cells and chains offers no difficulty. Since pairs of dual cells always intersect in cells, we have

$$(20)\quad E_i \times E'_j \cdot E^*_k \times E'^*_l = (-1)^{(p-i)(q-l)} E_i \cdot E^*_k \times E'_j \cdot E'^*_l,$$

the power of (-1) on the right being determined by the same considerations of indicatrices as before. From this follows

$$(21)\quad C_i \times C'_j \cdot C^*_k \times C'^*_l = (-1)^{(p-i)(q-l)} C_i \cdot C^*_k \times C'_j \cdot C'^*_l.$$

Suppose now that we consider *arbitrary* chains on K and K'. If N, N' are neighborhoods of L, L' on their respective complexes, $\mathfrak{N} = N \times K' + K \times N'$ is a neighborhood of Λ on $K \times K'$. Moreover if N, N' are the K- and K'-neighborhoods, $\mathfrak{N}$ is likewise the $K \times K'$ neighborhood. Using these facts we find readily that when the chains C_i, C_k on K are so disposed (in accordance with Ch. IV, § 3) as to yield an intersection cycle mod L, and similarly for C'_j, C'_l, mod L' on K', then the same holds for $C_i \times C'_j$ and $C_k \times C'_l$ on $K \times K'$ mod Λ. Similarly for the Kronecker index. From this follows that under the circumstances just stated, (19) holds whatever the chains, while (20) yields

$$(22) \quad C_i \times C'_j \cdot C_k \times C'_l \sim (-1)^{(p-i)(q-l)} C_i \cdot C_k \times C'_j \cdot C'_l \quad \text{mod } \Lambda.$$

26. **Application.** For the purpose of this application the factor complexes may be replaced by any subdivisions and we may therefore assume that they are both simplicial.

Let L, L^1, L^2, N, N^1, $\mathfrak{F}$ have the same significance for K as in Ch. III, Nos. 29, 30. Let N^2 be the K-neigborhood of $\mathfrak{F}$. Assuming as we may that N and N^1 are normal, it is a simple matter to show that N^2 is likewise normal. Denote the configurations analogous to L, $L^1, \cdots$, for K' by accents: L', $L'^1, \cdots$. Then from (2) and (5) follows that if we set

$$\Lambda^1 = K \times L'^1 + L^1 \times K',$$
$$\Lambda^2 = (K - L^1) \times L'^2 + L^2 \times (K' - L'^1)$$

the Λ's behave as regards $K \times K'$ like the L's as regards K.

Let a^i_μ, $b^i_{p-\mu}$ be associated canonical bases of $K - L^2 \bmod L^1$ and $K - L^1 \bmod L^2$, relatively to homologies $\approx$, or bases such that

$$(23) \qquad (a^i \cdot b^j) = \delta_{ij}.$$

We denote the elements of K' corresponding to a^i, b^i by b'^i, a'^i, the reverse order being particularly advantageous for the applications in the next chapter, and again

$$(24) \qquad (a'^i \cdot b'^j) = \delta_{ij}.$$

The number of cycles in each base of the first pair is $R_\mu(K - L^2; L^1)$, in each of the second pair $R_{q-\mu}(K' - L'^1; L'^2)$.

We now introduce two new subcomplexes of K:

$$K^1 = K - N^2, \qquad K^2 = K - N^1$$

and analogously, K'^1, K'^2 for K'. The complexes K^2, K'^2 are the same as the $\mathfrak{K}$ of Ch. III, No. 30. From what has been stated there and due essentially to the sliding properties of our neighborhoods, the sets a play exactly the same part for $K^1 \bmod L^1$ as for $K - L^2 \bmod L^1$ and likewise for the sets b' on K'. A similar remark applies to the sets b, a' with the indices 1, 2 interchanged and $\mathfrak{F}$, $\mathfrak{F}'$ in place of L^1, L'^1. Thus the sets a, a', b, b' are maximal independent sets for the complexes K^1, K'^2, K^2, $K'^1 \bmod L^1$, $\mathfrak{F}'$, $\mathfrak{F}$, L'^1, respectively. This is the one fact that we need to retain for the present.

27. Let us examine now the analogous situation as regards the Δ's and $K \times K'$. We denote the associated $N, \cdots, K^1, \cdots$ by $\mathfrak{N}, \cdots, \mathfrak{K}^1, \cdots$. It is a simple matter to show that

$$(25) \qquad \mathfrak{K}^1 = K^1 \times K'^1, \qquad \mathfrak{K}^2 = K^2 \times K'^2.$$

Since N and N^1 are normal neighborhoods of L, L^1, any cell, say of $K - L$ with vertices on L, is of the form $\sigma\sigma^*$ where σ is a cell of L, σ^* a cell of the boundary of N, and in particular $\sigma^* \neq 0$. Any point P of $\sigma\sigma^*$ has then a unique projection Q on σ^*, the end-point of the usual projecting segment. We will call Q the projection of P on $K - L$. If $P \not\subset N$, we will consider it as its own projection. Thus every point P of $K - L$ has a unique projection Q on $K - N$ which is on the closure of the cell of $K - L$ that carries P. A similar situation exists on K', the projection of the point P' of $K' - L'$ on $K' - N'$ being designated by Q'.

Given any point $P \times P'$ of $K \times K' - \Delta$, we now define $Q \times Q'$ as its projection on $K \times K' - \mathfrak{N}$. If $P \times P' \subset \mathfrak{N} - \Delta$, either $P \neq P'$ or $Q \neq Q'$, hence $P \times P' \neq Q \times Q'$. As there exists a closed cell of $K \times K'$ carrying both points, there is a unique segment joining them. These segments projecting the points of $\mathfrak{N}-\Lambda$ onto points of the boundary of $\mathfrak{N}$, play the same part as the usual segments associated with normal simplicial neighborhoods. They enable us to apply to the product complexes and the Λ's the same treatment as to the factors and the L's. Therefore a maximal independent set for the k-cycles of $\mathfrak{K}^1$ mod Λ^1 coincides with such a set for the k-cycles of $K\times K'-\Lambda^2$ mod Λ^1, and similarly with the indices 1, 2, interchanged.

28. A first consequence of what has just been said together with (25) and the theorem of No. 11, is that the cycles $a^i_\mu\times b'^j_{k-\mu}$, $\mu=0,1,2,\cdots,k$, constitute a maximal independent set for $K\times K'-\Lambda^2$ mod Λ^1, and $b^i_\mu\times a'^j_{k-\mu}$ for those of $K\times K'-\Lambda^1$ mod Λ^2. Hence, incidentally, this extension of (12):

$$\begin{aligned} &R_k(K\times K'-\Lambda^2;\Lambda^1)\\ (26)\qquad &=\sum_{\mu=0}^{k} R_\mu(K-L^2;L^1)\,R_{k-\mu}(K'-L'^2;L'^1),\end{aligned}$$

and similarly with the indices 1, 2, interchanged. *Since the manifold conditions were not utilized in establishing (26), it holds whatever the complexes involved.*

Let us now assume, however, that $K-L$ and $K'-L'$ are manifolds. Then (23) and (24) hold, and from (19) follows

$$(27)\quad (a^i_\mu\times b'^j_{k-\mu}\cdot b^h_{p-\mu}\times a'^l_{q-k+\mu})=(-1)^{(r+1-\mu)(k-\mu)}\,\delta_{ih}\,\delta_{jl}.$$

On the other hand,

$$\begin{aligned} &(a^i_\mu\times b'^j_{k-\mu}\cdot b^h_{p-\nu}\times a'^l_{q-k+\nu})\\ (28)\qquad &=\pm(a^i_\mu\cdot b^h_{p-\nu})\,(b'^j_{k-\mu}\cdot a'^l_{q-k+\nu})=0\end{aligned}$$

for $\mu\neq\nu$. For let, say, $\mu<\nu$. Approximate a^i_μ by a subdivision of K whose mesh is suitably fine and $b^h_{p-\nu}$ by the dual of that subdivision mod L. The two approximations do not meet since $\mu+p-\nu-p<0$. Hence their index, which is by definition that of $a^i\cdot b^h$, is zero. When $\mu>\nu$, $(b'^j\cdot a'^l)=0$. Hence in any case

$$(29)\qquad (a^i_\mu\times b'^j_{k-\mu}\cdot b^h_{p-\nu}\times a'^l_{q-k+\nu})=0\quad\text{for}\quad \mu\neq\nu.$$

Comparing (27) and (29) we see that the two associated bases for the k-cycles of $K \times K' - \varDelta^2 \bmod \varDelta^1$ and the $(p+q-k)$-cycles of $K \times K' - \varDelta^1 \bmod \varDelta^2$

$$(-1)^{(r-\mu)(k-\mu)} a^i_\mu \times b'^j_{k-\mu}; \quad b^h_{p-\nu} \times a'^l_{q-k+\nu}; \quad (\mu, \nu = 0, 1, 2, \cdots, k)$$

fulfill all conditions for associated canonical bases. This gives therefore a very simple rule for obtaining the canonical bases on $K \times K'$ when those on the factor complexes are known: *Except for the signs, the k-cycles have a canonical base consisting of all products of pairs of cycles of the related canonical bases on K, K' whose dimension $\leqq k$.*

29. **Important special cases.** For special choices of the complexes L^i, L'^i we have several noteworthy special cases.

I. $L^1 = L'^2 = 0$. Then $L^2 = L$, $L'^1 = L'$ and the manifolds are $K - L$, $K' - L'$. We have this time

$$\varDelta = K \times L' + L \times K'; \qquad \varDelta^1 = K \times L';$$
$$\varDelta^2 = K \times K' - \varDelta^1 = L \times (K' - L').$$

For the cycles of $K \times K' - \varDelta^2 \bmod \varDelta^1$ and $\approx$, we have minimum bases consisting of cycles of type

$$(\text{absolute cycle of } K - L) \times (\text{cycle of } K' \bmod L'),$$

while for the cycles of $K \times K' - \varDelta^1 \bmod \varDelta^2$, they will be cycles of type

$$(\text{cycle of } K \bmod L) \times (\text{absolute cycle of } K' - L').$$

The derivation of the canonical bases is exactly as before. If we denote by a, a' the absolute cycles, and by b, b' the relative cycles, the situation is formally as in No. 28.

II. $L^2 = L'^1 = 0$. This is essentially as I with the parts of K and K' interchanged throughout.

III. $L^2 = L'^2 = 0$, $L^1 = L$, $L'^1 = L'$, $\varDelta = K \times L' + L \times K'$. Then $a \times b'$ is a cycle mod $\varDelta$, and $b \times a'$ an absolute cycle of $K \times K' - \varDelta$.

IV. $L = L' = \varDelta = 0$. Both manifolds are absolute, and so are all cycles.

30. It is possible to solve an important intersection problem for $K \times K'$. Let Γ_k^1, Γ_{r-k}^2, $r = p + q$, be of the two respective types $a \times b'$, $b \times a'$. Let this time a_μ^i, $\cdots$, be merely minimal bases for $\approx$, and not necessarily canonical. We assume that we know the indices

$$\alpha_{ij}^\mu = (a_\mu^i \cdot b_{p-\mu}^j), \qquad \beta_{ij}^\mu = (a_\mu'^i \cdot b_{q-\mu}'^j).$$

The problem is to express $(\Gamma_k^1 \cdot \Gamma_{r-k}^2)$ in terms of the coefficients in

$$\begin{aligned} \Gamma_k^1 &\approx \sum \varepsilon_{ij}^\mu \, a_\mu^i \times b_{k-\mu}'^j, \\ \Gamma_{r-k}^2 &\approx \sum \eta_{hl}^{p-\nu} \, b_{p-\nu}^h \times a_{q-k+\nu}'^l. \end{aligned}$$

We have at once

$$(\Gamma_k^1 \cdot \Gamma_{r-k}^2) = \sum (-1)^{(r-\mu+1)(k-\mu)} \varepsilon_{ij}^\mu \, \eta_{hl}^{p-\mu} \, \alpha_{ih}^\mu \, \beta_{lj}^{q-k+\mu},$$

or in the matrix notation,

(30) $(\Gamma_k^1 \cdot \Gamma_{r-k}^2)$

$$= \sum (-1)^{(r-\mu+1)(k-\mu)} \, \mathit{trace} \; \eta^{p-\mu} \, \beta^{q-k+\mu} \, (\varepsilon^\mu)' \, \alpha^\mu.$$

When the bases are canonical $\alpha^\mu = 1$, $\beta^\mu = 1$, and we have the simpler relation

(31) $$(\Gamma_k^1 \cdot \Gamma_{r-k}^2) = \sum (-1)^{(r-\mu+1)(k-\mu)} \, \mathit{trace} \; \eta^{p-\mu} \, (\varepsilon^\mu)'.$$

The relations (30) or (31) solve the problem proposed.

Despite the simpler formula (31) it is on the whole preferable to take arbitrary bases and use (30). In the case of absolute manifolds that is a positive advantage since it enables us to take a unique set for each dimension and not two concurrent sets as necessitated by the conditions for canonical sets.

The problem just treated is of fundamental importance in the next chapter.

CHAPTER VI

TRANSFORMATIONS OF MANIFOLDS, THEIR COINCIDENCES AND FIXED POINTS

In the study of transformations of manifolds a fundamental problem is the determination of the number of *signed coincidences* or *fixed points.* It was first treated by Brouwer [1, 2, 7] and solved by him for cells, spheres and projective spaces. In recent papers [8, 9] we have studied transformations of absolute manifolds or manifolds with regular boundary, and arrived at explicit formulas for the numbers in question. The method there used, wholly unrelated to Brouwer's, is here applied to arbitrary manifolds, then to arbitrary complexes, with the result that we have a central formula, (22), which includes all known similar formulas as special cases. The extension to arbitrary compact metric spaces will be given in Ch. VII, § 6.

An elementary example will serve to illustrate the problem and its solution. A transformation T of the segment $0 \leqq x \leqq 1$ into the segment $0 \leqq y \leqq 1$ is represented by

$$f(x, y) = 0, \qquad 0 \leqq x,\ y \leqq 1,$$

that is, by a point-set in a certain square of the xy plane. Let T be so restricted that the point-set is an arc γ, and let T' be another transformation of same type with δ as its representative arc. The intersections of γ and δ correspond to the *coincidences* of T and T', or places x such that $T' \cdot x = T \cdot x$.

Their number is the number of crossings of the two arcs and it is clear that the slightest modification of the transformations may change it. Not so however for the Kronecker index $(\gamma \cdot \delta)$, and it is therefore that index, or the number of *signed coincidences*, which one will naturally seek to determine.

Replace the segments by two manifolds M, M', the square by $M \times M'$, the arcs by cycles, and there arises the exact situation studied at length in the present chapter.

When the "signed" numbers are $\neq 0$ there is at least one coincidence or fixed point, as the case may be. In certain important cases a greater lower bound may be assigned to the number of geometrically distinct fixed points. Regarding this question, not considered at all here, we shall merely refer the reader to the work of Birkhoff [1, 3], Nielsen [6] and Hopf [6].

§ 1. Position of the Problem

1. It is convenient for the sequel to start with two sets $A = \{a\}$, $B = \{b\}$, and a transformation T carrying over the subset A' of A into the subset B' of B. If $a \subset A'$ and b is one of the elements $T \cdot a$, the pair (a, b) determines a unique element $a \times b$ of the product set $A \times B$. The totality of pairs (a, b) is a subset G of the product set, uniquely determined by T. Conversely, given any subset $G = \{a \times b\}$ of $A \times B$, there is an associated transformation T between the set A' of all projections on A of the points $a \times b \subset G$ into the set B' of their projections on B. Thus there is a one-to-one correspondence between the subsets of $A \times B$ and the transformations T.

Let G, G' be the subsets of $A \times B$ corresponding to two different transformations T, T'. The *coincidences* or pairs (a, b) such that $b = T \cdot a = T' \cdot a$, are represented on $A \times B$ by the intersection $G \cdot G'$ of the two sets G, G'. For to each coincidence there corresponds an element $a \times b$ common to both, and conversely.

A similar representation holds for transformations T of subsets of A into subsets of A itself. We then take for B a duplicate of A, and if $a' = T \cdot a \subset A$, we take the image b of a' on B, and associate with the couple (a, a') the element $a \times b$ of $A \times B$. If G is the image of T on $A \times B$ and I that of the identity, the *fixed elements* of T correspond to the intersection $G \cdot I$ of G with I.

2. To leave the domain of generalities let us assume forthwith that *the two sets* A, B, *are connected, orientable manifolds* $K - L$, $K' - L'$ where the notations are as in Ch. V. Then, Δ being as in that Chapter, $K \times K' - \Delta$ is likewise a connected orientable manifold, and the problem of coincidences and fixed elements is reduced to a question already considered: to determine if two subsets of a manifold (here G, G', or G, I), do or do not intersect. We have given a set of necessary conditions in order that intersections be wholly absent (Ch. IV, § 7). Properly translated they will result in necessary conditions in order that fixed elements or coincidences do not occur. *When these conditions fail to be fulfilled we can affirm that coincidences of* T *and* T', *or fixed points of* T, *occur, as the case may be.*

REMARKS. I. The assumption that all manifolds are connected, or what is the same, that they are simple circuits is only mildly restrictive. For example, it is easy to pass from that case to that where every component circuit of one of the manifolds is transformed into a subset of a unique component circuit of the other. In fact even more general situations can be treated just as easily. For the present, however, there is something to be gained in making simplifying assumptions, the more so as all such restrictions are removed later (§ 7).

II. Since the part of L which is not on $\overline{K - L}$ does not play any role whatever in the theory, we may remove it, and similarly for L' and $K' - L'$. We make therefore the explicit assumption that $K = \overline{K - L}$, $K' = \overline{K' - L'}$, which will be frequently found advantageous. In short, we assume that K is the closure of a simple circuit which is an M_n, and similarly for K'.

§ 2. Representative Cycles of the Transformations. The Signed Coincidences and Fixed Points

3. The principle of the intersection method of Ch. IV is as follows: To determine whether two subsets G, G' of a manifold $K-L$ intersect or not, we look for a pair of cycles $\Gamma_h \subset G$, $\Gamma_k \subset G'$ whose intersection $\Gamma_h \cdot \Gamma_k \not\sim 0$, or, when $h+k-n = 0$, whose Kronecker index $(\Gamma_h \cdot \Gamma_{n-h}) \neq 0$. When such cycles can be found we are able to affirm that G and G' meet on $K-L$.

The situation is particularly simple when Γ_h, Γ_k may be chosen in some intrinsically unique way, for then the intersection cycle or the index are topological invariants of the pair of sets G, G' and the condition for intersection assumes the following form: *A certain invariant cycle (number) must be* $\not\sim 0$ $(\neq 0)$.

The application to our problem is obvious. The manifold is here $K \times K' - \Delta$ and the cycles must be cycles on the images G, G' of the transformations. The simplest possible situation is when $p = q = h = k = n$, and it is possible to assign unique n-cycles Γ_n, Γ'_n to G and G'. We then have a numerical topological invariant of the two transformations, the index $(\Gamma_n \cdot \Gamma'_n)$, called *the number of signed coincidences of the two transformations.* When we deal with transformations of $K-L$ into itself and $T' = I$, the identity, we have *the number of signed fixed points of* T.

In the sequel we shall consider exclusively transformations in the special situation just described, as it is only for these that anything like a general theory can be developed. We shall then treat the

Fundamental Problem. *To determine the number of signed coincidences of a pair of transformations of one manifold into another, or into itself. This includes, as a special case, the determination of the signed fixed points of a transformation of a manifold into itself.*

4. The problem as just proposed is still too general. For by reference to Ch. V, No. 30, it will be seen that a concrete

numerical solution can be obtained only when the cycles are considerably restricted. This is also quite clear when we remember that explicit formulas can only be expected for pairs of cycles taken out of some extensive classes. In view of the properties of the index (Ch. IV, §§ 3, 4), we must have pairs of classes such that *the cycles of one class never meet the boundaries of those of the other.*

As in Chapter V, § 3, let K, K' be our two closed complexes, whose dimensions are now both n. Let L, L' be closed subcomplexes of K, K' and L^1, L'^1 closed subcomplexes of L, L' with $L^2 = L - L^1$, $L'^2 = L' - L'^1$. We set again $\Lambda = K \times L' + L \times K'$, and consider the following three possible choices for Λ^1, Λ^2:

Case A. $\Lambda^1 = K \times L'^1 + L^1 \times K'$ $\Big\}$ $\Lambda^2 = \Lambda - \Lambda^1$
Case B. $\Lambda^1 = K \times L'$

Case C. $\Lambda^2 = L \times K'$, $\Lambda^1 = \Lambda - \Lambda^2$.

In the first two cases Λ^1 is closed and Λ^2 open; in the third case it is the reverse.

We now make the following

ASSUMPTIONS. I. *The representative image of any transformation under consideration coincides with an n-cycle mod* Λ^i *on* $K \times K' - \Lambda^j$, $i, j = 1, 2$; $i \neq j$; *where the choice of the* Λ*'s corresponds to one of the cases* A, B, C.

II. *All n-cycles on the particular representative image mentioned in I are* $\approx$ *to multiples of a unique cycle mod* Λ^i on $K \times K' - \Lambda^j$.

When both assumptions are satisfied the transformation is said to be of the first or second kind accordingly as $i = 1$ or 2. We will denote the transformations and cycles by T_{ia}, Γ_n^{ia}, $\cdots$ accordingly as they are of kind i and fall under case $A \cdots$.

We are now able to state the precise problem to be solved:

THE RESTRICTED PROBLEM. *To determine the number of signed coincidences of a pair of transformations* (T_{1a}, T_{2a}), (T_{1b}, T_{2b}), (T_{1c}, T_{2c}).

REMARKS. I. *For absolute manifolds there is only one type, as the six considered coincide.*

II. While we have been concerned so far solely with restricting the transformations we can avail ourselves of a very simple property to enlarge their range in an important direction. We are proposing to apply our intersection theory to pairs of cycles which do not meet each other's boundaries. Now a cycle mod Λ^i on $K \times K' - \Lambda^j$ is uniquely determined when we know it mod a certain neighborhood of Λ^i (Ch. II, § 5). *Therefore we can choose the transformations T_1, T_2 such that, after a suitable modification in sufficiently narrow neighborhoods of Λ^1, Λ^2, they become permissible associated pairs.* As we shall see this remark is of great practical value.

III. Referring to Ch. V, No. 29, it is seen that case B is a special case of A, obtained by making $L^1 = L'^2 = 0$, $L^2 = L$, $L'^1 = L'$. Similarly C corresponds to $L'^1 = L^2 = 0$, $L^1 = L$, $L'^2 = L'$ and interchanging Λ^1 and Λ^2. Thus A is all inclusive. It so happens, however, that the corresponding very general classes of transformations are not very interesting, and that the interesting classes all come under B or C. This is the justification for carrying them along throughout as special cases.

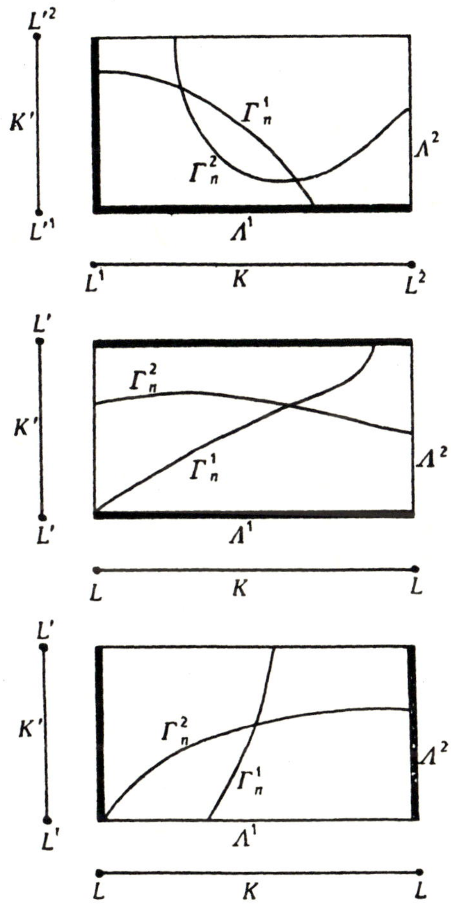

Fig. 15.

An important practical consequence is that we shall have a unique coincidence formula for all cases (formula (22)).

5. On the geometric side we have the following characterizations:

Case A. L'^1 carries no transform of a point of K by T_2. The set $T_2 \cdot L^1$ is vacuous.

Case B. $T_1 \cdot L \subset L'$ unless it is vacuous. L' carries no transform point of K by T_2.

Case C. $T_1 \cdot L$ is vacuous. The only points of $T_2 \cdot K$ on L' come from L.

The figure constructed with $n = 1$, gives a good schematic representation. The complexes are segments and their product a rectangle. Whenever a vertex is on a heavy line it is to be counted in with it, but not with the light line to which it may also belong.

REMARK. Here, as everywhere else, all our developments are valid for modular cycles. In particular, when the modulus is *two* orientability may be dispensed with.

6. While our assumptions will be adhered to until further notice, it is important to bear in mind that they are pure assumptions and nothing more. Moreover:

(a) Even under our assumptions we do not know as yet how to attach a unique cycle to T. For if Γ_n is a suitable cycle, so is $t\Gamma_n$, whatever the non-zero integer t.

(b) Assumption II is not fulfilled even by certain simple types of transformations. Thus, let the representative image carry several independent cycles Γ_n^i, $i = 1, 2, \cdots, r$, and let it coincide with the sum of their points. Then, for arbitrary t's, all cycles $\sum t_i \Gamma_n^i$ are cycles Γ_n that might be chosen as images for T. This case is easily realized: Take for K, K', absolute manifolds ($L = L' = 0$) and let T^1, T^2 be two distinct single-valued transformations of K into K', the corresponding images being two cycles Γ_n^1, Γ_n^2. Define T as the two-valued transformation assigning to every point x of K the two points $T_1 \cdot x$ and $T_2 \cdot x$. The image of T coincides with the points of any cycle of the double infinity $t_1 \Gamma_n^1 + t_2 \Gamma_n^2$.

7. As clearly shown by fig. 15, the six types selected include the most important transformations of manifolds. We consider a few special cases.

(a) *A single-valued continuous transformation T of $K-L$ into a subset of $K'-L'$, such that $\overline{T(K-L)} \subset K'-L'$ can be extended into a T_{2b}.*

The assumption is that we have a transformation of $K-L$ alone, nothing being said or known concerning the points of L. Since T is single-valued and continuous its image G on $K \times K'$ is a homeomorph of $K-L$, and by hypothesis $G \subset K \times (K'-L')$. Taking the situation as in case B, we have then $\overline{G} \subset K \times K' - \Delta^1$.

Let us suppose at first that G coincides with a complex and $\overline{G} - G = D$ with a subcomplex of G. Then since G is a homeomorph of a connected orientable M_n, it is itself such an M_n. It follows that $\overline{G}$ is an n-circuit and consequently it coincides with an n-cycle mod D, Γ_n, which is a base for all n-cycles mod D on G. Now when the point $A \times A'$ of G tends to D on G, its projection A can only tend to L, since otherwise $A \times A'$ would tend to some point of G. Therefore $D \subset L \times (K'-L') = \Delta^2$ and Γ_n is a cycle of $K \times K' - \Delta^1 \bmod \Delta^2$, i.e. it is a Γ_n^{2b} so that the given transformation has been extended to a T_{2b}.

Suppose now that we merely know that $\overline{G} \subset K \times K' - \Delta^1$. Let us replace K by a simplicial subdivision K^0, remove from K^0 the K^0-neighborhood N^0 of L, and let $B = T \cdot (K^0 - N^0)$. If the mesh η of K^0 is small enough, $F(B)$ will be as near as we please to Δ and also on $\overline{G}$. But since $\overline{G}$ does not meet Δ^1, that boundary is then as near as we please to Δ^2 and at a distance exceeding a certain $\xi > 0$ from Δ^1. Let us now deform the chain $F(B)$ onto Δ^2 after the manner of Ch. II, No. 25, add the deformation chain to B and thus change B into a cycle mod Δ^2, Γ_n. Owing to what has just been said, for η small enough, Γ_n will be on $K \times K' - \Delta^1$. Since all n-cycles on Γ_n differ from a multiple of B merely by a chain very near Δ^2, they are all multiples of $\Gamma_n \bmod \Delta^2$ on $K \times K' - \Delta^1$. Γ_n is then here also a Γ_n^{2b}. Therefore in the sense of the extension of No. 4, Remark II, the transformation is again a T_{2b}.

(b) *Under the same circumstances as in* (a), *if*

$$\overline{T\cdot(K-L)} \not\subset K'-L',$$

the transformation is a T_{2c}.

The treatment is exactly the same and we omit the details.

8. Let us suppose that we have identified certain cycles with our transformations. Then if we have a pair T_1, T_2 falling under the same case A, B or C, with Γ_n^i corresponding to T_i, we can find $(\Gamma_n^1 \cdot \Gamma_n^2)$ by (30) of Ch. V. We shall have here $p = q = n = r$,

$$(1) \qquad \begin{cases} \Gamma_n^1 \approx \sum \varepsilon_{ij}^s \, a_s^i \times b_{n-s}^{\prime j}, \\ \Gamma_n^2 \approx \sum \eta_{ij}^s \, b_s^i \times a_{n-s}^{\prime j}. \end{cases}$$

The matrices ε, η are evidently important characters of the transformations. It is interesting to see to what extent they depend upon the particular bases chosen. If we pass from the sets $\{a_s^i\}$, $\{b_{n-s}^{\prime i}\}$ to new sets such that

$$a_s^i \approx \sum \lambda_{ij} A_s^j,$$
$$b_{n-s}^{\prime i} \approx \sum \mu_{ij} B_{n-s}^{\prime j}; \ \lambda, \ \mu \text{ unimodular},$$

ε^s is replaced by $\lambda' \varepsilon^s \mu$. The invariants of ε^s under all these transformations are its rank and its invariant factors. These are then topological invariants of T_1 and similarly for T_2 and η^s.

By (30) of Ch. V we have then, for the number of signed coincidences of T_1 and T_2,

$$(2) \qquad \begin{gathered} (\Gamma_n^1 \cdot \Gamma_n^2) = \sum (-1)^{n(s+1)} \textit{ trace } \eta^{n-s} \beta^s (\varepsilon^s)' \alpha^s, \\ \alpha_{ij}^s = (a_s^i \cdot b_{n-s}^j); \qquad \beta_{ij}^s = (a_s^{\prime i} \cdot b_{n-s}^{\prime j}). \end{gathered}$$

When the bases on K and K' are canonical we have the simpler formula

$$(3) \qquad (\Gamma_n^1 \cdot \Gamma_n^2) = \sum (-1)^{n(s+1)} \textit{ trace } \eta^{n-s} (\varepsilon^s)'.$$

The two formulas (2), (3), are then explicit solutions of the problem proposed. They are however incomplete since, even for the simplest transformations, such as the identity, we have not shown how to calculate the matrices ε, η. The point is that, while these matrices are essential characters of the transformations, they are likely to be their least available characters.

What is then *a priori* the most natural information, of a topological nature, concerning a transformation? Since we are dealing with transformations of complexes, the basic information must be the effect on the cycle moduli, which are the simplest and the first topological attributes of a complex. There arises then the following problem: *To determine the matrices ε, η in terms of the transformations induced by T_1 and T_2 on the cycles of $K-L$.* This question will be taken up in § 4.

9. **The degree of a transformation (Brouwer).** Let for the present K also designate one of its own fundamental n-cycles mod L (sum of the n-cells normally oriented), and similarly for K' and the n-cycles mod L'. Let us confine our attention to a single transformation T. Among the product-cycles that appear in the expression of its fundamental Γ_n, there may be found a cycle $K \times A'$, where A' is a point of $K'-L'$. This can only occur when T is a T_{2b} or T_{2c}. Similarly, a possible cycle $A \times K'$ occurs only in a T_{1b} or T_{1c}. Let us write then generically,

$$\Gamma_n \approx \varepsilon^0 A \times K' + \varepsilon^n K \times A' + \cdots.$$

Since the manifolds $K-L$, $K'-L'$ are connected, no other terms than the two written will have a vertex for factor. Hence ε^0, ε^n are themselves the matrices ε^p for $p = 0$ or n. We now distinguish two cases:

(a) $\varepsilon^n \neq 0$. Then Brouwer's degree of T is defined as ε^0 if ε^n is positive, $-\varepsilon^0$ if ε^n is negative.

(b) $\varepsilon^n = 0$. Then the degree is defined as $|\varepsilon^0|$.

Another way of introducing the degree is as follows: Since Γ_n and $-\Gamma_n$ are equally suitable as representatives of T, when $\varepsilon^n \neq 0$ we choose the cycle such that $\varepsilon^n > 0$. When $\varepsilon^n = 0$ we take it such that $\varepsilon^0 > 0$. Then in all cases ε^0 is the degree.

The degree is then equal to the index $(\Gamma_n \cdot K \times A')$, or the number of signed points $T^{-1} \cdot A'$, the signature being understood in a reasonably obvious sense. We shall see later (No. 22) that it represents also the number of times that $T \cdot K$ covers K', i. e. the integer such that

$$T \cdot K - \varepsilon^0 K' \sim 0 \bmod L'.$$

When the manifolds are absolute, the degree is the number of times that $T \cdot K$ covers K' itself, and this is substantially the definition given for it by Brouwer [7]. The degree thus appears as the simplest and most important of the invariants connected with the matrices ε^p, and it is natural that it should have played a most important part in the investigations of Brouwer and others.

An entirely analogous interpretation can be given for ε^n and the inverse transformation T^{-1}. As a matter of fact, when K, K' are algebraic curves (Riemann surfaces), and T is an algebraic correspondence between them, ε^0 and ε^n are merely the noted *indices* of the correspondence, which appear, for example, in the Chasles-De Jonquières coincidence formula. Each represents the number of points of one of the curves that correspond to a single point of the other. In the theory of algebraic correspondences, these integers have been considered at least as far back as the middle of the nineteenth century.

10. Brouwer has defined a *class of transformations* of an absolute M_n into another M'_n, as a family of single valued transformations depending upon a certain number of continuous parameters. From the present point of view, a more or less equivalent definition of the class is as follows: It is a set of continuous single-valued transformations $\{T\}$ such that the representative cycles on $M \times M'$ are homotopic to one another. The problem arises of finding all the topological invariants of a class. Brouwer proved [7] that the degree is a class invariant.

Now if we refer to the classification of deformations in Ch. II, § 2, it becomes clear that the transformations may be subdivided into classes in different ways according to the type of deformation chosen. The type best adapted to the homology theory developed in the present book is not homotopy, but homological deformation.

DEFINITION. *Two transformations of one manifold into another are said to be in the same homology class, whenever they are of the same type* $T_{1a}, \cdots, T_{2c}$, *and moreover their representative cycles are* $\approx$ *to one another in the sense appropriate to the type.*

Thus if T, T' are both of type T_{2b}, and have representative cycles Γ_n, Γ'_n, they belong to the same homology class whenever

$$\Gamma_n \approx \Gamma'_n \text{ on } K\times K' - \varDelta^1 \bmod \varDelta^2.$$

This is the extension of the classification which we gave in [8] for transformations of an absolute manifold into itself. From the definition follows the important

THEOREM. *The arithmetical invariants of the matrices* ε *or* η, *as the case may be, are topological invariants of the homology classes, and so are the numbers of signed fixed points or coincidences. In particular, the degree of a transformation is an invariant of its homology class.*

§ 3. APPROXIMATIONS TO TRANSFORMATIONS OF COMPLEXES

11. In the present section the complexes need not be manifolds. We take K, K' simplicial, and consider a correspondence T between them. We assume concerning T that its image on $K\times K'$ coincides with a non-singular complex G. We shall now consider the approximation of T by a *polyhedral transformation*, i. e. with a polyhedral G.

Let $P\times P'$ be any point of G. The correspondences between that point and its projections P, P' on K, K', determine on them singular images g, g' of G. Since the meshes of the g's tend to zero with that of G, and since G

may be replaced by any subdivision, we can assume it such that g, g' are barycentrically deformable, without change of type, into subchains or subcomplexes of K, K' respectively.

Let $P \times P'$ be confined for the present to the cell E_h of G. Then P, P' are on cells e_h, e'_h on K, K'. Apply now the fundamental deformation theorem Ch. II, No. 15. P and P' go over into Q, Q' on simplexes σ, σ' of K, K' whose dimensions $\leqq h$. Since the correspondences between the point pairs $P \times P'$ and P or P', P and Q, P' and Q' are all barycentric, the simplexes σ, σ' are in a barycentric correspondence in which Q, Q' are associated pairs.

By the deformation theorem, Q, Q' are on the closures of the cells of K, K' that carry P, P'. Hence $Q \times Q'$ is on the closure of the cell of $K \times K'$ that carries $P \times P'$. Since that cell is convex there is a uniquely defined segment $\overline{P \times P' \; Q \times Q'}$ on $K \times K'$. The set of all these segments represents a deformation of G into G'. It coincides with a deformation complex for the deformation just mentioned, which I say is polyhedral. For let the simplexes

$$\sigma = A^0 A^1 \cdots A^h, \qquad \sigma' = B^0 B^1 \cdots B^h,$$

where the A's and B's are not necessarily all different, but those with the same superscript correspond to one another in the barycentric correspondence (Q, Q'). By passing to the coördinates in the spatial representations as in Ch. V, No. 4, we find that $Q \times Q'$ has for locus the simplex

$$\sigma^* = A^0 \times B^0 \; A^1 \times B^1 \cdots A^h \times B^h,$$

whose dimension is one less than the number of distinct vertices $A^i \times B^i$. We have therefore shown that G *may be barycentrically deformed into a simplicial* G', *in such manner that no point leaves its closed cell on* $K \times K'$ *throughout the deformation.*

As an example, when the factor cells are one-cells, we have for their product a rectangle, and σ^*, when its dimension is one, is a side or a diagonal of the rectangle.

A transformation T which carries over every simplex of K into a set of simplexes of K' is said to be *simplicial.* This is evidently the case with T' whose image is G'. Therefore

THEOREM. *A transformation T, whose image is a complex, can be indefinitely approximated by a simplicial transformation.*

Suppose in particular that g or g' have polyhedral parts. Then we may assume K, K', G, G', g, g' so subdivided that those polyhedral parts will be subcomplexes or subchains of the subdivisions. Referring again to the deformation theorem, these polyhedral parts (possibly subdivided) will appear unchanged in the complexes or chains derived from g, g'. As a special case, g or g' may coincide with K or K', and then they are fixed throughout the deformation.

APPLICATION. Let the transformation be single-valued. Then g coincides with K, so that we can take $Q = P$. The displacement from $P \times P'$ to $P \times Q'$ is in $P \times K'$, which insures that G', like G, meets every $P \times K'$ at a single point—namely Q'. This means that the approximating transformation is likewise single-valued. Therefore the theorem of No. 11 reduces to the following proposition due to Alexander [19]:

THEOREM. *A single-valued transformation can be indefinitely approximated by a single-valued simplicial transformation of a subdivision of K into K'.*

Observe that it was not necessary to subdivide K', for as a result of single-valuedness, a subdivision of K impresses one on G, whose mesh tends to zero with that of K.

REMARKS. I. As far as the preceding results are concerned it is immaterial whether K, K' are of the same dimension or not.

II. Suppose that T were single-valued only over $K - L$, L a locally connected subset of K. Then we could treat all but a neighborhood of $L \times K'$ on G as we have treated G itself, and apply Alexander's theorem. The new transformation would then be simplicial over all but any assigned neighborhood of L on $K - L$.

12. Conditions for the homeomorphism of two complexes (Alexander). This is the proper place to bring up a fundamental proposition proved by Alexander in the paper just quoted, [19], giving necessary and sufficient conditions for homeomorphism between two complexes K and K'. We shall merely describe the theorem without proof, first because it does not enter into any ulterior argument, next because it is really the first theorem in an alternate treatment of combinatorial topology.

The theorem is expressed in terms of *subdivision sequences* and *pseudo-identical* transformations which we first define.

A subdivision sequence $\{K^i\}$ of a complex K, is an infinite sequence of subdivisions of K, such that K^{i+1} is a subdivision of K^i, and that the mesh of K^i tends to zero with $1/i$.

Let K' be a simplicial subdivision of the simplicial complex K. A pseudo-identical transformation T of K' into K, is a single-valued simplicial transformation, such that every simplex of K' is transformed into a face of the simplex of K that carries it, or else into that simplex itself. It is easy to see that T is an approximation to the identical transformation in the sense of No. 11. We are now in position to state

ALEXANDER'S THEOREM. *In order that K, K' be homeomorphic, it is necessary and sufficient that there exist an infinite sequence of complexes $\{K^i\}$ and transformations T^i, such that: (a) $\{K^{2i}\}$ and $\{K^{2i+1}\}$ are respective subdivision sequences of K and K'. (b) T^i is a single-valued simplicial transformation of K^{i+1} into K^i. (c) $T^{2i}\, T^{2i+1}$ is a pseudo-identical transformation of K^{2i+2} into K^{2i}, and this for every i.*

§ 4. TRANSFORMATIONS OF THE CYCLES

13. We now return to the question as it was left at the end of No. 8. We confine our attention to transformations that satisfy our assumptions. For a T of that nature, there exists a characteristic Γ_n of appropriate type, such that all n-cycles in coincidence with it, are homologous to a multiple of it in a sense already defined. Given then T, we have to choose a definite cycle $t\Gamma_n$. It is natural to take $t = \pm 1$, and to consider more especially one of the two

cycles $\pm\Gamma_n$ as the cycle representing T. There still remains the choice of sign. We shall obtain a natural selection of it as a result of a closer study of the effect of T on the cycles of the manifolds.

14. **Single-valued transformations.** These transformations, highly interesting in themselves, give us the key to the whole situation.

We take again a manifold $K-L$, and subject it to a single-valued transformation T into a subset of $K'-L'$, whose closure $\subset K'-L'$, so that T is a T_{2b}.

We have seen that the image G of T on $K \times K'$, carries a characteristic fundamental Γ_n, the homeomorph of the Γ_n^0 by which the circuit $K-L$ is oriented. This Γ_n is then the representative n-cycle of T. We have the choice between Γ_n and $-\Gamma_n$, but of course in this case there is no hesitation: The cycle to be taken is the image of $T \cdot \Gamma_n^0$ itself, *orientation included*; Γ_n is a certain n-circuit, and its indicatrix is the image of an indicatrix of Γ_n^0. To be precise, we may assume that G is decomposed into the very cells which are the images of those of K or, here, are projected into those of K. Then the indicatrix of Γ_n has for projection that of Γ_n^0. Adopting permanently the convention of No. 9, we shall designate in the sequel by K, in chain and cycle relations, the circuit Γ_n^0 and by K' the similar circuit for the second manifold. Then if

$$K = \sum E_n^i, \tag{4}$$

we shall have

$$\Gamma_n = \sum \mathfrak{E}_n^i, \tag{5}$$

where $\mathfrak{E}_n^i$ has for projection E_n^i.

15. We have already considered a first polyhedral approximation in No. 11. It had the advantage of being simplicial, and above all, of representing likewise a single-valued transformation. Unfortunately the representative cycle, say Γ_n', was not a subcycle of $K \times K'$. For our immediate purpose it is imperative to have a representation by a subcycle.

As usual when we deal with that type of question, we may have to replace K, K' by simplicial subdivisions whose meshes are small. To avoid complications, let us agree to denote at any time by K, K' arbitrary simplicial subdivisions of the initial complexes. We shall designate by ε the corresponding mesh of $K \times K'$, and remember that under the circumstances it can be chosen arbitrarily small. We recall also that under our choice of metric for the product, the meshes of K and K' are also at most ε.

16. As we have seen then, we have for suitable K, K', the simplicial approximation

$$(6) \qquad \Gamma_n' = \sum \mathfrak{E}_n'^i,$$

where $\mathfrak{E}_n'^i$ is a simplex on a closed cell $\bar{E}_n^i \times \bar{E}_n'^i$ of $K \times K'$, projected on E_n^i into K. As for Γ_n', it is derived from Γ_n by a certain ε deformation, such that the projections of all points of Γ_n on K remain fixed. Let $A^i \times A'^i$ be one of the vertices of $\mathfrak{E}_n'^i$, and therefore of the carrying cell $E_n^i \times E_n'^i$ of $K \times K'$. The cell $E_n^i \times A'^i$ is then a face of $E_n^i \times E_n'^i$, and its projection is likewise E_n^i. If P is any point of $\bar{E}_n^i$, there correspond to it points $P \times A'^i$ and $P \times Q$ on $\bar{E}_n^i \times A'^i$ and $\bar{\mathfrak{E}}_n'^i$ projected into it, and therefore on the same $P \times K'$. Furthermore, as A'^i and Q are both on $\bar{E}_n'^i$, there is a unique segment $A'^i Q$ on K', and therefore a unique segment $\overline{P \times A'^i P \times Q}$ on $K \times K'$. The set of all these segments defines a deformation of $E_n^i \times A'^i$ into $\mathfrak{E}_n'^i$, again leaving the projections of all points undisturbed. The deformation chain D_n^i of the boundary of $E_n^i \times A'^i$, is on $F(E_n^i \times E_n'^i)$, and we have

$$\mathfrak{E}_n'^i \sim E_n^i \times A'^i + D_n^i,$$

$$(7) \qquad \Gamma_n' \sim \sum E_n^i \times A'^i + \sum D_n^i.$$

The cycle at the right is reducible to a subcycle of $K \times K'$, in accordance with the deformation theorem for combinatorial complexes, Ch. II, No. 45. This is done without displacing the cells in the first sum which are already cells of the product. As for the second sum, its elements undergo deformations during which they never leave their closed cells on $K \times K'$. Therefore they remain constantly on cells of types $\bar{E}_{n-1} \times \bar{E}_n'$.

The deformation introduces then no cell of type $E_n \times A'$, so that finally

(8) $$\Gamma'_n \sim \sum E^i_n \times A'^i + \sum_{k<n} t^k_{ij} E^i_k \times E'^j_{n-k} = \Delta_n.$$

Again let P be any point of $K - L$, $P \times Q$ the unique corresponding point of Γ' with the same projection P, $P \times R$ any point whatever of Δ with that projection. Owing to the mode of deriving Δ from Γ', R and Q are again on the closure of a cell of K', and there are uniquely defined segments RQ and $\overline{P \times R\, P \times Q}$ on K' and $K \times K'$. *Therefore Δ is likewise deformable into Γ', hence also into Γ, by an ε deformation $\mathfrak{D}^\varepsilon$ (ε for no point leaves its closed cell), leaving all projections on K invariant.* This is the basic property of Δ for our purpose.

17. Let now C_p be any chain on K. It is the projection of a chain $\mathfrak{C}_p$, which is the image on Γ_n of $T \cdot C_p$. The transform $C'_p = T \cdot C_p$ is the projection of $\mathfrak{C}_p$ on K'.

Let in particular C_p be an absolute cycle $\subset K - L$. Then C'_p is likewise an absolute cycle, and since the projection of a chain is in any case a singular image of it, C'_p is again an absolute cycle. We propose to determine both $\mathfrak{C}_p$ and C'_p in terms of the cycle C_p.

Consider now the two chains Γ_n and $C_p \times K'$. The first is an n-cycle on $K \times K' - A^1 \bmod A^2$, the second an $(n+p)$-cycle on $K \times K' - A^2 \bmod A^1$. Therefore their geometric intersection, which coincides with $\mathfrak{C}_p$, $\subset K \times K' - A$. Hence they have a well defined intersection Γ_p, which is an absolute cycle. That intersection is defined as a family of cycles homologous about the geometric intersection, that is about $\mathfrak{C}_p$ (Ch. IV, § 3).

Let $\{\Gamma^i_p\}$ be one of the constituent polyhedral sequences of the homologous set, and $\mathfrak{N}$ a neighborhood of $\mathfrak{C}_p$ on $K \times K'$. I say that for i above a certain value

(9) $$\Gamma^i_p \sim \pm \mathfrak{C}_p \text{ on } \mathfrak{N}.$$

More precisely: Except for its sign, $\mathfrak{C}_p$ is homologous on all its neighborhoods to cycles of the homologous family about $\mathfrak{C}_p$ which defines the intersection cycle $\Gamma_n \cdot C_p \times K'$.

18. Let us deform C_p homologically into a subcycle C_p^* of K^*, the dual of $K-L$. As a consequence we have the approximation $C_p^* \times K'^*$ to $C_p \times K'$,. and it is a subcycle of the dual K_{2n}^* of $K \times K' - \Lambda^1$. It is understood of course that K'^* is the dual of $K' - L'$. The deformation in this case is 2ε. It follows that $\Delta_n \cdot C_p^* \times K'^*$ is an appropriate approximation to $\Gamma_n \cdot C_p \times K'$. If the complexes K, K' run through a sequence of subdivisions of the initial complexes, whose meshes tend to zero, the corresponding sequence of the cycles $\Delta_n \cdot C^* \times K'^*$ is a sequence $\{\Gamma_p^i\}$ of our homologous set about $\mathfrak{C}_p$. Therefore the proof of (9) is reduced to that of

$$(10) \qquad \Delta_n \cdot C_p^* \times K'^* \sim \pm \mathfrak{C}_p \text{ on } \mathfrak{R},$$

for ε sufficiently small.

From (21) of Ch. V, we have now,

$$(11) \qquad \begin{aligned} &\Delta_n \cdot C_p^* \times K' \\ &= \sum E_n^i \cdot C_p^* \times A'^i \cdot K' + \sum t_{ij}^k E_k^i \cdot C_p^* \times E_{n-p}'^j \cdot K'^*. \end{aligned}$$

19. Let us apply to Δ the deformation $\mathfrak{D}^\varepsilon$ reducing it to Γ. As a consequence each chain at the right goes over into a chain on Γ whose projection on K is the same. The projections of the cells in the second sum are the cells $E_k^i \cdot C_p^*$, whose dimension $k+p-n<p$. Therefore the second sum is deformed into a zero-chain. As for the first, its projection on K is $K \cdot C_p^* = C_p^*$. Therefore it goes over into the chain $\mathfrak{C}_p^*$ on Γ_n whose projection is C_p^*. But associated with the reduction of C_p to C_p^* on K, there is a parallel one on Γ_n which amounts, say, to an η deformation, where η tends to zero with ε. Hence the cycle at the left in (10) is $(2\varepsilon+\eta)$-deformable into $\mathfrak{C}_p$. Since $2\varepsilon+\eta$ is as small as we please, (10) holds, with the sign $+$, and therefore likewise (9) with the sign $+$. The sign $-$ would prevail if Γ_n were projected into the circuit $-K$, i. e. if its indicatrix were projected into an inverted indicatrix of K. The sign is then in any case that of the Kronecker index $(\Gamma_n \cdot A \times K')$, where A is

any point of $K-L$. Let A^* be a vertex of K^* nearest to A, and let it be on E_n^i. Then, with Γ_n oriented as agreed,

$$(\Gamma_n \cdot A \times K') = (\Delta_n \cdot A^* \times K'^*) = (+E_n^i \times A'^i \cdot A^* \times K'^*) = +1.$$

If Δ_n is oriented in opposite fashion the index is -1. Therefore in all cases,

(12) $$\Gamma_n \cdot C_p \times K' \sim \eta\, \mathfrak{C}_p \text{ on } \mathfrak{N};$$

(13) $$\eta = (\Gamma_n \cdot A \times K'), \qquad C_p = Proj_K\, \mathfrak{C}_p.$$

REMARK. To simplify matters we have taken for C_p an absolute cycle. But it is evident that we might equally well have allowed it to be a cycle mod L. The attending modifications are insignificant, the chief one being that $\mathfrak{C}_p$, like Γ_n itself, is now a cycle mod Λ^2 on $K \times K' - \Lambda^1$.

20. **The general case.** When T is not single-valued, given C_p it is not possible in general to determine a unique $\mathfrak{C}_p$ of Γ_n whose projection on K is C_p. For there may exist points of K which are the projections of an infinite number of points of Γ_n. That happens whenever there are points of K having an infinity of transforms. Thus Δ_n of the preceding discussion is the image of a T transforming certain points into cells of dimension >0. T is single-valued on the n-cells of $K-L$ but not necessarily elsewhere.

For the general case then we define $\mathfrak{C}_p$ by means of a relation analogous to (12), with $\eta = \pm 1$ accordingly as $(\Gamma_n \cdot A \times K')$ is $+$ or $-$. The elements Γ, C, will have to be interpreted in suitable manner.

When the index $(\Gamma_n \cdot A \times K') \neq 0$, *we shall agree to orient* Γ_n *in such manner that this index be positive; when it is zero, in such manner that* $(\Gamma_n \cdot K \times A') \geqq 0$. Then $\mathfrak{C}_p$ is defined by means of

(12′) $$\Gamma_n \cdot C_p \times K' \sim \mathfrak{C}_p \text{ on } \mathfrak{N},$$

where $\mathfrak{N}$ is now a neighborhood of the geometric intersection of Γ_n and $C_p \times K'$.

A restriction that we impose on C_p is that it must be such as to enable us to determine *the intersection cycle at the left in accordance with the procedure* of Ch. IV, § 3. This restrictive condition merely requires that Γ_n and $C_p \times K'$ do not meet each other's boundaries on $K \times K' - \Lambda$. As in all applications to be made of (12′) in the sequel, Γ_n and $C_p \times K'$ will always be cycles of two dual types (on $K \times K' - \Lambda^1$ mod Λ^2 and $K \times K' - \Lambda^2$ mod Λ^1 or *vice versa*), the condition in question will always be naturally fulfilled.

We have just defined $\mathfrak{C}_p$ by analogy with the case of single-valued transformations. However, we can justify our definition on stronger grounds. Suppose that T is a single-valued transformation of a certain open subset G of $K - L$ into a subset of $K' - L'$. We can approximate G as closely as we please by a manifold $K - L^0 \subset G$, where $L \subset L^0$ and furthermore L^0 is a subcomplex of a subdivision of K. Then T is a single-valued transformation of the manifold $K - L^0$ into a subset of $K' - L'$. Its representative chain Γ_n^0, as a transformation of $K - L^0$, is a subchain of Γ_n. Let C_p^0, $\mathfrak{C}_p^0$ be the parts of C_p and $\mathfrak{C}_p$ on $K - L^0$ and Γ_n^0, respectively. $\mathfrak{C}_p^0$ is then the image of $T \cdot C_p^0$ on $(K - L^0) \times K'$. Given any neighborhood $\mathfrak{N}^0$ of $\mathfrak{C}_p^0$, we can find a representative of the left hand side in (12′) such that

$$(14) \quad \Gamma_n \cdot C_p^0 \times K' \sim \eta\, \mathfrak{C}_p^0 \bmod (K \times L' + L^0 \times K') \text{ on } \mathfrak{N}^0;$$

$$(14') \quad \eta = (\Gamma_n^0 \cdot A \times K').$$

Thus except for the sign of η, the parts of $\mathfrak{C}_p$ corresponding to the images of what may be described as "single-valued portions" of Γ_n, bear the correct relation to the projections on K.

We are not able to insure that $\eta = +1$ everywhere, and in fact, it is easy to show by an example that such a choice would be inadvisable. Suppose indeed that T is polyhedral, and let $\Gamma_n = \Gamma_n^0 + \Gamma_n^1 + \cdots$, the rest of Γ_n representing transforms of points not on $K - L^0$. Let Γ_n^0, Γ_n^1 have a common $(n-1)$-cell E_{n-1}, *not* on the boundary of Γ_n. Furthermore, let both represent single-valued transformations (as regards local elements) over $K - L^0$. It may then very well happen that they both carry cells E_n^0, E_n^1 with the same projection, E_n, on K. The two cells will have to be oppositely related to E_{n-1}, else Γ_n would have a boundary cell on $K\times K'-\Delta$. Hence their projections on E_n will be oppositely oriented. From this it is easy to deduce that if A is a point on E_n,

$$(\Gamma_n^0\cdot A\times K') = (E_n^0\cdot A\times K') = -(E_n^1\cdot A\times K')$$
$$= -(\Gamma_n^1\cdot A\times K').$$

As a consequence, if C_p^* is a subchain of K^* on E_n,

$$\Gamma_n^0\cdot C_p^*\times K' \sim \pm\mathfrak{C}_p^0,\quad \Gamma_n^1\cdot C_p^*\times K' \sim \mp\mathfrak{C}_p^1,$$

where $\mathfrak{C}^0$ and $\mathfrak{C}^1$ have the same projection C_p^* on K. The signs have to be taken oppositely in order that $\mathfrak{C}_p^0-\mathfrak{C}_p^1$ be without boundary cells on E_{n-1}. Fig. 16, which corresponds to $n=1$, illustrates the situation.

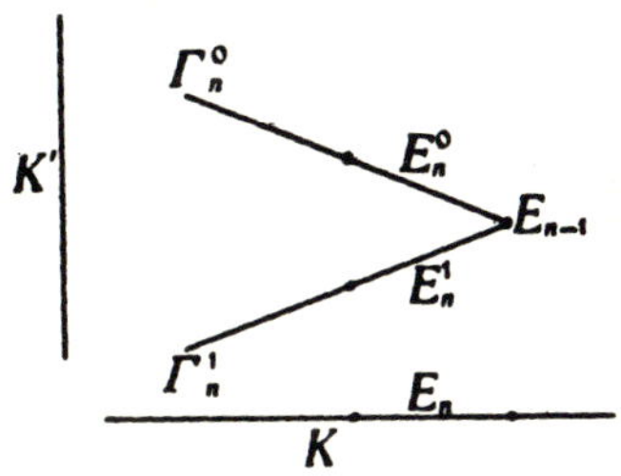

Fig. 16.

21. What is left from now on, is largely a matter of working out a few formulas.

Consider the cycles Γ_n^1, Γ_n^2 of No. 8. They stand for a pair of cycles Γ^{1a}, Γ^{2a}; $\cdots$, according to the interpretation given to the symbols $a, \cdots$. The formal relationship is the same for the three cases A, B, C. We propose to apply (12′) to the determination of the constants ε, η in (1). To determine the ε's we take $\Gamma_n=\Gamma_n^1$ and $C_p=b_p^h$; for the η's, on the contrary, we take $\Gamma_n=\Gamma_n^2$, $C_p=a_p^h$.

We have then from (1) and (21) of Ch. V,

$$\Gamma_n^1\cdot b_p^h\times K' \approx \sum \varepsilon_{ij}^{\mu}\,\overline{a_\mu^i\times b'^j_{n-\mu}}\cdot\overline{b_p^h\times K'}$$
$$\approx \sum \varepsilon_{ij}^{\mu}\,\overline{a_\mu^i\cdot b_p^h}\times\overline{b'^j_{n-\mu}\cdot K'}.$$

But $b'^j\times K'=b'^j$, hence

$$\Gamma_n^1\cdot b_p^h\times K' \approx \sum \varepsilon_{ij}^{\mu}\,\overline{a_\mu^i\cdot b_p^h}\times b'^j_{n-\mu}. \tag{15}$$

We have then from (12′), T_1 being the transformation whose image is Γ_n^1,

$$T_1 \cdot b_p^h \approx Proj_{K'} \sum \cdots.$$

But the term explicit in the sum in (15) is a chain whose projection on K' is $b'^j_{n-\mu}$. When $n-\mu<p$, that chain is not to be counted in $T_1 \cdot b_p^h$; when it is $>p$, $\mu+p-n<0$ and $a_\mu^i \cdot b_p^h = 0$. Therefore

$$T_1 \cdot b_p^h \approx Proj_{K'} \sum \varepsilon_{ij}^{n-p} \overline{a_{n-p}^i \cdot b_p^h \times b'^j_p}.$$

But

$$Proj_{K'} \overline{a_{n-p}^i \cdot b_p^h \times b'^j_p} = (a_{n-p}^i \cdot b_p^h)\, b'^j,$$

since the barred intersection consists merely of a certain number of points each taken with a $+$ or $-$ sign, the total number being the index that appears at the right.

From this follows finally,

$$T_1 \cdot b_p^h \approx \sum \varepsilon_{ij}^{n-p} (a_{n-p}^i \cdot b_p^h)\, b'^j_p. \tag{16}$$

As our problem has been formulated (No. 8), the effect of T_1 on the b's must be considered as known. Therefore we have a set of known homologies

$$T_1 \cdot b_p^h \approx \sum f_{hj}^p\, b'^j_p, \tag{17}$$

in the proper relative sense depending upon the case. Hence

$$\begin{gathered} \sum \varepsilon_{ij}^{n-p} (a_{n-p}^i \cdot b_p^h) = \sum \varepsilon_{ij}^{n-p}\, \alpha_{ih}^{n-p} = f_{hj}^p, \\ \alpha_{ih}^{n-p} = (a_{n-p}^i \cdot b_p^h). \end{gathered} \tag{18}$$

As for the α's, they are also known numbers—the indices for the cycles of two complementary bases $\{a_{n-p}^i\}$, $\{b_p^h\}$. Therefore α^{n-p}, f^p are known matrices. Furthermore, when the two bases just mentioned are not vacuous (i. e. when the Betti number R_{n-p} for the cycles a_{n-p} is $\neq 0$), $|\alpha^{n-p}| = \pm 1$, (Ch. IV, No. 17), hence α^{n-p} has an inverse. From (18) we find then,

$$\begin{gathered} (\alpha^{n-p})' \varepsilon^{n-p} = f^p, \\ \varepsilon^{n-p} = ((\alpha^{n-p})')^{-1} \cdot f^p. \end{gathered} \tag{19}$$

In the exceptional case there are no cycles of type $a_{n-p} \times b'_p$, nor cycles b_p. Therefore we may define ε^{n-p} and $\mathbf{f}^p$ as both zero. Hence the second relation (19) holds, provided that we make this convention, to be adhered to in the sequel: *Whenever the Betti numbers (both equal) for the types a_{n-p}, b_p are zero, α^{n-p} is to be replaced by the scalar multiplier unity, and likewise for its inverse.*

Similarly, if T_2 corresponds to Γ_n^2, and

$$T_2 \cdot a_p^h \approx \sum g_{hj}^p \, a_p'^j, \tag{20}$$

we find

$$\begin{aligned} (-1)^{(n+1)p} \, \alpha^p \, \eta^{n-p} &= \mathbf{g}^p \\ \eta^{n-p} &= (-1)^{(n+1)p} \, (\alpha^p)^{-1} \cdot \mathbf{g}^p. \end{aligned} \tag{21}$$

22. The extreme cases $p = 0$, n are highly instructive when we have a T_b or T_c. The extreme matrices $\alpha, \beta, \varepsilon, \eta, \mathbf{f}, \mathbf{g}$ are then of order zero or one. We designate the only term then present by α^0, β^n etc., respectively. The first two reduce to $+1$ whenever K, K' themselves can be taken as fundamental sets, absolute or mod L, L', for the dimension n.

We first observe that $K \times A'$ is an n-cycle of type Γ_n^{2b} or Γ_n^{2c}, while $A \times K'$ is a Γ_n^{1b} or Γ_n^{1c}. Therefore, in view of (19), $\varepsilon^0 = f^n$, hence

$$\Gamma_n^{1b} \text{ or } \Gamma_n^{1c} \approx f^n \, A \times K' + \cdots.$$

Similarly by (21), $\eta^n = g^0$, so that

$$\Gamma_n^{2b} \text{ or } \Gamma_n^{2c} \approx g^0 \, K \times A' + \cdots.$$

Referring to No. 9, and the conventions in No. 20, we find that f^n is Brouwer's degree. It is, in fact, the degree exactly as he defined it: the algebraic number of times that $T \cdot K$ covers K' mod L', which is precisely the definition of f^n. Let T' be the polyhedral approximation transforming simplexes of K (or of a subdivision of it still called K) into simplexes of K'. Any σ_n of $K' - L'$ will be covered with r transforms of positive n-simplexes of $K - L$ oriented like σ_n, and m oriented oppositely, and $r - m$ is the degree. In particular, for a T_{2b} or T_{2c}, $r = m$ and the degree is equal to zero.

When the degree > 0 the transformation is *sense-preserving*; when it is < 0 the transformation is *sense-reversing*.

Both cases are merged when we deal with transformations of absolute manifolds into one another. Since the cycles are all absolute, we take $b^i_p = a^i_p$; $b'^i_p = a'^i_p$, and write

$$\Gamma_n \approx \sum \varepsilon^p_{ij}\, a^i_p \times a'^j_{n-p}.$$

We have this time

$$\begin{aligned} T \cdot a^i_p &\approx \sum g^p_{ij}\, a'^j_p, \\ \mathbf{g}^p &= (\alpha^{n-p})'\, \varepsilon^{n-p}, \\ \varepsilon^n &= g^0, \quad \varepsilon^0 = g^n, \\ \Gamma_n &\approx g^n\, A \times K' + g^0\, K \times A' + \cdots, \end{aligned}$$

and the degree is g^n.

§ 5. The Numbers of Signed Coincidences and Fixed Points

23. Let us substitute the expressions found for the ε's and η's in (2), with s replaced everywhere by p. There results the following *fundamental coincidence formula:*

$$(22) \quad (\Gamma^1_n \cdot \Gamma^2_n) = \sum (-1)^{n-p}\, \mathit{trace}\, (\alpha^p)^{-1}\, \mathbf{g}^p\, \beta^p\, (\mathbf{f}^{n-p})'.$$

This very compact relation is valid for all choices of bases on K, K'. If the bases are *canonical*, that is, if $\alpha^p = 1$, $\beta^p = 1$, the coincidence formula, after replacing p by $n - p$, assumes the simpler form,

$$(23) \qquad (\Gamma^1_n \cdot \Gamma^2_n) = \sum (-1)^p\, \mathit{trace}\, \mathbf{g}^{n-p}\, (\mathbf{f}^p)'.$$

The coincidence formula can also be put in another form, which is sometimes more convenient. Instead of the matrices $\mathbf{f}$, $\mathbf{g}$, introduce the matrices φ, γ of the indices of the transformed cycles and the cycles b', a' or

$$\begin{aligned} \varphi^p_{ij} &= (a'^i_{n-p} \cdot T_1\, b^j_p), \\ \gamma^p_{ij} &= (T_2\, a^i_p \cdot b'^j_{n-p}). \end{aligned}$$

From (17) and (20) we find readily,

$$(\boldsymbol{f}^p)' = (\beta^{n-p})^{-1} \cdot \varphi^p; \qquad \boldsymbol{g}^p = (\gamma^p)(\beta^p)^{-1}.$$

Substituting in (22) we have this time

$$(24) \quad (\Gamma_n^1 \cdot \Gamma_n^2) = \sum (-1)^{n-p} \, trace\, (\alpha^p)^{-1} \gamma^p (\beta^p)^{-1} \varphi^{n-p}.$$

REMARKS. I. In comparing formulas of the preceding type with one another, it is well to keep in mind the different forms which they may assume, owing to the following properties:

(a) $trace\ \boldsymbol{a}\boldsymbol{b} \quad = trace\ \boldsymbol{b}'\boldsymbol{a}'$

(b) $trace\ \boldsymbol{a}\boldsymbol{b}\boldsymbol{a}^{-1} \quad = trace\ \boldsymbol{b}$.

From this follows

$$trace\ \boldsymbol{a}\boldsymbol{b} \cdots \boldsymbol{c} = trace\ \boldsymbol{b} \cdots \boldsymbol{c}\boldsymbol{a}, \text{ etc.}$$

Therefore in the matrix products at the right of all our formulas the cyclic order of the terms alone matters.

(c) Since the summations are from 0 to n for p, we may substitute everywhere $n-p$, for p.

II. In the derivation of our formulas we have assumed that $K-L$, $K'-L'$ are connected. When they are not connected, any continuous transformation of one into the other is equivalent to a set of transformations between the connected parts. From this follows readily that the formulas hold also in this case.

24. While the same formula holds for the three cases A, B, C, the symbols have different meanings. The following tables summarize the situation:

Cycle	a	a'	b	b'
Case A	on $K-L^2$ mod L^1	on $K'-L'^1$ mod L'^2	on $K-L^1$ mod L^2	on $K'-L'^2$ mod L'^1
Cases B, C	absolute cycle on $K-L$	absolute cycle on $K'-L'$	on K mod L	on K' mod L'

$$\Gamma_n^i = \text{a cycle on } K \times K' - \Delta^j \bmod \Delta^i, \qquad i \neq j = 1, 2.$$
$$\Delta = K \times L' + L \times K'.$$

Case A	$L^2 = L - L^1,\ L'^2 = L' - L'^1;$ $\Delta^1 = K \times L'^1 + L \times K'^1;\ \Delta^2 = \Delta - \Delta^1.$
B	$\Delta^1 = K \times L';\ \Delta^2 = \Delta - \Delta^1.$
C	$\Delta^2 = L \times K';\ \Delta^1 = \Delta - \Delta^2.$

25. The coincidence formulas which we have just derived include as special cases all those of our papers on the question [8, 9]. This will be shown explicitly below. The greater generality of our present result is largely due to the more general duality relations obtained in Ch. III. The method followed here is substantially the same as in our papers, with the improvement due to Hopf [7], that in deriving (19) and (21) we make use of projections. In our paper, we rested the derivation of the relations in question, on the following formula, quite interesting in itself:

$$(25) \qquad (\Gamma_n \cdot C_p \times C'_{n-p})_{K\times K'} = (-1)^p\,(TC_p \cdot C'_{n-p})_{K'},$$

where C_p, C'_{n-p} are suitably restricted chains. In particular, we may choose $C_p = a_p^h$ or b_p^h, with $C'_{n-p} = a'^j_{n-p}$, b'^j_{n-p} respectively. Except for the notations, (25) is the same as (59.2) of [8].

As for (24), it extends and includes the relation [40.5] of [9], which in turn generalized a similar formula derived for the two-dimensional case by Alexander [9].

26. **Transformations of a manifold $K - L$ into itself.** We first take cases B, C, and apply our results by choosing for K' another copy of K, with L' as the image of L. It is then natural to choose for the bases on K' the images of those on K, or a'^i_p, b'^i_p as the images of a^i_p, b^i_p. We shall have then $\alpha^p = \beta^p$, and in place of (22),

$$(26) \quad (\Gamma_n^1 \cdot \Gamma_n^2) = \sum (-1)^{n-p}\ \textit{trace}\ (\alpha^p)^{-1}\, \mathfrak{g}^p\, \alpha^p\, (\mathfrak{f}^{n-p})',$$

while (23) remains as before.

In case A, we can take for b'^{i}_{p}, a'^{j}_{p} the images of a^{i}_{p}, b^{j}_{p}. Then $\beta^p = (-1)^{p(n+1)}(\alpha^{n-p})'$, and we have this time:

$$(27)\ (\Gamma^1_n \cdot \Gamma^2_n) = \sum (-1)^{n(p+1)}\ trace\ (\alpha^p)^{-1}\,\mathbf{g}^p\,(\alpha^{n-p})'\,(\mathbf{f}^{n-p})'.$$

27. **Fixed point formulas.** By taking the identity I as one of our transformations we obtain the number of signed fixed points of a T in cases B or C.

Let Γ^0_n be the cycle corresponding to I. To have consistent definitions, if Γ_n corresponds to T, we shall define the *number of signed fixed points* as the index $(\Gamma_n \cdot \Gamma^0_n)$.

The identity unmodified can be considered as a T_{1b} or T_{2c}. Hence we have the possibility of obtaining the number of signed fixed points, or coincidences with I, for a T_{2b} or a T_{1c}, directly from our formulas.

By means of an additional device it is likewise possible to find the required number for T_{1b} or T_{2c}. It is based on the following observation in line with Remark II of No. 4: We are only interested in the case where T has a finite number of fixed points on $K-L$. Hence we may assume that there exists a neighborhood N of L, without fixed points of T. If we slightly modify I on N we shall add no new fixed points, and if we can thus turn I into a T_{2b} or T_{1c} we shall be able to apply our formulas once more.

Replacing K, if necessary, by a subdivision of sufficiently small mesh, still to be called K, we can assume that N is the K-neighborhood of L, and that it is normal.

Let Φ be the boundary of N. Since N is normal, through every point R of $N-L$, there passes a unique segment PQ, with P on L and Q on Φ. We define a single-valued transformation I^* of $K-L$, which is a T_{2b}, modification of I, as follows: On $K-N$ it coincides with the identity; any point R of $N-L$ is transformed into the point R' of the corresponding segment PQ such that $QR' = \frac{1}{2}QR$. The coincidences of this modified identity with T_{1b}, will yield the fixed points of the latter. For the fixed points of T_{2c}, we take I^{*-1} as the associated modification of the identity.

We verify at once that: (a) When T_{1b} or T_{1c} is I or I modified, every $\mathbf{f}^p = \mathbf{1}$. (b) When T_{2b} or T_{2c} is I or I modified, every $\mathbf{g}^p = \mathbf{1}$. Designating by Γ_n^0 the cycle corresponding to I or I modified, we have from (26) the number of signed fixed points:

For T_{1b} or T_{1c},

$$(28) \qquad (\Gamma_n^1 \cdot \Gamma_n^0) = \sum (-1)^p \text{ trace } \mathbf{f}^p.$$

For T_{2b} or T_{2c},

$$(29) \qquad (\Gamma_n^2 \cdot \Gamma_n^0) = \sum (-1)^p \text{ trace } \mathbf{g}^p.$$

We recall that the $\mathbf{g}$'s correspond to the transformations of the cycles of $K - L$, the $\mathbf{f}$'s to those of the cycles of K mod L.

APPLICATION. Let T be a deformation of $\overline{K - L}$ into a subset of $K - L$ (interior subset). Such a transformation is a T_{2b} whose effect on the absolute cycles is the same as the identity. Therefore, for the fixed points we have to apply (29) with $\mathbf{g}^p = \mathbf{1}$ for every p. But in that case the trace of $\mathbf{g}^p$ is equal to its order, which is $R_p(K - L) = R_{n-p}(K; L)$, (Ch. III, No. 34). Therefore this time

$$(30) \qquad (\Gamma_n^2 \cdot \Gamma_n^0) = (-1)^n \sum (-1)^p R_p(K; L).$$

This proves the

THEOREM. *The number of signed fixed points of a deformation of the orientable manifold $\overline{K - L}$ into an interior subset of itself, is equal to $(-1)^n$ times its Euler-Poincaré characteristic.*

28. **Absolute manifolds and manifolds with regular boundaries.** Our results include all those for fixed points obtained in our papers [8, 9] for these manifolds, as very special cases. When we have an absolute M_n, we may choose $a_i^p = b_i^p$; $a_i'^p = b_i'^p$, if the sets are not canonical, and then

$$(\alpha^p)' = (-1)^{p(n+1)} \alpha^{n-p}, \qquad (\beta^p)' = (-1)^{p(n+1)} \beta^{n-p}.$$

Hence (22) can be put in the form

$$(31) \quad (\Gamma_n^1 \cdot \Gamma_n^2) = \sum (-1)^p \text{ trace } \beta^p (\mathbf{g}^{n-p})' (\alpha^p)^{-1} \mathbf{f}^p$$

which, except for the notations, is the same as (39.1) of [9]. The coincidence formula (10.4) loc. cit., for the transformations of M_n into itself, is obtained by taking $\beta^p = \alpha^p$. As all the coincidence formulas for an absolute M_n, are reducible to these two, the identification is complete for the absolute M_n.

For an M_n with a regular boundary L, the identification of (35.3) of [9], the basic coincidence formula, with our (22) offers little more difficulty and need not be taken up here. The fixed point formula (36.2) of [9] is actually a special case of the present fixed point formula (29). It is true that the matrices which figure in (36.2) of [9] are the transformation matrices for the absolute cycles of K itself, whereas in (29) there appear only the same matrices for the absolute cycles of $K-L$. It follows however from Ch. III, No. 44, that when the boundary L is regular, every cycle of K is homologous to a cycle of $K-L$. Therefore, when the boundary L is regular, (29) is identical with the formula of our paper.

We have given as an expression for the number of signed fixed points of a deformation (loc. cit., formula (37.1)), the Euler-Poincaré characteristic of K itself, whereas above, we merely assert that it is $(-1)^n$ times the Euler-Poincaré characteristic of $K-L$ alone. However, when L is regular, we have shown in Ch. III, No. 45, that the characteristics of K and $K-L$ are equal. The extraneous factor $(-1)^n$ is due to a reversal in the order of the cycles Γ_n, Γ_n^0. Thus the discrepancy is only apparent.

29. **Cells and spheres.** Let T be a single-valued transformation of a cell E_n into itself, the cell being assumed to be a manifold with regular boundary. The formula to be applied is (29). Since the only absolute cycle $\not\sim 0$ is a point, $g^0 = 1$, $g^p = 0$ for $p > 0$. Hence

$$(\Gamma_n \cdot \Gamma_n^0) = g^0 = 1.$$

For a single-valued transformation T of an H_n which is an M_n, the situation is the same, save that now a point and H_n itself are $\not\sim 0$, hence the degree g^n also comes in; we still have $g^0 = 1$. Hence

$$(\Gamma_n \cdot \Gamma_n^0) = 1 + (-1)^n g^n.$$

We have therefore

THEOREM I. *The number of signed fixed points of a single-valued continuous transformation of a closed cell into itself, is one; for the n-sphere it is* $1 + (-1)^n g^n$, *where* g^n *is the degree of the transformation.*

THEOREM II. *The following single-valued continuous transformations leave at least one point invariant: (a) Transformation of* $\bar{E}_n$ *into itself. (b) Sense preserving (reversing) transformation of* H_{2n} (H_{2n+1}) *into itself. (c) Transformation of* H_n *into itself whose degree is* $\neq 1$ *in absolute value.*

30. The preceding results were given for ordinary (non-combinatorial) cells and spheres by Brouwer [7]. He also derived from them analogous propositions for projective spaces. A good exposition of his work along that line has been given by Feigl [2].

Brouwer [7] has shown that the fixed point theorem for $\bar{E}_n$ follows from Th. II c for the sphere. His reasoning is as follows: Take another copy $\bar{E}_n'$ of $\bar{E}_n$ and bring into coincidence the points of the boundaries that are the images of one another. The sum of the two closed cells is then an H_n which is an M_n (Ch. II, No. 49; Ch. III, No. 35). Let T be a continuous single-valued transformation of $\bar{E}_n$ into itself. We associate with it a continuous single-valued transformation T' of H_n, such that if A is a point of $\bar{E}_n$ and A' its image on $\bar{E}_n'$, $T' \cdot A' = T' \cdot A = T \cdot A$. If g^n is the degree of T', we must have $T' \cdot H_n \approx g^n H_n$. But $T' \cdot H_n \subset \bar{E}_n$, and hence ~ 0 mod H_n. Therefore $g^n = 0$, hence T' leaves at least one point of H_n invariant. But $T' \cdot \bar{E}_n' \subset \bar{E}_n$. Hence only points of $\bar{E}_n$ can be left invariant, and as T' coincides with T over $\bar{E}_n$, T has a fixed point on $\bar{E}_n$.

The preceding method is the one which we used in our paper [9] to reduce the coincidence and fixed point problem for an M_n with regular boundary to the same problem for an absolute M_n.

31. **Vector distributions.** Suppose that we have given a distribution $\mathfrak{D}$ of directed segments or vectors over $K - L$,

continuous and single-valued functions of their initial points. We can associate with $\mathfrak{D}$ a small deformation, such that if V is the vector applied at P, the point P undergoes a translation εV, where ε is a continuous function of P, which remains very small. If the vectors along L all point towards $K—L$, or if the reverse holds, we agree to choose ε such that εV points towards $K—L$ along L. Finally, where the vector is indeterminate, we take $\varepsilon = 0$, but it is not to be zero elsewhere. The deformation thus defined, has for fixed points the points of indetermination of the vectors. Therefore, the problem of finding the points of indetermination of a vector distribution on a manifold, falls under the category just treated. We can treat it by our method only when the vectors all point towards $K—L$, or away from it, along L. The deformation is then a T_{2c}. Therefore when the vectors along L point alike, the number of signed indeterminate positions of the vector distribution is equal to $\pm$ the Euler-Poincaré characteristic of $K—L$. When L is a regular boundary for the manifold M, the number in question is equal to $\pm$ the Euler-Poincaré characteristic of M.

Suppose that we have two vector distributions $\mathfrak{D}$, $\mathfrak{D}'$ such that: (a) The vectors of both $\mathfrak{D}$ and $\mathfrak{D}'$ point inward, or else outward, along L. (b) They are never collinear along L. (c) Their positions of indetermination are distinct. Then we can ask for the positions on $K—L$ where their vectors are parallel. To answer the question we choose the corresponding multipliers, such as ε above, equal at all points except those of small neighborhoods of the positions of indetermination and near L. Then the required positions are the coincidences of the associated deformations. Let us designate the deformations likewise by $\mathfrak{D}$, $\mathfrak{D}'$. Now, keeping $\mathfrak{D}$ fixed, we may modify $\mathfrak{D}'$ near L (taking its ε equal to zero along L) so as to reduce its representative cycle to the type Γ_n^{1b}. For evident reasons of continuity, the modified deformation $\mathfrak{D}''$ belongs to the homology class of the identity considered also as a T_{1b}. Therefore the number of signed coincidences of $\mathfrak{D}$, $\mathfrak{D}''$ is also given by (30). But the reduction of $\mathfrak{D}'$

to $\mathfrak{D}''$ can be accomplished without modifying the coincidences. Hence the number of signed positions of indetermination of the two given vector distributions is again $(-1)^n$ times the Euler-Poincaré characteristic.

32. In combination with the degree, systems of vectors have been found very useful in defining an index for the fixed points of a transformation. Let $K-L$ have the property that each point has for neighborhood an ordinary n-cell, and consider a single-valued transformation T of the manifold $K-L$ into itself, having an isolated fixed point P on $K-L$. We will take as a picture of an n-cell neighborhood on P, a region $\mathfrak{R}$ of S_n, and assume $P \subset \mathfrak{R}$. Let us then choose a very small (Euclidean) sphere H_{n-1} surrounding P. If Q is any point of H_{n-1}, its transform by T will have an image Q' $(\neq Q)$ on S_n. The ray from P parallel to the directed segment QQ', meets H_{n-1} at a point Q'', and the correspondence between Q and Q'' is a continuous single-valued transformation U of H_{n-1} into itself. We orient the E_n bounded by H_{n-1} as an indicatrix of $\mathfrak{R}$, hence of the initial manifold, then H_{n-1} in such manner that $E_n \rightarrow +H_{n-1}$. Under the circumstances, U has a uniquely defined degree τ, describing the algebraic number of times that Q'' covers H_{n-1} when Q covers it once. It is not difficult to show that τ is a topological character of T (Brouwer [7]). The number τ, introduced by Poincaré [1], is called the *index* of the fixed point P.

Suppose that T has a finite number of isolated fixed points $P^1, \cdots, P^r$ on $K-L$. Then we have an index τ_i for P^i, and we can calculate the sum

$$\zeta = \sum \tau_i.$$

If the fixed points are not isolated, we modify T slightly into a simplicial approximation T' whose ζ is then considered as the ζ of T itself. It turns out that *ζ is an invariant of the homotopy class of T,* (Brouwer [7]).

It is not difficult to identify ζ with the number of signed fixed points. From its mode of definition follows almost at once that we can assume that the cycles Γ_n, Γ_n^0, respective images of T and of the identity, intersect at isolated points on n-cells of the cycles and on $2n$-cells of $K \times K'$, each point counting only for ± 1 in the Kronecker index $(\Gamma_n \cdot \Gamma_n^0)$. But then if P is a fixed point, T behaves in the neighborhood of P like an affine transformation. If T is a repulsion from P, the transformation of H_{n-1} is the identity and $\tau = +1$. If T is an attraction towards P, each point of H_{n-1} is transformed into its antipode, hence $UH_{n-1} = (-1)^n H_{n-1}$ and $\tau = (-1)^n$. Taking coördinate systems, after the manner considered in more detail in Ch. VIII, § 2, or else as in our paper [8], No. 60, we find, by the simplest possible calculation, that the respective contributions to the Kronecker index, are respectively $(-1)^n$ and $+1$. Therefore

$$(-1)^n \zeta = (\Gamma_n \cdot \Gamma_{n,}^0)$$

or more accurately: *When the fixed points are isolated, the contribution of each to* $(\Gamma_n \cdot \Gamma_n^0)$ *is equal to* $(-1)^n$ *times its index.*

Suppose that we have a continuous vector distribution $\mathfrak{D}$ over $K-L$, and also a local coördinate system about any point P. Then if P is any point of indetermination, we can draw as above an H_{n-1} with center P. If Q is any point of H_{n-1}, we draw from P the ray parallel to the vector of $\mathfrak{D}$ through Q. If it meets the sphere at the point R, the transformation from Q to R is again a continuous single-valued transformation of H_{n-1} into itself. Its degree τ is called the *index* of the indetermination point P. We can again calculate the sum ζ of the indices as above. When the vectors along L all point inwards, ζ is $(-1)^n$ times the number of signed fixed points of the deformation associated with $\mathfrak{D}$, and by No. 28 $\zeta =$ the Euler-Poincaré characteristic of $K-L$. When the vectors point outwards along L, the distribution obtained by reversing them all, is of the preceding type, and has its ζ as just stated. But the reversal of vectors changes τ into $-\tau$, hence ζ into $-\zeta$, and therefore, in this instance, the sum of indices is minus the Euler-Poincaré characteristic. The indetermination in sign of No. 31 is thus removed.

Let T be a transformation into itself of an n-manifold $K—L$, which is a subset of S_n. If Q is the transform of any point P, the directed segments PQ determine a vector distribution whose ζ is precisely equal to $(—1)^n$ times the number of signed fixed points. With that particular type of manifold, the coincidence problem comes down to the same question of vectors. For if $Q = T\cdot P$, $Q' = T'\cdot P$, we have the field of all vectors QQ', and their indeterminations correspond to the coincidences of T and T'.

For further information concerning the properties of vector distributions, and their relation to the fixed point problem, see the summary by Feigl [2], where quite complete references will also be found, notably to some recent work by Hopf. See also Ch. VIII, No. 15.

§ 6. Special Properties of Single-Valued Transformations

33. In a recent note [8] Hopf has obtained a noteworthy relation between the matrices ε or η of a cycle Γ_n, image of a single-valued transformation T, from which certain highly interesting results follow.

Without committing ourselves for the present, let us observe that T is a T_{1b}, T_{2b} or T_{2c}. In any case the degree δ of T is the coefficient of $A \times K'$ in Γ_n, or the number of times that $T\cdot K$ covers K'. Let T' be a polyhedral approximation transforming simplexes of K or of a suitable subdivision (called K in any case) into simplexes of K'. When T is single-valued so is T', but for the present single-valuedness need not be assumed. $\delta = r—m$, where r, m are as at the end of No. 22. Any cell $\pm E_n'$ of $K'—L'$ is the projection of $r+m$ simplexes $\mathfrak{E}_n^2$ of Γ_n', the cycle representing T', of which, say the first r, are projected on K into $+E'$ and the others into $—E'$. It follows then, essentially from § 4, that if σ_p' is any simplex on E_n' it is the transform of $r+m$ simplexes σ^j on $K—L$, each on one of the cells transformed into E_n', and that among the σ^j's, r are transformed into $\xi\sigma_p'$ and m into $—\xi\sigma_p'$, where $\xi = \pm 1$ and depends solely upon

n and p. As we shall see, its value is determined without difficulty.

Let now K, K' be absolute manifolds and T single-valued. Then T' is likewise single-valued. Owing to the single-valuedness, σ'_p is the sole transform of all simplexes σ^j_p. Hence

$$T' T'^{-1} \cdot \sigma'_p = \xi \delta \sigma'_p,$$
$$T' T'^{-1} \cdot C'_p = \xi \delta C'_p.$$

When we pass to the cycles we may replace T' by T provided that we substitute $\approx$ for $=$. Hence

$$TT^{-1} \cdot \Gamma'_p \approx \xi \delta \Gamma'_p \text{ on } K'. \tag{32}$$

Let then

$$T^{-1} \cdot a'^i_p \approx \sum g^{*p}_{ij} a^j_p.$$

The sets a, b may be taken coincident, and similarly for a', b', so that we can write

$$\Gamma_n \approx \sum \varepsilon^p_{ij} a^i_p \times a'^j_{n-p}.$$

If we set

$$T \cdot a^i_p \approx \sum g^p_{ij} a'^j_p; \quad T^{-1} \cdot a'^i_p \approx \sum g^{*p}_{ij} a^j_p;$$
$$\alpha^p_{ij} = (a^i_p \cdot a^j_{n-p}); \qquad \beta^p_{ij} = (a'^i_p \cdot a'^j_{n-p}),$$

we find as in No. 21, and partly from (19) or (21):

$$\begin{cases} \mathbf{g}^p = (\alpha^{n-p})' \varepsilon^{n-p}, \\ \mathbf{g}^{*p} = (-1)^{n-p} (\beta^{n-p})' (\varepsilon^p)'. \end{cases} \tag{33}$$

Hence, from (32),

$$\mathbf{g}^{*p} \cdot \mathbf{g}^p = (-1)^{n-p} (\beta^{n-p})' (\varepsilon^p)' (\alpha^{n-p})' \varepsilon^{n-p} = \xi \delta, \tag{34}$$

where the numerical factor $\xi \delta$ at the right stands for its product by the unit matrix of the proper order. A similar remark holds for the matrix relations below.

The number ξ depends solely upon n and p. To determine it take $K = K'$, the bases on K' images of those on K, and for T the identical transformation. Then $\delta = 1$, $\alpha^p = \beta^p$; $\mathbf{g}^p = \mathbf{1}$. Hence, from (33),

$$(\alpha^{n-p})' \varepsilon^{n-p} = \mathbf{1}; \quad (\alpha^p)' \varepsilon^p = \mathbf{1}.$$

Under the circumstances, also

$$(\beta^{n-p})' = (-1)^{p(n+1)} \beta^p = (-1)^{p(n+1)} \alpha^p.$$

Substituting in (34) and passing to the determinant, we have, in view of $|\alpha^p| = \pm 1$, finally $\xi = (-1)^{n(p+1)}$. From this, after an obvious and elementary modification, we have the required relation

$$(35) \qquad (\varepsilon^p)' \alpha^p \varepsilon^{n-p} \beta^p = (-1)^{p(n+1)} \delta.$$

34. From the relation just obtained we can, with Hopf, deduce a number of interesting consequences, of which we list a few:

I. When there exists a single-valued transformation of K into K' whose degree $\delta \neq 0$, the rank of ε^p is equal to $R_p(K')$ for every p.

II. If T represents a transformation of K into itself the matrices ε are square, and their determinants satisfy

$$(36) \qquad |\varepsilon^p| \cdot |\varepsilon^{n-p}| = \delta^{R_p(K)}.$$

III. An H_n cannot have for single-valued transform by a T whose degree $\neq 0$, an absolute M_n with an $R_p \neq 0$, $p \neq 0, n$.

IV. The transformation T multiplies every Kronecker index by the degree δ. Hence when $\delta \neq 0$, intersecting cycles are transformed into intersecting cycles.

The proofs of I, II are immediate. Regarding III, the right hand side of (35), in this case, is $R_p(M_n)$, that of the left hand side zero, since $\alpha^p = 0$. Therefore $R_p(M_n) = 0$. As for IV, by a simple calculation we find:

$$\begin{aligned} \|(T a_p^i \cdot T a_{n-p}^j)\| &= g^p \beta^p (g^{n-p})' \\ &= (-1)^{(n+1)p} \alpha^p \varepsilon^{n-p} \beta^p (\varepsilon^p)' \alpha^p \\ &= \delta \alpha^p. \end{aligned}$$

Hence

$$(T a_p^i \cdot T a_{n-p}^j) = \delta (a_p^i \cdot a_{n-p}^j).$$

Then by the distributive law for indices, IV follows.

§ 7. Extension to Arbitrary Complexes

35. A most curious result was obtained recently by Hopf [7, 11]. Let T be a single-valued transformation of an arbitrary simplicial n-complex K into itself whose fixed points are on the n-cells of K. Its representation on $K \times K'$, where as previously, K' is a copy of K, is this time by a complex homeomorphic to K. Let Δ_n be the chain sum of the n-cells of that complex each oriented by means of the image of the indicatrix of the corresponding cell of K. Let Δ_n^0 correspond in similar fashion to the identical transformation I. The fixed points of T correspond to the intersections of Δ_n and Δ_n^0. If $P \times P'$ is any such point, and P is on E_n^i of K, then P' is on the image $E_n'^i$ of E_n^i on K'. If we reorient E^i we have to reorient E'^i, and hence we do not change the orientation of their product. On the other hand, this reversal of orientation is accompanied by a reversal of orientation for both Δ_n and Δ_n^0. It follows that it does not change the contribution of $P \times P'$ to $(\Delta_n \cdot \Delta_n^0)$. Hence this index has a value independent of the orientation of the cells of K.

Now if the effect of T on the cycles of K is still represented by (20), Hopf shows that

$$(\Delta_n \cdot \Delta_n^0) = \sum (-1)^p \, trace \, \mathfrak{g}^p.$$

In other words, (29) is applicable in the present instance, *as if K were an absolute manifold.* This appears, at first sight, as a most peculiar result, when we remember how essential a role the manifold conditions have played in deriving our coincidence and fixed point formulas. Hopf informed me some time ago, that the above result seems to stand by itself, and to hold only for single-valued transformations, and seems to break down for the more general type. This would explain why his proof is so very special, and seemingly not susceptible of being extended to other transformations.

We propose to attack the question from a wholly different angle from Hopf's. We shall show, following our recent papers [17, 18], that, properly interpreted, all the results of the present chapter can be extended to arbitrary complexes and indeed to arbitrary closed sets. At the present time we confine our attention to complexes; the complete extension will be taken up in the next chapter.

The treatment is based on a new class of cycles, called *pseudocycles*, associated with neighborhoods of the complex or set considered. Between cycles and pseudocycles there exist duality relations entirely analogous to those for manifolds. By means of them, the scheme of things leading to our formulas can be extended to complexes in the case of pairs of transformations of type B or C. However, the extension itself is based on the consideration of pairs of transformations of the general type A associated with certain auxiliary manifolds. Hence the type A plays here an essential, if indirect, part.

36. **The pseudocycles of a complex.** Let K be a given normal n-complex. We can always find an absolute simplicial orientable manifold $\mathfrak{K}_r$, having a subcomplex K', whose structure is the same as that of a simplicial subdivision of K. We may and shall impose upon $\mathfrak{K}$ the following additional conditions: (a) $r-n$ is even; (b) the $\mathfrak{K}$-neighborhood of K' is normal.

We can always find an S_r, such that $r-n$ is even, and that it carries a polyhedral image Π_n of a subdivision of K. Close S_r at infinity by a point, thus obtaining an H_r. The sphere can be decomposed into simplexes, and then by Ch. I, No. 61, Lemma I, there exists a subdivision $\mathfrak{K}_r$, of the complex covering H_r, having a subdivision K' of Π_n for subcomplex. Since $\mathfrak{K}$ can be replaced by its first derived, we may also assume that the $\mathfrak{K}$-neighborhood of K' is normal. Since $\mathfrak{K}$ is an absolute manifold, we have shown that a $\mathfrak{K}$ such as postulated at the beginning can be found for every K. Thus its existence implies no topological restriction on K. In particular, the $\mathfrak{K}$-neighborhood, N, of K, will be normal.

Since after all we are only concerned with topological properties of K, we may well replace it by K', or what is the same, assume that K is a subdivision of $\mathfrak{K}$ having the properties of K'.

REMARK. There is a good deal of arbitrariness as regards r. We can always replace S_r by any $S_{r+2k} \supset S_r$, and hence we may take for r an arbitrary number of its parity above a certain value. The condition that $r-n$ be even, is on the whole, essentially a convenience. It will result in an automatic extension of all formulas involving Kronecker indices and intersections of cycles.

37. Let C_{r-p} be any $(r-p)$-chain on $\mathfrak{K}_r$, and let us examine under what condition there can be found a chain C'_{r-p} not meeting K, such that

$$\lambda C_{r-p} \approx C'_{r-p} \text{ on } \mathfrak{K}; \qquad \lambda \neq 0. \tag{37}$$

If C'_{r-p} exists, we can find a subdivision $\mathfrak{K}'$ of $\mathfrak{K}$, such that if N' is the $\mathfrak{K}'$-neighborhood of K, $C'_{r-p} \subset \mathfrak{K}' - N' = \mathfrak{L}'$. Then we shall have, as a consequence of (37),

$$C_{r-p} \approx 0 \bmod \mathfrak{L}' \text{ on } \mathfrak{K}. \tag{38}$$

Therefore the boundary of C_{r-p} is on $\mathfrak{L}'$, and hence it does not meet K.

Let us apply now Th. I of Ch. IV, No. 17, with $K = \mathfrak{K}$, $L^1 = \mathfrak{L}'$, $L^2 = 0$. As a consequence, if Γ_p is an arbitrary absolute p-cycle of K, hence a cycle of $\mathfrak{K} - \mathfrak{L}'$, we shall have

$$(\Gamma_p \cdot C_{r-p}) = 0. \tag{39}$$

Conversely, if $F(C_{r-p})$ does not meet K, and (39) holds whatever the absolute cycle Γ_p of K, then C_{r-p} behaves as required. For we can choose $\mathfrak{L}'$ as above, such that $F(C_{r-p}) \subset \mathfrak{L}'$. Then C_{r-p} is a cycle of $\mathfrak{K}$ mod $\mathfrak{L}'$, and by the theorem already quoted,

$$C_{r-p} \approx 0 \bmod \mathfrak{L}' \text{ on } \mathfrak{K}.$$

Hence there exists a

$$C_{r-p+1} \to \lambda C_{r-p} - C'_{r-p}; \quad \lambda \neq 0, \quad C'_{r-p} \subset \mathfrak{L}'.$$

Since C'_{r-p} does not have any points on N', it does not meet K, and therefore (37) holds with C'_{r-p} as in that relation.

38. Let now $\{a^i_p\}$, $i = 1, 2, \cdots, R_p(K)$, be a minimum base for the p-cycles of K and $\approx$. Since N is normal the cycles a^i_p are independent on N. For if a linear combination of them were the boundary of $C_{p+1} \subset N$, we could deform C_{p+1} along the projecting segments of N onto K, without displacing the a's. Then the latter would have a linear combination ≈ 0 on K, which is ruled out. Given arbitrary integers t_i it follows then from Ch. IV, Nos. 17, 39, that we can find a chain B_{r-p}, which is a cycle of $\mathfrak{R}$ mod $\mathfrak{L}$ ($= \mathfrak{R} - N$), such that

$$(a^i_p \cdot B_{r-p}) = t_i, \qquad i = 1, 2, \cdots, R_p.$$

As above, $F(B_{r-p}) \subset \mathfrak{L}$, and hence it does not meet K. If B'_{r-p} is any chain whose boundary does not meet K, and such that

$$(a^i_p \cdot B'_{r-p}) = t_i$$

for every i, we shall have

$$(a^i_p \cdot B_{r-p} - B'_{r-p}) = 0.$$

But if Γ_p is any p-cycle of K, we have

$$\Gamma_p \approx \sum x_i a^i_p \text{ on } K, \text{ hence on } N.$$

Therefore,

$$(\Gamma_p \cdot B_{r-p} - B'_{r-p}) = \sum x_i (a^i_p \cdot B_{r-p} - B'_{r-p}) = 0$$

and hence, again by Ch. IV, No. 17, Th. I, $B_{r-p} - B'_{r-p}$ is $\approx$ on $\mathfrak{R}$, to a chain which does not meet K.

Conclusion. The integers t_i individualize, on every $\mathfrak{R}$, a unique family $\{B_{r-p}\}$ of chains whose boundaries do not meet K, and such that the difference of any two is $\approx$ on $\mathfrak{R}$ to a chain which does not meet K. The totality of all such families, for all choices of $\mathfrak{R}$'s, will be called an $(n-p)$-*pseudocycle* of K, and designated by b_{n-p}. A pseudocycle is uniquely determined by a set of integers t_i.

The chains B_{r-p}, upon which the existence of the pseudocycles rests, are after all merely the cycles of $\mathfrak{K}$ mod $\mathfrak{K}-K$. The situation is then the reverse of what it has been throughout this book so far. The analogue of our former L, which is the present K, is here replaced by its complement, and we have the cycles of the closed set mod the complementary open set. This is quite in line with the definition of Ch. II, § 1 or of our paper [14]. Incidentally the "cycles of a point" that Van Kampen [3] used in proving the invariance of manifolds are what the B's become when $\mathfrak{K}$ is any complex, and K a point of it.

It must be observed that it is only by taking the class of all B's for all $\mathfrak{K}$'s, that we obtain pseudocycles bearing a topologically invariant relation to the given complex K.

39. The pseudocycles just introduced may be considered as *absolute* pseudocycles. There is no difficulty in introducing the relative type. Let L be a closed subcomplex of K, and L^1 a closed subcomplex of L, with $L^2 = L - L^1$. Take for the a's the cycles on $K - L^1$ mod L^2. Then we have the pseudocycles of $K - L^2$ mod L^1, by taking for the B's cycles of $\mathfrak{K} - L^2$ mod $\mathfrak{L} + L^1$, and similarly when the indices 1, 2 are interchanged. If $L^2 = 0$, the pseudocycles of $K - L^1 = K - L$ are *absolute;* otherwise they are *relative.* Similarly with L^1 and L^2 interchanged.

Another, and our last extension of the pseudocycle, is associated with a product $K \times K'$. We designate all elements associated with K' by accents: $\mathfrak{K}'$, n', $\cdots$. Then the families of products $a_p \times B'_q$, $B_p \times a'_q$, define pseudocycles to be designated by $a_p \times b'_{q+n'-r'}$, $b_{p+n-r} \times a'_q$. It is always understood that the differences $n-r$, $n'-r'$ are even.

40. We are now in position to define homologies $\approx$, intersections, products, for combinations of cycles and pseudocycles. The meanings of the various symbols are always referred to the B's. Owing to the fact that $r-n$, $r'-n'$ are even, the behavior of the intersection or product symbols under permutation is as for manifolds. The following examples will serve to illustrate the situation.

I. By definition:

$$\begin{aligned}
(a_p^i \cdot b_{n-p}) &= (a_p^i \cdot B_{r-p}) = t_i,\\
(a_p \cdot b_{n-p}) &= (a_p \cdot B_{r-p}) = \sum x_i t_i; \qquad a_p = \sum x_i a_p^i,\\
(b_{n-p} \cdot a_p) &= (B_{r-p} \cdot a_p) = (-1)^{p(r+1)} (a_p \cdot B_{r-p})\\
&= (-1)^{p(n+1)} (a_p \cdot b_{n-p}).
\end{aligned}$$

II. Let K' and $\mathfrak{K}'$ be as in No. 39. The difference between the dimensions of $\mathfrak{K} \times \mathfrak{K}'$ and $K \times K'$ is $r + r' - n - n'$; here also > 0 and even. The product family $B_{r-p} \times B'_{r'-p'}$ is uniquely defined (up to zero divisors) on each $\mathfrak{K} \times \mathfrak{K}'$, and it determines an $(n+n'-p-p')$-pseudocycle of $K \times K'$ designated by $b_{n-p} \times b'_{n'-p'}$. Writing more simply q, q' for $n-p$, $n'-p'$, we verify by reference to the B's, that

$$a_p \times a'_{p'} \cdot b_q \times b'_{q'} \approx (-1)^{(n-p)(n'-q')} a_p \cdot b_q \times a'_{p'} \cdot b'_{q'},$$

which is essentially the extension of (22) of Ch. V.

41. The only homologies that are to be considered are $\approx$ and they suffice for all interesting applications.

To the pseudocycles correspond Betti numbers which we designate by ϱ in place of R, as for instance $\varrho_p(K; L)$. We have, then, by means of the indices, as in Ch. IV, § 2, also directly from the duality relations of Ch. III by proper choice of the sets L^i:

Duality Theorem for the Pseudocycles: *The duality relations are those obtained for the Betti numbers of manifolds, (Formulas (3), (7), (9) of Ch. III) when on any one side of them R is replaced by ϱ.*

42. **Application to transformations of arbitrary complexes.** Let the situation be as previously in this chapter—except that all manifold restrictions are dropped. We take then two distinct transformations of a subset of K into a subset of K', and examine their coincidences. We can assume K, K' to be subcomplexes of $\mathfrak{K}$, $\mathfrak{K}'$, where the latter are two complexes whose dimension may be taken to be the same, say r (No. 36, Remark).

We first introduce the *extension* $\mathfrak{T}$ of T. It is a transformation of a subset of $\mathfrak{R}$ into a subset of $\mathfrak{R}'$ which acts like T as between K and K'. A *contraction* is a $\mathfrak{T}$ associated with T as follows: Let PQ be one of the projecting segments of the normal neighborhood N of K, Q its end point on K. Then $\mathfrak{T} \cdot PQ = T \cdot Q$. In short, the transform by $\mathfrak{T}$ of any projecting segment is the same as the transform by T of its end point Q on K. An *expansion* is similarly related to T^{-1}. We now make the following restrictive assumption, analogous to those of No. 4 for manifolds:

ASSUMPTION: *Any transformation T to be considered possesses an extension $\mathfrak{T}$ with a unique representative r-chain except for the possible ambiguity in sign occurring in No. 20. The r-chain determines an n-pseudocycle representative of T, of type $a \times b'$ or $b \times a'$ (No. 39), which is again unique except for a sign ambiguity associated with that of the r-chain.*

We will say that two pseudocycles are *complementary* whenever they are of respective types $a \times b'$, $b \times a'$, where in each pair (a, b), (a', b'), the a's represent cycles of any particular type and the b's the associated type of pseudocycles, or *vice versa*. Here we no longer insist upon the a's being strictly cycles and the b's pseudocycles.

If we have a pair T_{1c}, T_{2b}, of transformations of L into L' whose extensions are an expansion $\mathfrak{T}_1$ and a contraction $\mathfrak{T}_2$, these extensions will be represented by complementary pseudocycles, Γ_n^1, Γ_n^2 formally given by homologies (1). They may be considered as a pair T_{1b}, T_{2b}, as between N, N', with $L + \Phi$, $L' + \Phi'$ in place of the former L's. The coincidences being the same as those of the given transformations T_{1c}, T_{2b}, the derivation of (22) can proceed exactly as before, yielding here also an invariant number which may be considered as the index of the coincidences. It is understood in each case, that the $\mathbf{f}$'s describe the effect of T_{1c} on the b's, and the $\mathbf{g}$'s that of T_{2b} on the a's. Therefore

THEOREM. *The fundamental coincidence formula (22) holds also for complexes under the interpretation just given for the various terms entering in it.*

43. REMARKS. I. When we have orientable manifolds everything reduces back to the earlier situation. For the B's now intersect K in cycles of same dimension as the pseudocycles. If b, b^* are a pseudocycle and the associated intersection cycle $K \cdot B$, the numbers $(a \cdot b)$, $(a \cdot b^*)$ are equal, from which follows readily that we can substitute everywhere b^* for b, i. e. go back to (22) in the strict sense.

II. The extensions can manifestly be replaced by any pair of transformations $\mathfrak{S}$, $\mathfrak{S}'$, of $\mathfrak{K}$ into $\mathfrak{K}'$, which have the following properties: (a) They transform K like T_1, T_2. (b) They transform $\bar{N}$ into a subset of N'. (c) Their coincidences in the vicinity of K, K', are those of T_1, T_2. We shall consider an example in the next number.

44. **Fixed point formulas.** Let L be a subcomplex of K, and T a transformation acting on $K-L$ as a T_b or T_c, which satisfies our basic assumption. Depending on circumstances, we have to have the identity or the identity modified, as the other transformation. As an example, suppose that T is a T_{2b}. Let its extension $\mathfrak{T}$ be a contraction and set $\mathfrak{L} = L + (\bar{N} - N)$. $\mathfrak{T}$ transforms every point of $\bar{N}$ into one of $K - L \subset \bar{N} - \mathfrak{L}$. Therefore it is likewise a T_{2b} as acting on $(\bar{N}, \mathfrak{L})$ in place of our customary pair (K, L). The assumption of No. 42 means, in this case, that $\mathfrak{T}$ has a representation

$$\Gamma_r \approx \sum \eta_{ij}^p \, B_{p+r-n}^i \times a_{n-p}'^j,$$

where a_{n-p}^j is an absolute cycle of $K-L$, and $a_{n-p}'^j$ its image on the copy K' of K. In an evident sense, then, T_{2b} is represented by the pseudocycle

$$\Gamma_n \approx \sum \eta_{ij}^p \, b_p^i \times a_{n-p}'^j.$$

As to the identity, we take for its extension the identical transformation for $\mathfrak{K}$. It is representable as

$$\Gamma_r^0 \approx \sum \varepsilon_{ij}^p \, a_p^i \times B_{r-p}'^j,$$

and determines the unique pseudocycle

$$\Gamma_n^0 \approx \sum \varepsilon_{ij}^p \, a_p^i \times b_{n-p}'^j.$$

The effect of the identity on the pseudocycles is represented by $\mathfrak{f}^p = 1$ for every p. Therefore (29) gives here also the number of signed fixed points, exactly as it stands, the $\mathfrak{g}$'s representing again the effect of T upon the absolute cycles of $K-L$. Similarly for the case T_{2c}. Regarding T_{1b} and T_{1c}, the application of our method requires that we define them as follows: A T_{1b}, or T_{1c}, for the pair (K, L), is a transformation which possesses some extension $\mathfrak{T}$, which is a T_{1b} or T_{1c} for the pair $(\bar{N}, L+\Phi)$. In this case we take for the extension of the identity, the transformation reducing every projecting segment on $\bar{N}$ to its end-point on K. Under these circumstances, (28) gives the number of signed fixed points, with the $\mathfrak{f}$'s representing, this time, the effect of T on the relative pseudocycles. We have, then,

THEOREM. *The number of signed fixed points of a transformation T_b or T_c, of a subset of $K-L$ into a subset of itself, is given by (28) or (29). It is understood that the $\mathfrak{f}$'s and $\mathfrak{g}$'s represent the action of T upon the cycles or pseudocycles, according as Γ_n is of type $\sum a\times b'$ or $\sum b\times a'$, with the first factor a pseudocycle or a cycle.*

45. REMARK. Suppose that all fixed points of T are on the n-cells of K. We can approximate T, as closely as we please, by a polyhedral transformation T' whose fixed points behave likewise. Let P be one of them, on E_n of K, and let Δ_n be the n-chain image of T' on $K\times K'$, and Δ_n^0 the analogue for the identity. Then P is an intersection of Δ_n and Δ_n^0, as well as of Γ_r and Γ_r^0. By an elementary consideration of indicatrices or else by the vector method of No. 32, very convenient here, it is then ascertained that the contributions of P to $(\Delta_n\cdot\Delta_n^0)$ and $(\Gamma_r\cdot\Gamma_r^0)$ are equal. Under the circumstances

$$(\Delta_n\cdot\Delta_n^0) = (\Gamma_r\cdot\Gamma_r^0) = (\Gamma_n\cdot\Gamma_n^0).$$

In other words *the number of signed fixed points is here again the actual Kronecker index of two n-chains on $K\times K'$.*

The argument goes through in the case of the fixed points, but not of coincidences, due to the reason brought out in

No. 35: If $P \subset E_n$ and E'_n is the image of E_n on K', the orientation of $E \times E'$ is independent of that of E.

46. **Single-valued transformations.** Let T be a contínous single-valued transformation which is a T_{2b} or T_{2c}. The contraction defined in No. 42 is then likewise continuous and single valued, its type being the same for $\bar{N}$, $\mathfrak{L}\,(= L + \mathfrak{K} - N)$, as that of T for K, L. Therefore in the representation of T, in the form $a \times b'$ or $b \times a'$, the first factor is a pseudocycle. Hence from the proposition of No. 44:

THEOREM. *For a single valued T_{2b} or T_{2c}, the formulas (29), (28) hold respectively in the strict sense, i. e. with the matrix $\boldsymbol{g}$ representing the action upon the cycles and not upon the pseudocycles.*

When $L = 0$ we have to use (29) in its strict form. *Coupling this result with the Remark of No. 45 this is precisely the result of Hopf, mentioned at the beginning of the section.*

APPLICATION. Let K be a projective plane. Then $R_0(K) = 1$, $R_1(K) = R_2(K) = 0$. Also $f^0 = 1$. Hence the number of signed fixed points $= 1$. Therefore *every continuous single-valued transformation of the projective plane into itself has at least one fixed point* — a well known result due to Brouwer [2].

CHAPTER VII

INFINITE COMPLEXES AND THEIR APPLICATIONS

If we wish to extend any combinatorial result obtained for finite complexes, for example the duality theorems, we usually face two difficulties: a) Passage from arbitrary singular chains on the finite complex K to subchains of K. b) Passage from a subcomplex L of K to an arbitrary closed subset L of K. The first is taken care of as a rule by a deformation theorem of some sort, the second by the intervention of an infinite complex. This becomes clear when we observe that the most natural way of correlating the topology of K and L is by means of a sequence of polyhedral neighborhoods of L each contained in its predecessor. The sum of the differences of pairs of consecutive neighborhoods is an infinite complex approximating $K-L$.

Infinite complexes arise also in another important connection. For Alexandroff has shown [10] that the topology of a compact metric space L is intimately connected with the properties of a certain infinite sequence of finite complexes, whose dimension is equal to the Urysohn-Menger dimension of L. Here also we have an infinite complex.

The preceding considerations justify the investigation of infinite complexes and manifolds for their own sake. This investigation constitutes the subject matter of §§ 1–3. The most elusive task proved to be the determination of the bases for the various types of cycles considered and the extension of the duality theorems of Ch. III for open manifolds (§ 3), the chief difficulties being caused by the infinite Betti numbers that may occur everywhere. Following this we have applied the general theory of the infinite complex to the homology theory of compact metric spaces, the separation of Euclidean spaces by their closed subsets: Alexander's duality relation and its consequences, (§§ 4, 5). Finally in § 6 we extend to all compact metric spaces the coincidence and fixed point formulas of Ch. VI, obtaining what we believe to be a basic and final theorem in that direction (No. 61). Our general coincidence formula is thus one of the most important and most general formulas in topology.

The only explicit contribution to the theory of the infinite complex in the literature are the investigations of Kerékjártó [10] on the infinite M_2 for which in particular he has given a classification similar in many respects to that of the finite M_2. This is an interesting question, but like the classification of the finite M_2, it stands largely by itself, and practically none of the questions taken up here were considered by Kerékjártó.

§ 1. General Properties of Infinite Complexes

1. The infinite complexes under investigation are n-complexes, with a finite number of cells incident with any given cell. They are defined as the finite types (Ch. I, Nos. 5, 51; Ch. II, Nos. 43, 50) except that the polyhedral image is now assumed on the parallelotope of Hilbert space

$$\mathfrak{H}:\quad 0 \leqq x_i \leqq \frac{1}{i}, \qquad i = 1, 2, \cdots, +\infty.$$

$\mathfrak{H}$ is a compact metric space, the distance function being the usual Euclidean distance function. We have thus a metric attached to K.

The set of cells of K will be assumed enumerable. Regarding the dimension n of K, it is important to bear in mind that $n = +\infty$ is not excluded; i. e. while the dimensions of the cells must be finite, they need not have an upper bound.

It will be sufficient to consider only normal complexes (Ch. II, No. 50), the boundary of a q-cell being a combinatorial $(q-1)$-sphere which is a finite $(q-1)$-subcomplex of K. The complex has incidence matrices infinite both ways but the number of terms $\neq 0$ in any row or column is finite.

Since K is normal, it possesses a simplicial first derived K' by means of which it is possible to define regular and semi-regular cells, and hence also manifolds.

Since we have infinite complexes, we will naturally have also infinite chains. Let $\{E_p^i\}$ be an infinite sequence of cells on K, singular or otherwise, but with the double property that every $\bar{E}_p^i$ meets at most a finite number of closed cells of K, every closed cell of K at most a finite number of cells $\bar{E}_p^i$. The set of all the points on the closed cells $\bar{E}_p^i$ associated with a symbol

$$C_p = \sum t_i E_p^i,$$

is a p-chain on K. It is infinite when the number of t's not zero is infinite. In the non-modular case the t's are allowed to be arbitrary rational numbers. It has the important consequence that *when C_p is infinite it may not have any multiple which is an integral chain* (chain with integral coefficients). This is decidedly at variance with the case of finite chains.

We shall consider only the non-modular case and rational chains. The treatment will apply directly to mod 2, and, with unimportant modifications, to mod m as well.

The boundary relations and homologies, the cycles and also the homology characters, are defined as in the finite case. However with rational chains allowed, we cannot define torsion which is essentially a diophantine property. Therefore in the domain of rational chains the sign $\approx$ must prevail throughout.

The definition of orientable or non-orientable circuits is again as before.

Where a subdivision of a complex or chain occurs, it is understood that each cell of the complex or chain must be subdivided into a finite number of cells. In short, locally the behavior on K must be as on a finite complex.

2. The first and simplest question before us, is the extension of the first deformation theorem (Ch. II, No. 15). The second deformation theorem (Ch. II, No. 45) is then proved as in the finite case.

We consider a K_p (singular or not) on a simplicial K having the same property as the infinite chain: every closed cell of K_p meets at most a finite number of closed cells of K and *vice versa*.

Let K_{n-1}, E_n^j, P^j, H^j be as in the proof for finite complexes. Denote by $\mathfrak{E}$ a generic cell of K_p. By modifying slightly the treatment loc. cit. we shall show that the theorem holds provided that the diameter of any $\bar{\mathfrak{E}}$ which meets $\bar{E}_n^j$, is less than a certain ζ_j, that may depend upon j. We may again proceed here by recurrence. We have then an aggregate of ζ's for K_{n-1}. Let ζ_j' denote the smallest of them for all $(n-1)$-faces of E_n^j. The theorem holds, by assumption, for any complex on K_{n-1} such that every $\bar{\mathfrak{E}}$ having a point on the boundary of E_n^j is of diameter $< \zeta_j'$, for every j. In order that the proof for the finite case be applicable, it is merely necessary that the following two conditions be fulfilled:

(a) Any $\bar{\mathfrak{E}}$ that meets K_{n-1} is exterior to the spheres H.

(b) The projection $\bar{\mathfrak{E}}'$ of such an $\bar{\mathfrak{E}}$ from the P's onto K_{n-1}, is of diameter less than the ζ' corresponding to any E_n which it meets.

In order that (a) be satisfied we merely need to take ζ_j less than the distance from $F(E_n^j)$ to the sphere H_j.

The diameter β of the projection from the P's onto K_{n-1} of any set of diameter α that meets $F(E_n^j)$ tends to zero with α. Hence there exists an α, say α_j, such that every corresponding $\beta < \zeta_j'$. If we take $\zeta_j < \alpha_j$, condition (b) will hold, and the theorem will be applicable.

When the cells of K_p do not possess properties (a), (b), it has a subdivision K_p' whose cells do. For any $\bar{\mathfrak{E}}$ of K_p meets at most a finite number of cells $\bar{E}_n^j$. The least of the corresponding numbers ζ_j is a certain δ, and we take for K_p' a subdivision such that its cells on $\bar{\mathfrak{E}}$ are of diameter $< \delta$, this for every $\bar{\mathfrak{E}}$ — a condition that can obviously be fulfilled.

The extension of the deformation theorems is therefore completed.

As in the finite case we have the following two important consequences.

(a) The homology characters determined by means of the subchains of K are the same as when all singular chains are allowed.

(b) The homology characters calculated in accordance with (a) (combinatorial characters), are topological invariants, and similarly for the circuit properties.

§ 2. Ideal Elements

3. The more interesting and difficult part of the theory of infinite complexes, especially in view of the applications, is found in the relations between a complex and its ideal elements. We must first introduce these elements.

An infinite K is *locally* compact: every point possesses a neighborhood N, here its K-neighborhood, such that $\bar{N}$ is compact. On the other hand K is not compact as a whole, for the set of its vertices has no limit point.

The most general category of ideal elements is to be obtained by sequences of neighborhoods, analogous to those which are used in turning the non-compact set of rational numbers into a compact set, through the addition of mutually inclusive neighborhoods. In topology such ideal elements have been considered explicitly by Kerékjártó [10] in his work on the infinite M_2. The *Primenden* introduced by Carathéodory [1] in his investigations on the boundaries of two-cells, are also very closely related to them.

By an *ideal element*, L, of K we understand then a sequence $\{N^i\}$, $i=1,2,\cdots,+\infty$, of sets open on K such that

$$\text{I.}\quad N^i \supset \bar{N}^{i+1}; \qquad \text{II.}\quad \prod N^i = 0.$$

If $L=\{N^i\}$, $L'=\{N'^i\}$ are two ideal elements we define $L \subset L'$ by the condition that for every j there is an i such that $N^i \subset N'^j$. We say that L' *includes* L. When both $L \subset L'$ and $L' \subset L$ we say that $L = L'$, and call the sequences *equivalent*.

The sequences $\{N^i\}$ have two simple properties, the first of which is obvious:

(a) Every sequence is equivalent to any one of its subsequences.

(b) $\{N^i\}$ is equivalent to an $\{N'^i\}$ such that every N' is an exact sum of cells of a fixed subdivision K' of K.

To prove (b), observe that in view of conditions I, II, the boundaries of at most a finite number of sets N^i meet a given cell $\bar{E}$ of K. The numbers $d(F(N^i)\cdot \bar{E},\ F(N^{i-1})\cdot \bar{E})$ corresponding to all pairs $F(N^{i-1})$, $F(N^i)$ meeting $\bar{E}$, have a positive lower bound δ for that cell. If there are no such pairs for a given E we take as δ the diameter of E. Subdivide K into K' such that the cells of K' on E are of diameter $<\delta$, and this for every cell of K. The set $\{N'^i\}$ of all cells of K' whose closures meet $\bar{N}^i$ is open. Moreover $\{N'^i\}$ satisfies I and the condition of equivalence with $\{N^i\}$ by construction, hence also II. Therefore it is a sequence defining the same ideal element, and of the desired nature.

Since we are free to replace K by any simplicial subdivision, whenever we shall deal with a finite number of ideal elements—and this is the situation throughout—we shall legitimately confine our attention to the neighborhoods that are exact sums of cells of K.

4. Let E^h, $h=0,1,2,\cdots,+\infty$, be the cells of K and consider the sequence of closed finite subcomplexes, K^i, of K formed

as follows: $K^1 = \bar{E}^1$; $K^{i+1} =$ the closure of the K-neighborhood of $K^i + \bar{E}^i$. If we set now $\mathfrak{N}^i = K - K^i$, it is a simple matter to verify that $\Lambda = \{\mathfrak{N}^i\}$ is an ideal element. Therefore *every infinite complex possesses ideal elements.* More important however, is the fact that Λ *includes every other ideal element.* For if $L = \{N^i\}$, since $K - \mathfrak{N}^j$ is a closed subcomplex of K, there must be an N^i without points on it, and hence $\subset \mathfrak{N}^j$, so that $L \subset \Lambda$. We will call Λ *total ideal element* or *set.*

We will now consider the former ideal elements such as L, as *closed* elements, and *define* the open elements in terms of Λ. Essentially they are the complements of the closed elements, but we must describe them in terms of neighborhoods. When $L = \{N^i\}$ is closed and $L \neq \Lambda$ (i. e. there is at least one element $\not\subset L$) there exists an N^i such that $K - N^i$ is an infinite subcomplex of K. For otherwise, by the very construction of $\{\mathfrak{N}^i\}$, there would exist some $K^j = K - \mathfrak{N}^j \supset K - N^i$ and hence $N^i \supset \mathfrak{N}^j$, $L \supset \Lambda$. Since also $L \subset \Lambda$, we would have $L = \Lambda$, contrary to assumption. Since $K - N^i$ is infinite it possesses some closed ideal element $L' = \{N'^i\}$. We have $N'^i \subset K - N^i$. The totality of all elements L' so defined is said to be an *open* ideal element, and that open element is called the *complement* $\Lambda - L$ of L relatively to Λ. A closed element $\{N'^i\} = L'$ is said to be on $\Lambda - L$, $(L' \subset \Lambda - L)$, whenever there exists i, j such that $N'^j \subset K - N^i$.

5. **The cycles mod ideal elements.** We shall now introduce cycles of two dual types relatively to two complementary ideal elements. They may be defined in various ways, as is frequently the case when passing from the finite to the infinite K. One will want to have definitions that are as natural as possible, and also such as to yield a duality theory which comes nearest to that of Ch. III. These conditions are fulfilled by those to be given presently.

We will designate the two complementary ideal elements by L^1, L^2, with $L^1 = \{N^i\}$ as the closed element, and we will set $K^i = K - N^i$.

(a) The chain C_p is defined as on $K - L^2$, whenever it has at most a finite number of cells on every K^i. It will be considered as on $K - L^1$ whenever it is on at least one K^i.

(b) A chain C_p on $K - L^i$, will be considered as a cycle mod L^j $(i, j = 1, 2; i \neq j)$ whenever it has no boundary.

Observe that we do not exclude the possibility that L^1 or $L^2 = 0$. In particular when $L^2 = 0$, $L^1 = \varLambda$, the cycles on $K - L^2 \bmod L^1$ are the cycles on $K \bmod \varLambda$, i. e. the infinite cycles of K, while those on $K - L^1 \bmod L^2$ are the cycles on $K - \varLambda$, i. e. the finite cycles of the complex. When $L^1 = \varLambda$, K^i is a finite complex, when $L^1 \neq \varLambda$, it is infinite.

(c) The homology

$$(1) \qquad C_p \approx 0 \bmod L^1 \text{ on } K - L^2,$$

signifies that for every i there exists a chain C^i_{p+1} on $K - L^2$, and a chain C^i_p on $\bar{N}^i$, such that

$$(2) \qquad C^i_{p+1} \to C_p - C^i_p.$$

(d) The homology

$$(3) \qquad C_p \approx 0 \bmod L^2 \text{ on } K - L^1,$$

signifies that there exists at least one complex K^i, such that if $\varLambda^i$ is its total ideal set,

$$(4) \qquad C_p \approx 0 \bmod \varLambda^i \text{ on } K^i.$$

Since $\varLambda^i \subset L^2$, we may also conveniently write this in the form

$$C_p \approx 0 \bmod L^2 \text{ on } K^i.$$

When $L^2 = 0$, $L^1 = \varLambda$, (1) becomes

$$(1') \qquad C_p \approx 0 \bmod \varLambda \text{ on } K.$$

It means that for every i,

$$(2') \qquad \bar{\mathfrak{N}}^i \supset C^i_p \approx C_p \text{ on } K.$$

We may think of (1′) as implying that $C_p \approx$ a chain on the total ideal set, or "ideal chain", or according to (2′), that it is $\approx$ to a chain on any assigned "neighborhood" $\overline{\mathfrak{N}}^i$ of $\varDelta$. This homology is different from

(5) $$C_p \approx 0 \text{ on } K,$$

which implies that C_p is the boundary of some C_{p+1} on K. The sequel will show however that the homologies (1′) are the more important of the two. The distinction between the two types must be kept clearly in mind.

When we make $L^2 = 0$, $L^1 = \varDelta$ in connection with (3), we have

(3′) $$C_p \approx 0 \quad \text{on} \quad K - \varDelta,$$

whose interpretation is that for some i,

(4′) $$C_p \approx 0 \quad \text{on} \quad K - \mathfrak{N}^i.$$

In other words (3′) implies that C_p is a finite cycle ≈ 0 on a finite subcomplex of K.

The homologies that we have just introduced are "finitely" additive, that is if a finite number of chains $C_p^1, \cdots, C_p^r$, are ≈ 0 in accordance with (c) or (d), so is any linear combination

$$C_p = \sum_{j=1}^{r} t_j C_p^j.$$

Since our cycle moduli may be expected to include linear combinations of an infinite number of chains, we make the additional convention:

(d′) Let the chains of the infinite sequence $\{C_p^j\}$ be all on the same fixed K^i, and let each $C_p^j \approx 0$ in the sense of (d). Then if

$$C_p = \sum t_j C_p^j$$

represents a true chain on K it is likewise $\approx 0 \bmod L^2$ on $K - L^1$.

Thus (d), (d′) represent the conditions in order that (3) be satisfied. When $L^1 = \varDelta$, $L^2 = 0$, since a finite chain has only a finite number of cells on $K - \mathfrak{N}^i$, the sum just written must be finite. Hence in that case (d′) does not go beyond (d).

If L is a subcomplex of K we may introduce the cycles of $K - \mathit{\Delta} - L$, $K \bmod L$, $K \bmod (\mathit{\Delta} + L)$, $K - \mathit{\Delta} \bmod L$. The formal modifications are insignificant. Thus for the last type which corresponds to the finite cycles mod L, we must merely add the mention "mod L" in (3′) and (4′). These new types of cycles are not important in the applications, but they will be found useful in developing the theory of the two initial types.

6. To the various cycles ≈ 0 just introduced there correspond relations of dependence for their types. Since we allow rational chains throughout, we have to give up zero-divisors and hence torsion coefficients, but we may introduce the Betti numbers, which may however be finite or infinite. In particular, the numbers corresponding to (1′) will be designated by $R_p(K; \mathit{\Delta})$, those corresponding to (5) by $R_p(K)$.

The Betti numbers may serve in defining the circuit as in the finite case. It is then immaterial whether we use $R_n(K; \mathit{\Delta})$ or $R_n(K)$, for when $p = n$, the boundary relations are merely the identical linear relations between the cycles, so that $R_n(K; \mathit{\Delta}) = R_n(K)$.

Since the various relations introduced are intrinsic relations between K and its chains, preserved under a homeomorphism of K alone, the corresponding Betti numbers and the circuit properties are topological invariants of K.

7. **The bases for the cycles mod ideal elements.** Given any particular type of cycle and the associated homologies, if the corresponding Betti number is finite, any maximal independent set of the cycles is a base for the type. When the Betti number is infinite, however, the problem is much more complex, and it is that problem which we shall now consider. Besides its intrinsic interest, its solution will enable us to extend greatly the scope of later results.

Let $\mathfrak{M}$ be a modulus of p-chains, $\mathfrak{M}^0$ its zero-modulus. The term *modulus* is to be understood as in Ch. I, No. 21, except that rational coefficients are allowed. A base $\{C_p^i\}$ is defined as a set of elements of $\mathfrak{M}$ independent mod $\mathfrak{M}^0$ such that for every element C_p of $\mathfrak{M}$, we have

(6) $$C_p \approx \sum t_i C_p^i \quad \mathrm{mod}\, \mathfrak{M}^0.$$

A base may be finite or infinite. We distinguish the following types of infinite bases:

I. *Semi-finite.* Then for every element of $\mathfrak{M}$ the sum in (6) is finite.

II. *Totally infinite.* The sum in (6) may be infinite.

III. *Semi-infinite.* In this type the base is of the form $\{C_p^{ij}\}$, with both indices running from 1 to $+\infty$. Every element is represented in the form

(6′) $$C_p \approx \sum t_{ij} C_p^{ij} \quad \mathrm{mod}\, \mathfrak{M}^0,$$

one of the indices i, j remaining finite but not necessarily the other, and we have the following two possibilities: (a) One of the indices, say i, remains finite for every C_p, independently of j. (b) i remains finite for every j, but its range varies with j. We will say that the corresponding bases are semi-infinite respectively of the *first kind* and *second kind.*

In each of the types described, all the independent elements mod $\mathfrak{M}^0$ are obtained by making the t's take arbitrary values consonant with the particular type considered. Thus in IIIa we have an element by assigning arbitrary values to the rational numbers t_{ij}, $i \leqq h$, and taking $t_{ij} = 0$ for $i > h$.

Suppose that every C_p^i of a totally infinite base is a subchain of K:

$$C_p^i = \sum x_{ij} E_p^j.$$

Then, every linear combination

$$C_p = \sum t_i C_p^i = \sum t_i x_{ij} E_p^j$$

being a proper element of $\mathfrak{M}$, the coefficient of the cells in the last sum are all finite, whatever the numbers t_i. This can occur only when each cell is found in at most a finite number of elements C_p^i. That property will characterize totally infinite bases.

Suppose that $\{E_p^j\}$ is the set of all cells of K in some fixed order. Then if

$$C_p = \sum t_j E_p^j$$

is any subchain of K, with the indices j in ascending order, the cell E_p^i corresponding to the first coefficient $t_i \neq 0$, is called the *first cell* of C_p. If $\{C_p^i\}$ is a set of chains, a sufficient condition that no E_p^j occur in more than a finite number of chains C_p^i, is that their first cells be all different.

8. The cycles that primarily interest us are those of the two basic dual types on $K - L^i \bmod L^j$. As a preliminary step it is necessary, however, to consider first the simpler types involving only the total ideal set, and also perhaps a subcomplex L of K, as at the end of No. 5.

The various moduli $\mathfrak{M}$ corresponding to the cycles and homologies of No. 5 are of three distinct types, according as the cycles are infinite on $\varDelta$, L^2 or L^1, and we will treat the bases in that order. Owing to the deformation theorem, we may assume that we have only subchains throughout.

I. *The cycles are infinite on* $\varDelta$. The mode of becoming infinite is then unrestricted. The cycles may be absolute cycles of K, cycles mod L, cycles of $K - L$. In the latter two cases $\{E_p^i\}$ consists only of the cells of $K - L$, but except for that the treatment is the same. We will designate by Γ_p, $\varDelta_p$ generic elements of $\mathfrak{M}$, $\mathfrak{M}^0$. The homologies will all be understood mod $\mathfrak{M}^0$.

Assuming that $\mathfrak{M}$ is not vacuous, the aggregate of cells that are first cells of subcycles of K in $\mathfrak{M}$, is a subsequence $\{E_p^{h_i}\}$ of $\{E_p^i\}$. Consider the cycles whose first cell is $E_p^{h_i}$. We select some p-cycle of $\mathfrak{M}$ with $E_p^{h_i}$ as its first cell, preferably a $\varDelta$, if one is to be found. Dividing the cycle by the coefficient of $E_p^{h_i}$ in it, we will have a cycle

$$\Gamma_p^i = E_p^{h_i} + t_1 E_p^{h_i+1} + \cdots. \tag{7}$$

9. The sequence $\{\Gamma_p^i\}$ has the following properties:

(a) No p-cell occurs in more than a finite number of elements Γ_p^i. Therefore

$$\Gamma_p = \sum x_i \Gamma_p^i \tag{8}$$

is a p-cycle of $\mathfrak{M}$ whatever the rational numbers x.

(b) Every p-cycle of $\mathfrak{M}$ is representable in the form (8), to be precise with $=$ replaced by $\approx$, when it is not a subcycle.

We have to prove this statement only for subcycles. Now if

$$\Gamma_p = \sum y_i E_p^i$$

the cycle

$$\Gamma_p' = \Gamma_p - \sum y_{h_i} \Gamma_p^i \subset \mathfrak{M},$$

and in its expression there enter no cells $E_p^{h_i}$. This is impossible however, since its first cell is of that type. Hence $\Gamma_p' = 0$, and Γ_p is of the required form.

(c) If in the expression (8) there is a single Γ_p^i, not a Δ, $\Gamma_p \not\approx 0$.

For let Γ_p be a Δ, and Γ_p^j the first cycle in the sum not a Δ. Omitting a finite number of Δ's in the sum, we have a Δ whose first cell is $E_p^{h_j}$. But, by assumption, if there are Δ's with $E_p^{h_j}$ as their first cell, one of them is to be chosen as Γ_p^j, hence Γ_p^j is a Δ, and this contradiction proves (c).

We will now remove the Δ's from $\{\Gamma_p^i\}$ and continue to designate the remaining sequence by $\{\Gamma_p^i\}$. If $\{\Delta_p^i\}$ is the removed sequence we have for every Γ_p of $\mathfrak{M}$:

$$\Gamma_p = \sum x_i \Gamma_p^i + \sum y_i \Delta_p^i.$$

The second sum is a cycle having the property that the order of its first cell can be raised indefinitely by suppressing a finite number of Δ's. With the present type of modulus $\mathfrak{M}$, this means that the second sum is homologous to a cycle on an $\mathfrak{N}^i$ with i as high as we please, and hence ≈ 0. Therefore

$$\Gamma_p \approx \sum x_i \Gamma_p^i.$$

Thus every cycle of $\mathfrak{M}$ depends upon the cycles Γ_p^i and since these are independent by (c), $\{\Gamma_p^i\}$ is a base for $\mathfrak{M}$. This base is totally infinite.

10. II. *The cycles are infinite on L^2.* The important point is that the ideal element L^2 is an open ideal element ($L^2 = A - L^1$),

and we do not exclude $L^2 = 0$, in which case the cycles are finite.

Let $K^i = K - N^i$ with $N^0 = K$, $K^0 = 0$, and let $\{E_p^{i\alpha}\}$ designate the p-cells of $K^i - K^{i-1}$, where i is fixed for the present. Let Γ_p, Δ_p be as before. We will select a set of cycles $\{\Gamma_p^{i\beta}\}$ on K^i in the same manner as previously, except that the cells $\{E_p^{i\alpha}\}$ are alone taken into consideration: Γ_p^{i1} is a Γ having in its expression a first cell of the set $\{E_p^{i\alpha}\}$, whose order α is least, etc. We break up, here also, the sequence $\{\Gamma_p^{i\beta}\}$ into two, the one consisting of cycles that are not Δ's, which we henceforth designate alone as $\{\Gamma_p^{i\beta}\}$, the other a sequence $\{\Delta_p^{i\beta}\}$.

Let us denote by E_p^{γ} the cells $E_p^{j\alpha}, j < i$, regrouped in some order. Suppose that $E_p^{\gamma_1}$ is the first cell of the set found in an infinity of cycles $\Gamma_p^{i\beta}$. Whatever t, β', the cycle $\Gamma_p^{i\beta} + t\Gamma_p^{i,\beta+\beta'}$ is a $\Gamma \subset K^i$, and with the same first cell of the set $\{E_p^{i\alpha}\}$ as $\Gamma_p^{i\beta}$. Furthermore we can always choose t, β' such that the new cycle will lack $E_p^{\gamma_1}$. This yields a new base $\{\Gamma_p^{i\beta}\}$ with the same first cells of $\{E_p^{i\alpha}\}$ as before but in whose cycles every E_p^{δ}, $\delta < \gamma_1$, will be found at most a finite number of times, and $E_p^{\gamma_1}$ not at all. Let $E_p^{\gamma_2}$, $(\gamma_2 > \gamma_1)$, be the first E_p^{γ} found in an infinity of new cycles $\Gamma_p^{i2}, \Gamma_p^{i3}, \cdots$. We treat the latter as before for suppression of $E_p^{\gamma_2}$, and so on. Thus there will be obtained a final sequence $\{\Gamma_p^{i\beta}\}$ such that every cell of K^i appears in at most a finite number of cycles of the sequence, and that the first cells of $\{E_p^{i\alpha}\}$ in these cycles are as before.

A similar treatment may be applied to $\{\Delta_p^{i\beta}\}$. This being done, every finite or infinite linear combination of the cycles $\Gamma_p^{i\beta}$, $\Delta_p^{i\beta}$ with rational coefficients is a $\Gamma \subset K^i$, and in fact a Δ if it includes only cycles of the second category.

11. Consider now any cycle Γ_p infinite on L^2. It will be on some K^i. Let K^i correspond to the least i such that $\Gamma_p \subset K^i$, and let x_β, y_β be the coefficients in Γ_p of the first cells of $\Gamma_p^{i\beta}$, $\Delta_p^{i\beta}$, when only cells $E_p^{i\alpha}$ are considered. Then

$$\Gamma_p^1 = \Gamma_p - \sum x_{i\beta}\, \Gamma_p^{i\beta} - \sum y_{i\beta}\, \Delta_p^{i\beta}$$

is a cycle on K^i which has among its cells $E_p^{i\alpha}$ none that is first cell of a $\Gamma_p^{i\beta}$ or $\Delta_p^{i\beta}$. Therefore in the expression of Γ_p^1 there enter no cells $E_p^{i\alpha}$, hence $\Gamma_p^1 \subset K^{i-1}$ Repeating the reasoning we find in general

$$\Gamma_p^{h+1} = \Gamma_p^h - \sum_\beta x_{i-h,\beta}\,\Gamma_p^{i-h,\beta} - \sum_\beta y_{i-h,\beta}\,\Delta_p^{i-h,\beta} \subset K_{i-h-1};$$
$$\Gamma_p^0 = \Gamma_p; \qquad \Gamma_p^i = 0.$$

Hence, by addition,

$$\Gamma_p = \sum_{j \leqq i;\,\beta} x_{j\beta}\,\Gamma_p^{j\beta} + \sum_{j \leqq i;\,\beta} y_{j\beta}\,\Delta_p^{j\beta}, \tag{9}$$

and therefore, by (d′) of No. 5,

$$\Gamma_p \approx \sum_{j \leqq i;\,\beta} x_{j\beta}\,\Gamma_p^{j\beta}. \tag{10}$$

The cycles at the right are independent, for $\Gamma_p \approx 0$, implies that the cycle is a Δ and by the first cell argument repeatedly used, applied this time to $\{E_p^{i\alpha}\}$, this may be ruled out. Therefore the double sequence $\{\Gamma_p^{i\alpha}\}$ has all the properties of the base. When $L^2 \neq 0$, the base is in general semi-infinite of the first kind. When $L^2 = 0$, the cycles of $\mathfrak{M}$ are finite, the complexes K^i are finite, and so are the partial bases $\{\Gamma_p^{i\alpha}\}$. Hence in that case the base is semi-finite.

REMARK. We may replace throughout K^i by $K^i - \Phi^i$, where Φ^i is the boundary of N^i. This is equivalent to substituting everywhere $\overline{N}^i$ for N^i.

12. III. *The cycles are infinite on* L^1. This time $\mathfrak{M}$ consists of the absolute cycles of $K - L^2$, that is, the cycles mod L^1 on $K - L^2$. The treatment of this case will be based on the notion of *cut* of a subchain. The cut of the subchain C_p by Φ^i, is the part of the subchain which is on K^i, or sum of its p-cells not on N^i. If C_p is infinite on L^1, its cuts are all finite and conversely.

Consider in particular those cuts, $C_p^{i\alpha}$, of the p-cycles mod L^1, which are on $\overline{K^i - K^{i-1}}$. They constitute a special class of finite cycles of $\overline{K^i - K^{i-1}}$ mod Φ^i. By the same scheme as in the preceding case, we obtain a base $\{C_p^{i\alpha}\}$ for such cycles, the preferred cycles Δ being now the boundaries of finite chains on K^i, and $\{E_p^{i\beta}\}$ the set of p-cells of $K^i - \Phi^i - (K^{i-1} - \Phi^{i-1})$.

Let then $\Gamma_p^{i\alpha}$ be the cycle mod L^1 having $C_p^{i\alpha}$ for its cut. If Γ_p is any cycle whatever, and $t_{i\alpha}$ the coefficient of the first cell of the cycle $\Gamma_p^{i\alpha}$ in its expression, when only cells E_p^i are taken into consideration,

$$\Gamma_p \approx \sum t_{i\alpha} \Gamma_p^{i\alpha}. \tag{11}$$

For in the first place, the cut C_p^i of Γ_p by K^i being finite, the number of cells of K^i in Γ_p is finite. It follows that in the sum at the right, the number of coefficients $t_i \neq 0$ with given i is finite, and therefore the sum represents a true subchain of K, whose cuts are cycles mod the respective Φ^i, hence a cycle of the type considered. The difference of the two sides of (11) has for cut by Φ^1,

$$C_p^1 - \sum t_{1\alpha} C_p^{1\alpha} \approx 0 \quad \text{mod } \Phi^1 \text{ on } K^1 - L^2,$$

for it includes no first cell of the cut of any cycle $\not\approx 0$. Therefore,

$$\Gamma_p^1 = \Gamma_p - \sum t_{1\alpha} \Gamma_p^{1\alpha} \approx 0 \quad \text{mod } \bar{N}^1 \text{ on } K - L^2.$$

We may therefore replace Γ_p by a cycle on $\bar{N}^1 - L^2$, to which the same reasoning may be applied, etc. Hence (11) holds mod $\bar{N}^i$ on $K - L^2$ for every i, therefore also mod L^1 on $K - L^2$. Due to the very mode of constructing the cycles $\Gamma_p^{i\alpha}$, we cannot replace Γ_p in (11) by zero unless every $t = 0$. Hence $\{\Gamma_p^{i\alpha}\}$ is a base which is in general semi-infinite.

13. The preceding results prove:

THEOREM. *There exist bases for all the types of cycles introduced and related homologies. The base is infinite for Type I, semi-infinite of the first kind for Type II, of the second kind for Type III when the cycles are infinite, semi-finite for the same three types when the cycles are finite.*

REMARK. A base for ordinary linear dependence and all subchains is obtained by adding to anyone of the preceding bases a certain set $\{\varDelta_p^i\}$ consisting of cycles ≈ 0, and also the set of all cells $\{E_p^i\}$ not first cells of cycles of $\mathfrak{M}$, whether relatively to all cells ordered in a specific way or only to a certain subset $\{E_p^{i\alpha}\}$, as the case may be. An important property is that *the $\varDelta$'s may always be chosen finite.* For an infinite $\varDelta$ is a linear combination of finite $\varDelta$'s, one of which will have the same first cell, and hence can take its place in $\{\varDelta_p^i\}$.

14. **Product of two infinite complexes.** The results of qualitative nature obtained in Ch. V, are extended without difficulty. The question of the base and the relations between the Betti numbers require further consideration.

We shall use the notations of Ch. V, § 3, except that the L's are now ideal elements and no manifold conditions are imposed. It is more in keeping with the question to call λ, λ' the total ideal elements of K, K', reserving $\varLambda$ for the total ideal element of $K \times K'$. We shall denote the analogues of $N^i, \cdots$ for K' by $N'^i, \cdots$. Then $\varLambda = \{K \times \mathfrak{N}'^i + \mathfrak{N}^i \times K'\}$. It has for closed subset $\varLambda^1 = \{K \times N'^i + N^i \times K'\}$, its complement being $\varLambda^2 = \varLambda - \varLambda^1$.

Let $\{\varGamma_\mu^i\}$ be the base for the μ-cycles of $K - L^2 \bmod L^1$, obtained in the preceding numbers, and merely rewritten as a single ordered sequence. As a by-product of the construction of the minimum base, we have a sequence $\{\varDelta_\mu^i\}$ of boundaries of finite $(\mu+1)$-subchains, which is a maximal linearly independent set for the cycles of type $\varDelta$. To have a base for all μ-subchains and ordinary linear dependence, we merely have to add to $\{\varGamma_\mu^i\}$, $\{\varDelta_\mu^i\}$, the set $\{E_\mu^i\}$ of all μ-cells which are not first cells of a $\varGamma$ or $\varDelta$. The same configurations for K' are designated by the same symbols with accents.

15. Since no μ-cell of K is found in more than a finite number of the elements $\varGamma$, $\varDelta$, E, and similarly for K', no cell of $K \times K'$ is found in more than a finite number of the elements

$$(12) \qquad \varGamma_\mu^i \times \varGamma_\nu'^j,\ \varGamma_\mu^i \times \varDelta_\nu'^j,\ \cdots,\ E_\mu^i \times E_\nu'^j.$$

Hence every linear combination with rational coefficients of those for which $\mu+\nu=h$, is a proper h-subchain of $K\times K'$. On the other hand, since every μ-cell of K is a linear combination of the elements Γ_μ, Δ_μ, E_μ, and similarly for the ν-cells of K', every h-cell of $K\times K'$ is a linear combination of the product elements of dimension h. Since the factor elements of a given dimension are linearly independent, this is also the case for the product elements. Therefore these product elements constitute a base for all h-subchains of $K\times K'$ and linear dependence.

Let then γ_h be an h-cycle on $K\times K'-\varLambda^2 \bmod \varLambda^1$. We first reduce it to a subcycle and continue to call the latter γ_h. According to the above,

$$(13)\qquad \gamma_h=\sum \Gamma_\mu^i\times\Gamma'^j_{h-\mu}+\cdots+\sum E_\mu^i\times E'^j_{h-\mu},$$

where we have nine sums corresponding to the same number of types of product elements. To simplify we have omitted the coefficients throughout.

We observe first that, since the cycles Δ, Δ' are boundaries of finite chains, we can choose for their bases some of the cycles $\Delta_\mu^i=F(E_{\mu+1}^i)$, $\Delta'^i_\mu=F(E'^i_{\mu+1})$. Next, it is easy to show that the Γ's and Δ's may be so chosen that the first cell of the set $\{E_p^{i\beta}\}$ of No. 12 associated with any one Δ is found in no other Δ and in no Γ, and similarly for the cycles Γ', Δ'. Hence if $C\times C'$ is any term in one of the sums in (13) with C or C' a Δ, some point $A\times A'$, $A\subset C$, $A'\subset C'$, will actually be a point of γ_h. As a consequence, if we pick out any group of terms such as just considered in (13) the chain which is their sum can only be infinite on $\varLambda^1$ like γ_h itself.

Referring now to the boundary formula (6) of Ch. V we find: (a) Since $F(E)$, $F(E')$ are linear combinations of a finite number of cycles Δ, Δ', the boundaries of the terms $E\times E'$ in (13) do not cancel out, as they must since γ_h is a cycle. Therefore these terms are altogether absent in (13). (b) Owing to the form of γ_h, and in view of the remark made above concerning the infinite parts of its subchains, the terms with one factor a Δ and the other a Γ or Δ, make up a C_h which is the boundary of a C_{h+1} infinite on $\mathcal{A}^1$, and hence $C_h \approx 0$ mod $\mathcal{A}^1$ on $K \times K' - \mathcal{A}^2$. Therefore the terms in question may be suppressed in (13) with $\approx$ in place of $=$. (c) For similar reasons, and in view of the relations

$$E \times E' \rightarrow E \times \Delta' \pm \Delta \times E' \approx 0,$$

we may replace all terms $E \times \Delta'$ by terms $\Delta \times E'$.

Owing to these properties then, γ_h is homologous to a similar cycle in whose expression there are only terms $\Gamma \times \Gamma'$, $\Gamma \times E'$, $E \times \Gamma'$ and $\Delta \times E'$. The consideration of the boundaries shows however that the sum of the boundaries of the terms $\Delta \times E'$ must be $\equiv 0$. Hence, if the reduced γ_h contains a term $\Delta^i_\mu \times \sum t_j E'^j_{h-\mu}$, its boundary must be $\equiv 0$, and therefore $\sum t_j E'^j_{h-\mu}$ must be a cycle of K'. But that cycle must have its infinite part on L^1. Hence it is a cycle on $K - L^2$ mod L^1, which is excluded since every cycle of that nature is a linear combination of cycles Γ, Δ. Thus there are no terms $\Delta \times E'$. We can show similarly that there are no terms $\Gamma \times E'$, $E \times \Gamma'$. Therefore the reduced cycle includes only terms $\Gamma \times \Gamma'$, or explicitly,

(14) $\quad \gamma_h \approx \sum t^\mu_{ij} \Gamma^h_\mu \times \Gamma'^j_{h-\mu}$ mod $\mathcal{A}^1$ on $K \times K' - \mathcal{A}^2$.

Let K^i, Φ^i be as before and K'^i, Φ'^i the same for K'. The assumption $\gamma_h \approx 0$ implies dependence of the products of the cuts of the cycles on $K^i \times K'^i$ mod $K \times \Phi'^i + \Phi^i \times K'$. Dependence of the latter, on the other hand is reducible to dependence of such products on a product of finite complexes, whereas we already know by Ch. V, No. 28, that independence of the factors implies that of the products. Hence the cycles $\Gamma^h_\mu \times \Gamma'^j_{h-\mu}$ are independent.

As the elements of $\{\Gamma^h_\mu \times \Gamma'^j_{h-\mu}\}$ are independent, and each cycle of $K \times K' - \mathcal{A}^2$ mod $\mathcal{A}^1$ is expressible in terms of them in the form (14), the set is a base for the cycles of this type. Thus the base is obtained in the present instance as in Ch. V, No. 28, regardless of whether the Betti numbers are finite or not. As a consequence, formula (26), Ch. V, holds here also.

The same results are now readily deduced for the cycles of dual type. For let γ_h be a cycle of $K \times K' - \Lambda^1 \bmod \Lambda^2$. There is a $K^i \times K'^i \supset \gamma_h$. Hence by the result just proved, applied to $K^i \times K'^i$, γ_h depends upon the products of infinite cycles on $K^i \times K'^i$, and hence upon the products of cycles of the bases for $K - L^1 \bmod L^2$ and $K' - L'^1 \bmod L'^2$. A similar reasoning shows that the latter products are independent and hence those of dimension h constitute a base for the cycles γ_h.

16. The preceding choice of complementary ideal sets Λ^1, Λ^2, is the analogue of the choice of such sets in Ch. V, No. 26. Another one more important for the applications, is the less symmetrical choice similar to that of Ch. V, No. 29. This time L, L' are replaced by the total ideal elements λ, λ' of K, K'; we take $\Lambda^1 = \{K \times \mathfrak{N}'^i\}$, which might be designated as $K \times \lambda'$, and, as before, $\Lambda^2 = \Lambda - \Lambda^1$. The reasoning is exactly the same as above with the following interpretation of the elements: The cycles Γ^i_μ, Δ^i_μ are cycles of $K - \lambda$ (finite cycles of K), the corresponding accented cycles those of $K' \bmod \lambda'$ (infinite cycles of K'). The result is again as before. We have then proved

THEOREM. *When Λ^1 is one of the two ideal sets $\{K \times N'^i + N^i \times K'\}$, $\{K \times \mathfrak{N}'^i\}$ and Λ^2 is its complement, the bases for $K \times K' - \Lambda^i \bmod \Lambda^j$ are obtained as in the similar situations of Ch. V.*

17. **Application of ideal elements.** We now return to the single infinite complex K, so that Λ, L^1, L^2 are again as in No. 5.

A case of great importance in the applications is when K is an open subset of a compact metric space $\mathfrak{R}$. With each ideal element L there is associated a unique closed set $l \subset \mathfrak{R} - K = \lambda$, the total ideal element Λ being now the closed set λ, that is the complement of K relatively to $\mathfrak{R}$. Let L^1 correspond to l^1; then L^2 corresponds to $\lambda - l^1 = l^2$, and the infinite chains on $K - L^i$ are actual chains on the set $\mathfrak{R} - l^i$, and similarly for the cycles. In the calculation of the homology characters an important observation is to be made: In considering the cycles on $\mathfrak{K}-l^i \bmod l^j$, we may take in only the finite cycles, or else all the cycles. *Accordingly as we do the one or the other, we may obtain different values for the homology characters.*

Let for example C_p be a finite chain on $\mathfrak{K}-l^2$. We may apply the deformation theorem to the part of the chain on K, and if C_p has points on l^1 that will automatically turn C_p into an infinite subchain of $K-L^2 \bmod L^1$. However there may exist infinite chains of the same type not derived from any finite chain, and the cycles of such nature may affect the Betti numbers as mentioned above.

It will be understood that *in all similar cases we agree to take the largest class open to our choice.* Here then all the cycles are to be included.

When λ, l^1 and $\mathfrak{K}$ are locally connected, the finite and infinite chains give rise to the same homology characters. For in that case, if $\{N^i\} = L^1$, and C'_p is the cut of any infinite cycle C_p on $\mathfrak{K}-l^2$ by an N^i, i sufficiently large, $C_p - C'_p$ can be deformed onto l^1, and C'_p into a finite cycle on $\mathfrak{K}-l^2 \bmod l^1$. By considerations similar to those of Ch. II, § 5, we may then show that the finite cycles can take the place of the infinite cycles, without affecting the homologies between them, and similarly with l^1 and l^2 interchanged.

18. **Application to finite complexes.** In the preceding chapters the central configuration under discussion has been practically throughout an open set $K-L$, difference between the closed complex K and its closed subcomplex L. As we have observed on several occasions, L might as well have been replaced by any closed polyhedral configuration on K. The theory developed has then been the topology of open complexes, or open polyhedral sets. The infinite complex enables us to extend most of it to the case where L is any closed subset of K, that is *to arbitrary open subsets of a finite complex.*

Let indeed the $\mathfrak{K}$ of the preceding number be a finite n-complex, with λ as a closed subset. Let $\mathfrak{K}^i$ be the ith derived of $\mathfrak{K}$, $(\mathfrak{K}^0 = \mathfrak{K})$, $\mathfrak{N}^i$ the $\mathfrak{K}^i$-neighborhood of λ. Then the infinite complex

$$K = \mathfrak{K} - \mathfrak{N}^0 + \sum (\mathfrak{N}^i - \mathfrak{N}^{i+1})$$

coincides with $\mathfrak{K} - \lambda$. If l^1 is a closed subset of λ, and $l^2 = \lambda - l^1$, the results of the preceding numbers are directly applicable here.

In various parts of Chs. II, V, we have obtained a number of propositions regarding a finite complex and certain subsets L, L^1, L^2 restricted at the time to subcomplexes. *What we have just said implies the extension to similarly related arbitrary closed sets, of the part or all of these propositions in which* $\approx$ *alone enters.* As an example, the theorems of invariance of Ch. II, § 5, hold for a perfectly arbitrary open subset $K - L$ of a complex K. The infinite complex appears to be the ideal tool for all extensions of the properties of $K - L$, L a subcomplex, to the same with L an arbitrary closed set on K.

The power of the method is perhaps well illustrated by the proofs which it yields for *Brouwer's theorem on invariance of dimensionality.* To avoid special considerations of boundaries, we will take for $\mathfrak{R}$ a region on an (ordinary) n-sphere H_n. Then $\lambda = H_n - \mathfrak{R}$ is closed, and we can construct an infinite complex K coincident with $\mathfrak{R}$. If K' is a maximal connected part of K, it is a circuit mod λ, and hence coincides with an n-cycle mod $(\lambda, 2)$. The number of such circuits is $R_n(K; \lambda, 2)$, since no two have any points in common. Since all cycles of dimension $n+i$ are $\approx$ to subcycles of K mod λ, all numbers $R_{n+i}(K; \lambda, 2)$ vanish. Hence n is the largest p such that $R_p(K; \lambda, 2) \neq 0$, and $R_n(K; \lambda, 2)$ is the number of connected parts whose sum is $\mathfrak{R}$. Owing to the invariance of the Betti numbers of the infinite complex mod its total ideal set, if $\mathfrak{R}$ and $\mathfrak{R}'$ are homeomorphic their dimensions are the same.

§ 3. Infinite Manifolds and their Duality Theory

19. Infinite n-manifolds, (n finite), are defined as in the finite case. We propose to extend to them as much as possible of the theory of finite manifolds, and in particular to prove the duality relations for the Betti numbers mod ideal elements.

On the qualitative side there is nothing new. In particular the theory of cycle intersections goes through without modification. We may take advantage of the generalized polyhedral intersections of Ch. IV, § 6, and define the more general type by means of these. The treatment of Ch. IV, §§ 3, 4, is applicable with unimportant modifications, mostly in the nature of replacing distance conditions by neighborhood conditions. The intersecting chains are now subjected to the restriction that they do not meet each other's boundaries on K. As a consequence, the proposition of Ch. IV, No. 29, is replaced by the following: Owing to the restriction on the boundaries, and in the notations loc. cit., the fundamental complex K may be so chosen that if C_p, C_q are respectively deformed over their K- and K^*-neighborhoods, into $C_p^\alpha \cdot C_q^{*\alpha}$, all cycles $C_p^\alpha \cdot C_q^{*\alpha}$ thus obtained, are homologous on $\mathfrak{M}$. The rest of the treatment is similarly modified, wherever necessary, and the results are as in Ch. IV.

Regarding $(C_p \cdot C_{n-p})$, it is to be observed that the number of isolated intersections and the index may be infinite.

For the sake of clearness, whenever we have cycles of two dual types, we shall designate the cycles of one type by Γ, say those of $K - L^1 \bmod L^2$, the cycles of the other type by G. Then Γ or $G \approx 0$, will be understood in each case in the sense appropriate for the type. Thus, for the G just considered, it is mod L^1 on $K - L^2$.

As regards all indices $(\Gamma_p \cdot G_{n-p})$, it is to be observed that they are always finite. For there exists in every case a readily detected submanifold M of K carrying one of the two cycles, and which meets only a finite number of cells of the other, hence their intersection is as if they were finite chains on a finite manifold.

We shall establish the following propositions, in which the two dual types are as in § 2, relatively to a pair of complementary ideal elements L^1, L^2 with L^1 closed. They constitute further extensions of the intersection theorems of Ch. IV, § 2, (Poincaré-Veblen-Weyl), and of the duality relations for the Betti numbers (Ch. III).

THEOREM I. *In order that one of the cycles* Γ_p, $G_{n-p} \approx 0$, *it is necessary and sufficient that*

$$(\Gamma_p \cdot G_{n-p}) = 0 \tag{15}$$

whatever the other cycle.

THEOREM II. *The bases* $\{\Gamma_p^\alpha\}$, $\{G_{n-p}^\alpha\}$ *may be so chosen that*

$$(\Gamma_p^\alpha \cdot G_{n-p}^\beta) = \delta_{\alpha\beta}. \tag{16}$$

The cycles Γ_p, G_{n-p} *have then the representations*

$$\begin{cases} \Gamma_p \approx \sum (\Gamma_p \cdot G_{n-p}^\alpha)\, \Gamma_p^\alpha, \\ G_{n-p} \approx \sum (\Gamma_p^\alpha \cdot G_{n-p})\, G_{n-p}^\alpha. \end{cases} \tag{17}$$

From the representations follows:

THEOREM III. *When the base* $\{G_{n-p}^\alpha\}$ *is totally infinite, (general case), there exists a* G_{n-p} *such that the indices* $(\Gamma_p^\alpha \cdot G_{n-p})$ *have arbitrarily assigned values, and similarly for the cycles* Γ_p. *When the bases are semi-finite or semi-infinite, the choice of indices is arbitrary only in so far as the resulting representations (17) are of the type demanded by the bases.*

As an example if $\{G_{n-p}^\alpha\}$ is semi-finite we can assign arbitrarily a finite number of indices $(\Gamma^\alpha \cdot G)$, but the rest must be zero.

As a consequence of the pairing off of the cycles of the bases we have:

THEOREM IV. *Fundamental duality relation for the Betti numbers:*

$$R_p(K - L^1; L^2) = R_{n-p}(K - L^2; L^1). \tag{18}$$

In particular when $L^1 = \Lambda$, $L^2 = 0$,

$$R_p(K - \Lambda) = R_{n-p}(K; \Lambda). \tag{19}$$

Formula (19) is the duality relation for the finite and infinite cycles as the associated dual types. It extends (9) of Ch. III.

THEOREM V. *The numbers* r *of Ch. III satisfy*

$$r_p(K - \Lambda) = r_{n-p}(K - \Lambda). \tag{20}$$

THEOREM VI. *The preceding propositions hold when any one of the ideal elements L^i, Λ is increased by a subcomplex L of K whose infinite part, if it has one, is on that element.*

The addition of L does not appreciably modify the proofs, so that this proposition need not be considered separately. It is stated explicitly in view of an important application later.

Referring to Nos. 17, 18, 19, we can translate our theorems into theorems on certain compact spaces, and in particular on manifolds $K-\lambda$, where K is a finite complex and λ an arbitrary closed subset of K. They imply:

THEOREM VII. *The intersection properties of Ch. IV, § 2, in particular the extended theorem of Poincaré-Veblen-Weyl, likewise the duality relations for the Betti numbers given in Ch. III (formulas (7), (20)), are valid when the sets L, L^1 are arbitrary closed sets on K, with $L^1 \subset L$.*

The best definition of the manifold $K-\lambda$ is that when $K-\lambda$ is turned into an infinite complex, as in No. 18, the latter is a manifold. It implies that every cell of a subdivision of K which is on $K-\lambda$ is regular for that subdivision. The invariance of the manifold conditions (Ch. III § 4) shows then that the definition does not depend upon the particular mode of constructing the infinite complex.

As the sequence of our propositions shows the proof of the fundamental duality relation rests on the extended Poincaré-Veblen-Weyl theorem—a reversion to Poincaré's first attempted proof of his own duality theorem. In this we are in agreement with all recently published proofs of extended duality relations, notably our own [9, 12, 14], those of Alexandroff [6, 10], Pontrjagin [1]. Frankl [1], Van Kampen [1, 3], exception being made however for Alexander's proof of his duality relation [8] (the first beyond Poincaré's), and its extension by Alexandroff [8].

I am advised by Alexander that he has developed the duality theory for a closed subset of S_r and its residual set, likewise for a closed subset of a Hilbert parallelotope. I take this occasion to state that we have both worked entirely independently.

20. The crux of the whole question, is the proof of Theorems I, II. Almost all the difficulties are already found in the treatment of an analogous but somewhat simpler case, which we have to take up as a preliminary step. It refers to a manifold $K-L$, L a subcomplex of K, the dual types being the cycles G_{n-p} of $K-L \bmod \Lambda$ (infinite cycles of $K-L$), and Γ_p of $K-\Lambda \bmod L$ (finite cycles of $K \bmod L$). The first three propositions of No. 19 hold for them and will be proved together. After that the theorems for the case that interests us primarily will offer very little difficulty.

Let here $K^i = K - \mathfrak{N}^i$, and let Ψ^i be the sum of the cells of the boundary Φ^i of $\mathfrak{N}^i$ not on L, and Ω^i the subcomplex common to L and K^i. The complex K^i is finite and closed, with Ω^i, $\Psi^i + \Omega^i$ as closed subcomplexes, and $K^i - \Psi^i - \Omega^i$ is a manifold.

The same argument as in Ch. IV No. 39, shows that the condition in Th. I, is necessary in order that Γ or $G \approx 0$, and its sufficiency alone needs to be proved.

Let then (15) hold for a fixed G_{n-p} and arbitrary Γ_p. The cut C^i_{n-p} of G by Ψ^i is a cycle of $K^i - \Omega^i \bmod \Psi^i$. Any p-cycle of $K^i - \Psi^i \bmod \Omega^i$ is a Γ_p and since it intersects G only on C^i_{n-p},

$$(\Gamma \cdot C^i) = (\Gamma \cdot G) = 0.$$

Hence (Ch. IV, No. 17),

$$C^i \approx 0 \bmod \Psi^i \text{ on } K^i - \Omega^i,$$

for every i, and therefore $G \approx 0$. Thus the condition in Th. I is sufficient to make $G \approx 0$.

21. It is much more difficult to derive the same result for Γ, and we can obtain it only in a roundabout way after proving Th. II.

Consider then a base $\{\Gamma^\alpha_p\}$ obtained as in No. 10, but with K^i replaced by $K^i - \Phi^i$, as described in No. 11, Remark, and let us construct the other base so as to satisfy (16).

From No. 11 and the same Remark follows that if $\{\Gamma^{i\beta}_p\}$ is a base for $K^i - \Psi^i \bmod \Omega^i$,

$$\Gamma^{i\beta}_p \approx \sum t^i_{\beta\varepsilon} \Gamma^\varepsilon_p, \tag{21}$$

where the only numbers $t \neq 0$ are those for which $\Gamma^{\varepsilon} \subset K^i - \Psi^i$, and the homologies are mod L on $K - \Lambda$, as for all cycles Γ.

It follows from Ch. IV, Nos. 20, 39 (Th. III for finite complexes), that there exists a cycle C^i_{n-p}, of $K^i - \Omega^i$ mod Ψ^i, such that

$$(22) \qquad (\Gamma^{i\beta}_p \cdot C^i_{n-p}) = t^i_{\beta\alpha}.$$

Let then $\Gamma_p \subset K^i - \Psi^i$. We shall have

$$(23) \qquad \Gamma_p \approx \sum x_\beta \, \Gamma^{i\beta}_p \quad \text{mod } \Omega^i \text{ on } K^i - \Psi^i,$$

$$(24) \qquad \Gamma_p \approx \sum y_\varepsilon \, \Gamma^{\varepsilon}_p,$$

where by No. 11, $y_\varepsilon = 0$ when $\Gamma^{\varepsilon}_p \not\subset K^i - \Psi^i$. Thus the homologies (21), (23), (24) involve only a finite number of cycles throughout. In fact, they are homologies between finite chains on some finite complex. We can therefore eliminate the cycles $\Gamma^{i\beta}$ between (21) and (23), the latter being taken mod L on $K - \Lambda$, which is weaker than mod Ω^i on $K^i - \Psi^i$. As a consequence, we find

$$(25) \qquad \Gamma_p \approx \sum x_\beta \, t^i_{\beta\varepsilon} \, \Gamma^{\varepsilon}_p.$$

Since $\{\Gamma^{\varepsilon}_p\}$ is a base, the expression of every Γ_p in terms of it is unique. Therefore the coefficients in (24) and (25) must be the same, or

$$(26) \qquad \sum x_\beta \, t^i_{\beta\varepsilon} = y_\varepsilon.$$

But from (22) and (23) we find

$$(27) \qquad (\Gamma_p \cdot C^i_{n-p}) = \sum x_\beta \, t^i_{\beta\alpha} = y_\alpha,$$

and in particular

$$(27') \qquad (\Gamma^{\beta}_p \cdot C^i_{n-p}) = \delta_{\alpha\beta}.$$

As the index (27) is independent of i provided that $\Gamma_p \subset K^i - \Psi^i$, we may replace in it i by $i+1$. But if D_{n-p} is the cut of C^{i+1}_{n-p} by Ψ^i, it includes all intersections of C^{i+1}_{n-p} and Γ_p. Hence

$$(\Gamma_p \cdot D_{n-p}) = (\Gamma_p \cdot C^{i+1}_{n-p}) = y_\alpha.$$

From this follows that $C^i_{n-p} - D_{n-p}$, which is a cycle of $K^i - \Omega^i \bmod \Psi^i$, has the property that

$$(\Gamma \cdot C^i - D) = 0,$$

whatever the cycle Γ of $K^i - \Psi^i \bmod \Omega^i$. By Ch. IV, No. 17 (Th. I for finite complexes),

$$C^i_{n-p} - D_{n-p} \approx 0 \quad \bmod \Psi^i \text{ on } K^i - \Omega^i.$$

There exists then a subchain D'_{n-p} of Ψ^i, which is also a subchain of $\overline{\mathfrak{N}}^i - \mathfrak{N}^{i+1} - L$, such that

$$C^i_{n-p} - D_{n-p} \approx D'_{n-p} \quad \text{on } K^i - \Omega^i.$$

Therefore

$$C^{i+1}_{n-p} \approx C^i_{n-p} + (C^{i+1}_{n-p} - D_{n-p} - D'_{n-p}) \quad \text{on } K^{i+1} - \Omega^{i+1}.$$

Consequently we can replace C^{i+1}_{n-p} by the chain at the right, and (22) will continue to hold with C^i_{n-p} replaced by C^{i+1}_{n-p}. Let D^i_{n-p} denote the chain in parenthesis, which is a subchain of $\overline{\mathfrak{N}}^i - \mathfrak{N}^{i+1} - L$, and let h be the least integer such that $K^h - \Psi^h \supset \Gamma^\alpha_p$. We will now construct C^i_{n-p} for every i as follows: (a) For $i < h$ we take $C^i_{n-p} = 0$. This is in accordance with (22), since the numbers $t^i_{\beta\alpha}$ are then all zero. (b) For $i \geqq h$ we take

$$C^{i+1}_{n-p} = C^i_{n-p} + D^i_{n-p}.$$

Since $D^i_{n-p} \subset \overline{\mathfrak{N}}^i - \mathfrak{N}^{i+1}$, $\overline{\mathfrak{N}}^{i+1} \subset \mathfrak{N}^i$, only two consecutive D's intersect. Hence

$$G^\alpha_{n-p} = C^h_{n-p} + \sum_{i=h}^{+\infty} D^i_{n-p}$$

represents a proper subchain of $K - L$, for when it is written as a linear combination of the cells of K, all coefficients are finite. Furthermore, since the cut C^i_{n-p} has no boundary cells on K^{i-1}, G^α_{n-p} has none whatever on $K - L$, hence it is properly a cycle of type G. Finally the D's do not meet Γ_p^α, since they are separated on K by the boundary of some $\mathfrak{N}^{h+i}$. From this and (27') follows

$$(\Gamma_p^\beta \cdot G_{n-p}^\alpha) = (\Gamma_p^\beta \cdot C_{n-p}^h) = \delta_{\alpha\beta},$$

which is (16). Similarly from (27) we have

$$(\Gamma_p \cdot G_{n-p}^\alpha) = y_\alpha. \tag{28}$$

Now if $\Gamma_p \not\approx 0$, there is at least one $y \neq 0$, and therefore at least one index $(\Gamma \cdot G) \neq 0$. This is the part of Th. I which was still left unproved for the present dual types Γ, G.

22. The sets $\{\Gamma_p^\alpha\}$, $\{G_{n-p}^\alpha\}$ satisfy then (16), and the first is a base by assumption. It is easy to show that this holds likewise for the second. In the first place, by construction G_{n-p}^α has cells on K^i only for i above a certain value; hence any cell of K occurs in at most a finite number of cycles G_{n-p}^α. As a consequence $\sum y_\alpha G_{n-p}^\alpha$ is a G whatever the rational numbers y. Let now G_{n-p} be arbitrary. We first reduce it to a subcycle, still to be called G_{n-p}. Since every Γ^α is on some K^i, and the cut of G by Φ^i (its part on K^i) is finite, the indices $(\Gamma^\alpha \cdot G)$ are all finite. Therefore

$$G'_{n-p} = G_{n-p} - \sum (\Gamma_p^\alpha \cdot G_{n-p}) G_{n-p}^\alpha$$

is a G. Now any particular Γ^β intersects G, G', and the cycles G^α which it meets, on a finite submanifold of K, and it meets only a finite number of the cycles G^α. It follows that

$$(\Gamma_p^\beta \cdot G'_{n-p}) = (\Gamma_p^\beta \cdot G_{n-p}) - \sum (\Gamma_p^\alpha \cdot G_{n-p}) (\Gamma_p^\beta \cdot G_{n-p}^\alpha) = 0,$$

and hence $(\Gamma \cdot G') = 0$ for every Γ, since each $\Gamma \approx$ a linear combination of a finite number of cycles Γ^α. By Th. I then, already proved, $G' \approx 0$; consequently G is representable in the form (17). If a linear combination of the cycles G^α is ≈ 0, we find, by means of Th. I, that every coefficient $= 0$, hence the cycles G^α are independent, and $\{G_{n-p}^\alpha\}$ is a base.

To complete the proof of Th. II, all that remains to be done, is to show that every Γ admits of the representation (17).

But that is implied by the fact that the coefficients y of the representation (24) satisfy (28).

Thus Theorems I, II are completely established in the present instance and therefore Theorems III, IV follow as in No. 19. We have, of course, in place of (18), the duality relation

$$R_p(K-\Lambda; L) = R_{n-p}(K-L; \Lambda). \tag{29}$$

23. The ground is now prepared for the theorems of No. 19 concerning the cycles Γ of $K-L^1 \bmod L^2$ and their dual type G, the cycles of $K-L^2 \bmod L^1$. We will set here $K^i = K-N^i$, $\Phi^i =$ the boundary of N^i, so that $K^i-\Phi^i$ is a manifold.

The necessity of the condition in Th. I is established, as regards a fixed G_{n-p}, in the same manner as in Ch. IV. For if $G_{n-p} \approx 0$, it bounds a C_{n-p+1} which meets Γ on some finite submanifold of K. Since Γ is an infinite cycle, (20) of Ch. IV holds and hence (15) is verified. Conversely, let (15) hold for a fixed G and arbitrary Γ. Then, if C^i_{n-p} is the cut of G by Φ^i, $(\Gamma \cdot C^i) = 0$ for every $\Gamma \subset K^i-\Phi^i$. Hence by Th. I, proved for the dual cycles of No. 20, applied to $K^i-\Phi^i$,

$$C^i_{n-p} \approx 0 \quad \bmod \Phi^i \quad \text{on } K^i-L^2.$$

Since this is true for every i, $G \approx 0$. Thus Th. I is completely proved for a fixed G.

24. When Γ_p is the fixed cycle the proofs are here also more difficult. Let K^* be the dual of K constructed as in Ch. III, and let us reduce Γ_p as usual to a subcycle of K, but G_{n-p} to a subcycle G^*_{n-p} of K^*.

Supposing now $\Gamma_p \approx 0$, we will take advantage of the fact that some $K^i-\Phi^i \supset \Gamma_p$. By (9), and in its notations,

$$\Gamma_p = \sum_{j \leq i;\, \beta} y_{j\beta}\, \Delta_p^{j\beta}. \tag{30}$$

Now any p-cell of K^i belongs at most to a finite number of Δ's. Furthermore G^*_{n-p} has only a finite number of cells with points on K^i. Hence G^* meets at most a finite number of cells of Γ, or cycles $\Delta^{j\beta}$, present in (30). As a consequence, if we substitute for Γ, G^*, and the Δ's, their expressions in terms of the cells of K or K^*, we verify directly the relation

$$(\Gamma \cdot G^*) = \sum y_{j\beta} (\Delta^{j\beta} \cdot G^*).$$

Hence $(\Gamma \cdot G^*) = 0$ if

$$(31) \qquad (\Delta^{j\beta} \cdot G^*) = 0.$$

But every Δ is either $\approx 0 \bmod L^2$ on some $K^h - \Phi^h$, or else it is a linear combination of Δ's of that type. By repeating the argument everything is reduced to proving that $(\Delta \cdot G^*) = 0$, when $\Delta \approx 0 \bmod L^2$ on some $K^h - \Phi^h$. But if

$$K^h - \Phi^h \supset C_{p+1} \to \Delta_p,$$

C_{p+1} and G^*_{n-p} intersect on some finite submanifold of K; hence the intersection properties of Ch. IV, § 1 hold, and in particular, formula (20) of that chapter is verified. It follows $(\Delta \cdot G^*) = 0$, and hence $(\Gamma \cdot G^*) = 0 = (\Gamma \cdot G)$, that is (15) holds. In other words $\Gamma \approx 0$ implies (15) whatever G, or the condition of Th. I is also necessary when Γ is the fixed cycle.

25. From this point on the treatment is practically as in No. 21. We construct the bases $\{\Gamma_p^{i\alpha}\}$, $\{\Gamma_p^{i\beta}\}$, by means of a fixed numbering of the p-cells of K. We have again (21), (23), (24). Here also in (21) only those coefficients t are $\neq 0$ for which $\Gamma_p^\varepsilon \subset K^i - \Phi^i$. Furthermore by No. 11, for a given ε, $t^i_{\beta\varepsilon} = 0$ when β exceeds a certain value. On the strength of this we justify passing from (23) to (27) as follows: Owing to the property of the t's the coefficients in (27) are all finite. Hence (25) represents a true $\Gamma_p \subset K^i$. Since each p-cell of K occurs only in a finite number of cycles Γ_p^α, $\Gamma_p^{i\beta}$, from the expressions of the cycles Γ_p^α in terms of the cells, together with (21) and (23), we can compute the coefficient of any one cell at the right in (23) and in (25), and we find the same coefficient in both. Therefore (23) and (25) represent the same

p-cycle. The rest is exactly as before, except that in place of Theorems I, II, III for finite complexes we must apply the same theorems for the infinite manifold $K - L$ of No. 21. Our Theorems I, II, III of Nr. 19 are therefore completely proved.

26. Regarding Th. V (formula (20)), a proof of a similar nature might be used, but actually it may be obtained by a mild modification of Poincaré's matrix proof of the duality relations for finite absolute manifolds, which we will not take up here. As for Th. VII, it is a direct consequence of the considerations of Nos. 16, 17, 18.

REMARK. The chief difficulty offered by the basic case just considered arises from the more general homology condition (d′) of No. 5. On the other hand, the primary justification for this condition, is that it is the most general condition which preserves the fundamental property: $(\Gamma \cdot G) = 0$ when $\Gamma \approx 0$.

27. **Formulas for coincidences and fixed points.** The machinery is now all set up for extending the formulas for the coincidences and fixed points of Ch. VI to transformations of infinite manifolds when the L's of Ch. VI are two complementary ideal elements, and similarly for Λ^1, Λ^2 and $K \times K'$. When the Betti numbers are all finite, so are the matrices α^p, β^p, and the fundamental formula (22) holds.

Suppose that some or all the Betti numbers are infinite. With arbitrary bases we are unable to carry through the computations of Ch. VI, No. 21, since we do not know how to find the inverse of an arbitrary infinite matrix. We have shown however how to construct associated pairs of bases satisfying (16), and hence such that $\alpha^p = 1$, $\beta^p = 1$. This time the computation goes through, and we obtain (23) of Ch. VI. This is then the formula to be used for infinite manifolds.

The resulting fixed point formulas (28), (29) of Chap. VI, also hold here, and by passing from the special bases to arbitrary bases, it is easily verified that they hold regardless of the bases chosen. Here however we are not at liberty to introduce a modified identity, and we can only apply

(29), Ch. VI, to a transformation T_{2b} and (28), Ch. VI, to a T_{1c}. In particular (29) gives the number of signed fixed points of a single-valued transformation of K into a finite subset of itself.

Referring to No. 18, we have

THEOREM. *The formulas (23), (28), (29) of Ch. VI, for coincidences and fixed points hold when the L's are arbitrary closed subsets of finite complexes* K, K', *with* $K-L, K'-L'$ *manifolds, and* $L^1 \subset L$, $L'^1 \subset L'$.

Thus (28) gives the number of signed fixed points of a single-valued transformation T of an arbitrary open subset A of a finite complex K, when A is a manifold, in case $\overline{T \cdot A} \subset A$. The generality of these results need not be emphasized. We still have left the extension of the results of Ch. VI, § 7, but it demands a study of the cycles and homology characters of general metric spaces, and this constitutes our next topic.

28. CONCLUSION. Referring to No. 18, and taking into account what we have obtained since, we have this: In all questions regarding a finite K and its closed subsets L, L^1, $(L \subset L^1)$ treated in the previous chapters, all restrictions concerning the L's may be dropped provided the field is restricted to homologies $\approx$. Thus the minimum base must be replaced by the base, torsion disappears, etc. On the other hand, there is but little weakening of the fundamental formulas for coincidences and fixed points.

§ 4. COMPACT METRIC SPACES AND THEIR HOMOLOGY CHARACTERS

29. THEOREM. *Every compact metric space L has a homeomorph on the Hilbert space parallelotope*

$$\mathfrak{H}: \quad 0 \leqq x_i \leqq \frac{1}{i}, \quad i = 1, 2, \cdots, +\infty.$$

This fundamental theorem was first proved by Urysohn [1, 2] in his work on metrisation. He actually proved it for all separable metric spaces but in the case of compact spaces it can be established very quickly. See on this question Chittenden [1].

Since L is also separable there exists on it an everywhere dense enumerable set of points $\{P^i\}$, $i = 1, 2, \cdots, +\infty$. If P is any point of the space we will set

$$x_i(P) = \frac{1}{i} \frac{d(P, P^i)}{1 + d(P, P^i)}$$

and call Q the point of $\mathfrak{H}$ whose coördinates are $x_i(P)$.

Let $\mathfrak{L}$ be the subset of $\mathfrak{H}$ consisting of all points Q thus obtained. To every P of L corresponds a unique Q of $\mathfrak{L}$. Conversely given Q on $\mathfrak{L}$ there is at least one corresponding point P and there can be only one, for if P' were any other we would have $d(P, P^i) = d(P', P^i)$, and since $\{P^i\}$ is everywhere dense on L, this can only occur if $P = P'$. Thus we pass from L to $\mathfrak{L}$ by a transformation T which is one-to-one. From the difinition of distance it follows readily that $d(P, P^i)$ is a continuous function of P, hence T is continuous. Since L is compact, by a well known theorem in real variables T^{-1} is likewise continuous, hence T is a homeomorphism, and the proof of the theorem is complete.

From this point on we identify L with its image $\mathfrak{L}$, i. e. take L on $\mathfrak{H}$.

It may be observed that in the present section $\mathfrak{H}$ plays merely the role of background. We could very well dispense with it, and reason directly on L, after the manner of Vietoris [1], or Alexandroff [10], but by making use of the elementary linear properties of $\mathfrak{H}$, we shall be able to proceed more speedily.

30. We understand by an *infinite p-chain, C_p, on L,* an infinite chain on $\mathfrak{H}$, with simplicial, convex or other cells,

$$C_p = \sum t_i E_p^i,$$

such that the diameter and distance of E_p^i from L on $\mathfrak{H}$ tend both to zero with $\frac{1}{i}$. If $F(C_p)$ is a finite chain, C_p is said to define a $(p-1)$-*cycle*, γ_{p-1}, *on* L. When C_p differs from an $F(C_{p+1})$ by a finite chain, we write $\gamma_{p-1} \approx 0$ on L. Furthermore, every finite p-cycle defines a $(p-1)$-cycle ≈ 0 on L. We add the condition that every infinite sum of finite cycles which is an infinite p-chain on L, likewise defines a $\gamma_{p-1} \approx 0$ on L. This is entirely analogous to the condition (d′) of No. 5 for infinite cycles on an infinite complex.

Since we have cycles and homologies between them, we have a definition for the Betti numbers of L. It follows from our definitions that in the homology theory we may consider solely simplicial chains with vertices *strictly* on L, the more general type having been introduced more as a matter of convenience than anything else. Hence if L, L' are two homeomorphic compact metric spaces which are both on $\mathfrak{H}$, the p-cycles on L, L' are in one to one correspondence with preservation of the homologies between them. Therefore the associated Betti numbers are the same. Hence they are topological invariants of L.

REMARK. There is considerable latitude in the definition of the homology characters. Thus, if we consider only integral chains, we have one type, whereas rational chains yield another. The proper choice to be made, is the one which yields a theory in closest analogy with the known invariant theory of complexes. Thus here we take rational chains, and not integral chains, for they are the type that enables us to extend the duality and separation theorems of Ch. III. See in this regard our paper [14].

The operations of subdivision, section, deformation, on the C_p's, are to be understood in the same sense as for finite chains on an S_r. The only criterion is that these operations transform an infinite chain on L into an infinite chain on L, and involve only infinite chains on L as deformation chains.

31. **A fundamental complex for L.** We shall apply a most fruitful idea, recently introduced by Alexandroff [2, 10], to obtain a fundamental infinite complex for L, to whose homology theory we shall reduce that of L. A concept similar to Alexandroff's was communicated to us several years ago

by Alexander, but it never appeared in print, and it is scarcely necessary to state that Alexandroff's result was obtained independently.

The basis lies in the following immediate corollary of the Borel theorem: *For any* $\varepsilon > 0$, L *is the sum of a finite number of open sets of diameter* $< \varepsilon$. Denoting such a covering by $\{G^\alpha\}$, $\alpha = 1, \cdots, s$, and putting $F^\alpha = \bar{G}^\alpha$, we say that $\{F^\alpha\}$ is an *ε-covering* of L; ε is called the *mesh* of the covering. A new covering $\{F^{*\alpha}\}$ of mesh ε can be constructed from $\{F^\alpha\}$ as follows:

$$F^{*1} = F^1; \qquad F^{*\alpha} = \overline{G^\alpha - G^\alpha \sum_{\beta=1}^{\alpha-1} F^\beta}.$$

Such a covering has the property that no two of the F^*'s have an inner point in common.

A covering $\{F^{*\beta}\}$ is a *subdivision* of $\{F^\alpha\}$, if each F^* is a subset of some F, and each F is a sum of F^*'s. If $\{F^\alpha\}$ is an ε_1-covering of L, and $\{F^{*\beta}\}$ is an ε_2-covering, there is a covering of L which is a subdivision of both $\{F^\alpha\}$ and $\{F^{*\beta}\}$. Put $G^{\alpha\beta} = G^\alpha G^{*\beta}$, and then $\{F^{\alpha\beta}\}$, where $F^{\alpha\beta} = \overline{G^{\alpha\beta}}$, is the required covering. The mesh of the new covering is the least of the two numbers ε_1 and ε_2. There may be constructed, therefore, a sequence of coverings Σ^i such that each is a subdivision of the preceding, no two sets of any Σ have an inner point in common, and the mesh ε_i of Σ^i decreases monotonely to zero with $\frac{1}{i}$.

32. Let $\{F^{i\alpha}\}$ be the closed sets that make up Σ^i. We shall choose points $A^{i\alpha}$, one for each $F^{i\alpha}$, at a distance $< \varepsilon_i$ from it and in such an order that no A^i precedes an A^j when $j < i$. Furthermore, if at any time we have already selected r of the points, the $(r+1)$st is to be so chosen that together they are the vertices of a non-degenerate σ_r on $\mathfrak{H}$. All these conditions can manifestly be satisfied. We will now construct the simplex $A^{i\alpha} \cdots A^{i\beta}$, wherever $F^{i\alpha} \cdots F^{i\beta}$ have at least one point in common. The set of all these simplexes includes the faces of any simplex of the set. Therefore, owing to the mode of selecting the points $A^{i\alpha}$, the sum of the simplexes

is a closed complex Φ^i. The sequence $\{\Phi^i\}$ was introduced by Alexandroff [10], who made a thorough study of its properties in relation to L. It constitutes the point of departure for the construction of our fundamental complex.

We introduce with Alexandroff a mapping μ_i of Φ^i on Φ^{i-1} defined as follows: For each $F^{i\alpha}$ we choose one $F^{i-1,\beta} \supset F^{i\alpha}$ and have μ_i map $A^{i\alpha}$ on $A^{i-1,\beta}$. Owing to the construction of the Φ's, the vertices of any simplex of Φ^i is mapped on those of a single simplex of Φ^{i-1}. We extend μ_i to the complexes by making it barycentric.

Suppose that by our procedure σ_p of Φ^i is mapped into σ_q of Φ^{i-1}, and consider the segments joining corresponding points. Their sum is a $(p+1)$-cell which is not flat, called the *joining cell* of the two simplexes. By construction $\sigma_p\,\sigma_q$ is a non-degenerate σ_{p+q+1} which carries the joining cell and no two such σ's intersect, nor do they meet the cells of the Φ's. Hence the joining cells are non-singular and no two intersect. As any face of a joining cell is a joining cell or else a cell of a Φ, the sum of the Φ's plus the joining cells is a complex, which is our fundamental infinite complex K for L.

33. We shall assume that $\{\Sigma^i\}$ begins with Σ^0 and call the initial complex Φ instead of Φ^0. The sum of the cells of K consisting of all Φ^j's, $j > i$, the cells joining them, and also Φ^i to Φ^{i+1}, will be designated by N^i, $(N^0 = K)$. It is an infinite open subcomplex of K whose boundary is Φ^i, and $\{N^i\}$ defines the total ideal element of K. On $\mathfrak{H}$ we have $\bar{K} = K + L$ so that L is a closed subset of the closed set $\bar{K}$. The sets $\{N^i\}$ constitute a sequence of sets, open in K, whose closures in $\mathfrak{H}$ have for common part precisely L, so that in a sense L may be identified here with the total ideal element of K.

An important property of K, evident from the construction of the joining cells, is that through every point P of N^i there passes a unique arc projecting it onto any Φ^j, $j \leqq i$. It is a sum of the segments that arise in the construction of the joining cells one of which, PQ passes through P, projecting it on Q of a certain Φ^h, $h \geqq i$, the next one projecting Q onto Φ^{h-1}, etc.

REMARK. As we shall not utilize the internal properties of the cells of K, the fact that they are not flat does not matter. However if it be desired to have simplicial cells, all that is necessary is to replace μ_i by a succession of barycentric deformations into intermediary complexes each of which has one vertex not belonging to its predecessor, the first being Φ^i and the last Φ^{i-1}.

34. DEFORMATION THEOREM. *Every infinite chain on the compact metric space L is reducible by subdivion, section, deformation to a subchain of the fundamental complex K.*

We shall need the following property of $\{\Sigma^i\}$, readily obtained by elementary metric considerations (Alexandroff [10]): For every Σ^i there exists a $d_i > 0$ tending to zero with $\frac{1}{i}$, such that if a simplicial E_p, whose vertices are on L, is of diameter $< d_i$, the sets $F^{i\alpha}$ which carry its vertices have a point in common. The cell E_p will then be called *proper* for Σ^i. When it is proper for Σ^i it is also proper for any preceding Σ. The number of improper cells for Σ^i, in any infinite chain on L, is always finite.

By subdivision and deformation we reduce the initial chain to a simplicial chain with vertices on L, then discard all cells not proper for Σ^0. We can express the new chain in the form

$$C_p = \sum C_p^i,$$

where: (a) the chains C_p^i are finite and no two have any common p-cells; (b) C_p^0 consists of all p-cells of C_p proper for Σ^0 but not for Σ^1; (c) C_p^i, $i > 0$, consists of all p-cells proper for Σ^i but not for Σ^{i+1}, not found in its predecessors, plus all p-cells having a vertex in common with C_p^{i-1}.

Any cell of C_p^i can be put into one of three classes accordingly as (a) its vertices are all on C_p^{i-1}; (b) some but not all are on C_p^{i-1}; (c) none are on C_p^{i-1}.

Now it is an immediate consequence of the definition of proper cells, that C_p^i can be mapped into a subchain of Φ^i, each vertex B going over into a vertex $A^{i\alpha}$ whose $F^{i\alpha} \supset B$. The mapping of the vertices of C_p^i thus obtained will be designated by λ_i. If a vertex B of C_p is common to C_p^{i-1} and C_p^i, it is not a vertex of any other C_p^j, hence it receives exactly two assignments by the λ's, one by λ_{i-1}, the other by λ_i. If E_p.of C_p^i is of type (a) or (c), we leave the mapping of its vertices as fixed by λ_{i-1} or λ_i, respectively.

Suppose now that E_p is of type (b), so that $E_p = A^0 \cdots \cdots A^q B^0 \cdots B^r = \sigma_q \sigma_r$, $q + r + 1 = p$, where σ_q includes all vertices also on C_p^{i-1}. Then we have unique simplexes $\lambda_{i-1} \sigma_q$, of Φ^{i-1}, $\lambda_i \sigma_q$, $\lambda_i \sigma_r$, $\lambda_i \sigma_q \sigma_r$ of Φ^i, also a joining cell $\mathfrak{E}$ of $\lambda_{i-1}\sigma_q$ and $\lambda_i \sigma_q$. We make a section of E_p separating σ_q and σ_r and introducing new vertices P^{jk} on $A^j B^k$. Such a section can always be made since the simplex E_p is the join of σ_q and σ_r. As a consequence of the section there results a decomposition of E_p into two convex cells E_p', E_p'', whose vertices are the A's and P's for the first, the B's and P's for the second. Map A^j on $\lambda_{i-1} A^j$, B^k on $\lambda_i B^k$, P^{jk} on $\lambda_i A^j$ for every k, then extend the mapping barycentrically to the cells E' and E'' There results a mapping of E'' into $\lambda_i \sigma_q \sigma_r$, and E' into the joining cell $\mathfrak{E}$. The total mapping thus described is consistent throughout C_p. This proves the theorem.

35. Let C_{p+1} be an infinite chain representing a p-cycle, γ_p, of L. By the preceding theorem we can reduce it to a subchain of K still representing a cycle $\approx \gamma_p$, and with a finite boundary Γ_p. Let C_{p+1}^i be the cut of C_{p+1} by Φ^i, and Γ_p^i its boundary on Φ^i, where i is sufficiently high to have $\Gamma_p \subset K - N^i$ We can remove C_{p+1}^i from C_{p+1} without affecting γ_p. Then, by sliding the points of Γ_p^i along the projecting lines of K, it can be deformed successively into subcycles of Φ^{i-1}, Φ^{i-2}, $\cdots$, Φ, and the resulting deformation is represented by a subchain $C_{p+1}'^i$ of K, under our very construction of K. As a consequence, γ_p will be represented by an infinite cycle of $K \bmod \Phi$, $\Gamma_{p+1} = C - C^i + C'^i$.

THEOREM. *The two homologies $\gamma_p \approx 0$ on L, $\Gamma_{p+1} \approx 0$ mod Φ on K, are equivalent.*

The second homology implies the first by definition, whether Γ_{p+1} bounds an infinite C_{p+2}, or is an infinite sum of such cycles or is a sum of finite cycles, each of the latter being $\approx 0 \bmod \Phi$, as deformable along the projecting lines of K onto Φ. When the first holds, Γ_{p+1} may be a sum of finite cycles. By the observation just made each of these $\approx 0 \bmod \Phi$, hence likewise $\Gamma_{p+1} \approx 0 \bmod \Phi$. Suppose that Γ_{p+1} is not such a sum or else Γ_{p+1} is an infinite sum of such cycles of the preceding type and we need merely carry out the following argument for one of them. There exists then on L a $C_{p+2} \to \Gamma_{p+1} +$ a finite C_{p+1}. From the deformation theorem, we can reduce the new chains to subchains of K at the cost of replacing perhaps Γ_{p+1} by a homologous cycle mod Φ, still to be called Γ_{p+1}, which is clearly permissible. Now for a certain i, and all chains being subchains,

$$\bar{N}^i \supset C'_{p+2} \to \Gamma_{p+1} - C^i_{p+1} + D_{p+1};\ D_{p+1} \subset \Phi^i.$$

Let D_{p+2} be the deformation chain corresponding to sliding D_{p+1} along the projecting lines down to Φ. We have

$$C'_{p+2} + D_{p+2} \to \Gamma_{p+1} - C^i_{p+1} + C'_{p+1};\ C'_{p+1} \subset K - N^i.$$

Hence $$\Gamma_{p+1} \approx C^i_{p+1} - C'_{p+1}.$$

But the last cycle is a finite cycle mod Φ, hence reducible to Φ by sliding it down along the projecting lines of K. It follows that it is $\approx 0 \bmod \Phi$, and so is Γ_{p+1}.

COROLLARY. *The homology relations for the p-cycles of L are the same as for the infinite $(p+1)$-cycles mod Φ of K. It follows, in particular, that there is a totally infinite base for the p-cycles of L (No. 13).*

36. **The Vietoris cycles.** The homology theory of L may be built up along somewhat different lines (Vietoris [1], Lefschetz [14]). We consider with Vietoris ε-chains and cycles on L. They are finite chains or cycles such that the diameters of their cells as well as their maximum distances from L are $< \varepsilon$. An ε-cycle Γ_p is $\varepsilon \approx 0$ if there exists an ε-chain $C_{p+1} \to \Gamma_p$. A Vietoris cycle γ_p on L is a sequence $\{\Gamma^i_p\}$ such that all cycles Γ^i_p whose i exceeds a certain value are ε-homologous to one another. By definition $\gamma_p \approx 0$ whenever for every ε the cycles Γ^i_p whose i is large enough are $\varepsilon \approx 0$.

From the deformation theorem applied to $\sum \Gamma_p^i$, we infer that every Vietoris cycle can be represented by a sequence $\{\Gamma_p^i\}$, where Γ_p^i is a subcycle of Φ^i. Furthermore, the ε-homologies between the Γ's can be replaced by bounding relations for finite subchains on an N^i whose i is large enough.

We will now compare the two representations for the cycles on L. Let γ_p be a cycle under our previous definition, and Γ_{p+1} its representative infinite cycle mod Φ. The notations being as before, we have

$$D_{p+1}^i = C_{p+1}^i - C_{p+1}^{i+1} \rightarrow \Gamma_p^i - \Gamma_p^{i+1} \approx 0 \quad \text{on } \overline{N^i - N^{i+1}}.$$

Since the cells of $\overline{N^i - N^{i+1}}$ tend to zero with $\frac{1}{i}$, $\{\Gamma_p^i\}$ is a Vietoris cycle.

Conversely, let $\{\Gamma_p^i\}$ be a Vietoris cycle already reduced to the type with Γ_p^i a subcycle of Φ^i. We have boundary relations

$$D_{p+1}^i \rightarrow \Gamma_p^i - \Gamma_p^{i+1},$$

where the D's are again subchains, and $D^i \subset N^j$ such that j increases indefinitely with i. By extending each D down along the projecting lines of K, it is a simple matter to replace $\{\Gamma_p^i\}$ by a cycle $\{\Delta_p^{h_i}\}$, $\Delta_p^{h_i} \subset \Phi^{h_i}$, such that the corresponding

$$D'^{h_i}_{p+1} \rightarrow \Delta_p^{h_i} - \Delta_p^{h_{i+1}}$$

is on $\overline{N^{h_i} - N^{h_{i+1}}}$. Then

$$\Gamma_{p+1} = \sum D'^{h_i}_{p+1}$$

is a cycle mod Φ associated in the same manner as previously with a Vietoris cycle $\{\Delta_p^i\}$, $\Delta_p^i \subset \Phi^i$, having $\{\Delta_p^{h_i}\}$ as a subsequence. From this follows readily that both define Vietoris cycles homologous on L.

37. We have thus paired off the two types of representations of cycles of L, in such manner that with every Γ_{p+1} mod Φ defining a cycle γ_p on L, there is associated the Vietoris cycle $\{\Gamma_p^i\}$ consisting of the parts of the boundaries of its cuts

on the complexes Φ^i. This pairing off associates with one another the representations of arbitrary linear combinations of any given set of cycles. To show that the resulting homology theories are the same, it is sufficient to prove the

THEOREM. *Let Γ_{p+1} be a cycle mod Φ representing γ_p of L in the sense previously defined, $\{\Gamma_p^i\}$ the associated Vietoris cycle. The two homologies $\Gamma_{p+1} \approx 0$ mod Φ, $\{\Gamma_p^i\} \approx 0$, are equivalent.*

If the first homology holds Γ_{p+1} is a finite or infinite sum of bounding cycles any one of which can replace it in the argument. For such a cycle we have immediately a

$$C_{p+2} \to \Gamma_{p+1} \quad \text{mod } \Phi.$$

Hence, passing to the cuts by Φ^i,

$$C_{p+2}^i \to C_{p+1}^i + C_{p+1}'^i \quad \text{mod } \Phi; \qquad C_{p+1}'^i \subset \Phi^i.$$

Since Φ and $C_{p+1}'^i$ do not meet, and $C_{p+1}^i + C_{p+1}'^i$ is a cycle mod Φ, the boundary cells of C_{p+1}^i on Φ^i and all those of $C_{p+1}'^i$ must cancel one another; hence

$$-F(C_{p+1}'^i) = \Gamma_p^i \approx 0 \text{ on } \Phi^i$$

for every i, or the second homology follows from the first.

Conversely the second homology implies that we have, for every i, a finite chain

$$D_{p+1}^i \to \Gamma_p^i,$$

where if j is the highest integer such that $N^j \supset D^i$, j increases indefinitely with i. Now

$$\Gamma_{p+1}^i = C_{p+1}^i - C_{p+1}^{i+1} - (D_{p+1}^i - D_{p+1}^{i+1}) \to 0,$$

so that Γ_{p+1}^i is a finite cycle. The number of such cycles having cells on any $K - N^i$ being finite, their sum, for $i = 1, 2, \cdots, +\infty$, is a proper infinite chain on K. If we add up however we obtain precisely Γ_{p+1}. Hence Γ_{p+1} is an infinite sum of finite cycles, and therefore it is ≈ 0 mod Φ.

Since the Vietoris representation and the representation by infinite cycles yield the same homology theory under the correspondence between them, we are justified in defining as identical the p-cycles defined by Γ_{p+1} or by the associated sequence $\{\Gamma_p^i\}$, the two constituting two different representations of the same γ_p on L.

38. REMARKS. I. When L is locally connected the present homology theory does not differ from the old. For the characteristic property of L is that chains sufficiently near it can be deformed onto it without leaving its vicinity, the argument being exactly as in Ch. II, § 5. Therefore if $\{\Gamma_p^i\}$ is a Vietoris cycle for L, above a certain i all the Γ's are deformable onto L. Furthermore, when on L they are homologous to one another on any assigned neighborhood of L. The chains that they bound in pairs are then deformable onto L for i large enough, hence the cycles imbedded in L (on L in the strict sense) whose i is large enough, all belong to the same family of homologous cycles in the earlier sense of the term. For the same reason, if $\{\Gamma_p^i\} \approx 0$, the $C_{p+1}^i \rightarrow \Gamma_p^i$ for i above a certain value, is deformable with Γ_p^i onto L, hence $\Gamma_p^i \approx 0$ on L. Thus the homologies are preserved when we pass from the Vietoris cycles to the cycles on L. But if Γ_p is a cycle on L, all that is necessary is to subdivide it indefinitely and to take the sequence $\{\Gamma_p^i\}$ of its subdivisions as the Vietoris cycle. The correspondence in the reverse direction manifestly preserves the homologies. A similar argument has been applied recently to complexes by Vietoris [3].

II. Brouwer [13] and Alexandroff [7, 10] have defined homology characters without having recourse to infinite chains or sequences. If $\varrho(\varepsilon)$ is any particular homology invariant defined for (finite) ε-chains, they set $\varrho(L) = \lim_{\varepsilon \rightarrow 0} \varrho(\varepsilon)$. An equivalent precedure is to take $\varrho(L) = \lim_{i \rightarrow +\infty} \varrho(\Phi^i)$. We have shown [14] that the Betti numbers thus obtained are the same as those derived from rational (not integral) chains by infinite chains or Vietoris cycles, whenever L is a subset of an S_r.

III. If L^1 is a closed subset of L, we can define the Betti numbers of $L \bmod L^1$ as follows: Considering L as the total ideal element of K, L^1 is a closed ideal element which is a subset of L. A C_{p+1} subchain of K, such that $F(C_{p+1})$ is a cycle $\mod L^1$ defines a p-cycle $\mod L^1$, γ_p. If C_{p+1} differs from an $F(C_{p+2})$ by a cycle $\mod L^1$ we write $\gamma_p \approx 0 \mod L^1$ on L. This enables us to introduce the Betti numbers $\mod L^1$ in the usual manner.

IV. We have limited the treatment to compact spaces. It can be readily extended, however, to a locally compact metric space L. The number of sets in every $\{F^{i\alpha}\}$ is then in general enumerable but not finite. As a consequence the Φ's are infinite complexes, and the space L plays the part of a closed ideal element of K, but it is not its total ideal element. In place of the cycles considered, we have then those of $K-L'$ $\mod L$, where L' is the open ideal element complement of L. These cycles define here also the absolute cycles of L, but otherwise the rest is as before.

39. **The dimension of a space L and its bearing upon the homology theory of L.** We owe to Urysohn [3] and Menger [1] a definition of the *dimension* of an arbitrary separable metric space L which has become classic.

Both authors first define the dimension of L at a point P of the space, basing it upon the notion of *ε-separation*. The point P is *ε-separated* by the subset B of L if we have

$$L = A + B + C;\ A \supset P:\quad d(A+B) < \varepsilon;\ A \cdot \bar{C} + \bar{A} \cdot C = 0.$$

We have then the following recurrence for the dimension of L: (a) L is of dimension -1 when and only when it is vacuous. (b) It is of dimension n at P when it is not of dimension $< n$ and furthermore, is ε-separated by a set of dimension $< n$ whatever ε. (c) L is of dimension n if it is of dimension at most n at every point, and of dimension n for some point.

Around the preceding definition there has been built up a far reaching analysis of metric spaces. The following theorems regarding the dimension of a compact metric space L are particularly closely related to the homology theory.

I. The least dimension of any fundamental complex K attached to L is $n+1$, where n is the dimension of L (Alexandroff [2, 10]).

II. Every n-dimensional compact metric space L has a homeomorph in S_{2n+1} (Menger [1]).

From Alexandroff's result, in conjunction with the definition of the relative Betti numbers (No. 38), follows:

III. If n is the dimension of L all Betti numbers absolute or relative, for the cycles of dimension $> n$ are zero.

Let $\mu(L)$ designate the highest dimension for which there are non-zero Betti numbers (absolute or relative) attached to L. In other words there exists a closed subset L^1 of L, such that some Betti number $R_\mu(L; L^1) \neq 0$, and μ is the highest integer for which this holds. Of course $L^1 = 0$ is not ruled out, i. e. the Betti number $\neq 0$ may be $R_\mu(L)$ itself. Now we already know that when L is an n-complex, $\mu(L) = n$. The possibility of extending this proposition to an arbitrary compact metric space had much to do with our introducing the relative Betti numbers some years ago [12, 14] (1927–28). That this proposition does not hold in general has just been established by Pontrjagin [2] as follows: From the result of No. 15, applied to the approximating complexes of the spaces L, L', we derive the relation

$$\mu(L \times L') = \mu(L) + \mu(L'). \tag{32}$$

Hence

$$\mu(L) = \dim L, \tag{33}$$

implies

$$\dim L \times L' = \dim L + \dim L'. \tag{32'}$$

Both (32), (32′) hold for the dimensions 0, 1, and for these dimensions (33) holds. However Pontrjagin has constructed an example for which $\dim L = \dim L' = 2$; $\dim L \times L' = 3$, and hence (32′) fails.

It seems probable that (33) holds for a locally connected L. See also on this question Alexandroff [11].

§ 5. Closed Subsets of S_r

40. We shall now investigate the closed subsets of S_r, particularly as regards the relations between their topology and that of their residual sets. A considerable part of our results holds for subsets of any finite complex, but the extensions are fairly obvious and there is much to gain in adopting the simpler situation at the outset. On the questions to be treated in the present section, and as regards recent work by Alexander, see also our observation at the end of No. 19.

Let then L be a closed subset of S_r. We take $r > 1$ in order to avoid uninteresting and well known special cases. Let E_r be a convex polyhedral cell $\supset L$, Φ its boundary, K an infinite simplicial complex coincident with $\bar{E} - L$. K is constructed by means of certain neighborhoods N^i of L which are turned into subcomplexes of K (No. 18). We designate by Φ^i the boundary of N^i, and take $N^0 = K$, $\Phi^0 = \Phi$.

We shall first correlate the cycles of L with those of K, which will lead us rapidly to the duality relation of Alexander and the separation properties of S_r by L or of a sphere by a closed subset.

As regards the infinite chains and cycles on L in the sense of No. 29, we assume them all immersed in S_r. That this is not a restriction as far as homologies on L are concerned, is shown as in No. 30, when we compared two homeomorphic spaces immersed in the Hilbert parallelotope.

41. Given C_p on E, with a finite boundary $-\Gamma_{p-1}$, we may first reduce it to a subchain. Preserving the same notations for the reduced elements, C_p will determine, in the sense just considered, a cycle, γ_{p-1}, of L, and it will be a natural extension of our conventions to write

$$C_p \rightarrow \gamma_{p-1} - \Gamma_{p-1},$$

or if $C_p = \Gamma_p$, a cycle mod Φ,

$$\Gamma_p \rightarrow \gamma_{p-1} \quad \text{mod } \Phi. \tag{34}$$

We add the convention that if $\gamma_{p-1} \approx \delta_{p-1}$ on L, we also write

$$C_p \to \delta_{p-1} - \Gamma_{p-1},$$

$$\Gamma_p \to \delta_{p-1} \mod \Phi. \tag{35}$$

We now consider the reverse process: Given δ_{p-1} we shall find Γ_p satisfying (35).

Assuming first $p > 1$, if D_p is the chain defining δ, we may write

$$D_p = \sum D_p^i,$$

where D_p^i is finite and on N^{i+1}. There exists then an associated Vietoris sequence $\{\varDelta_{p-1}^i\}$ such that

$$D_p^i \to \varDelta_{p-1}^{i+1} - \varDelta_{p-1}^i.$$

Since E_r is a cell it contains a $C_p^j \to \varDelta_{p-1}^j$. From this follows

$$C_p^{i+1} - C_p^i - D_p^i \to 0.$$

Since $p > 1$ the absolute cycle at the left ≈ 0 on E_r, or

$$E_r \supset C_{p+1}^i \to C_p^{i+1} - C_p^i - D_p^i. \tag{36}$$

We can take the $\varDelta$'s and C's polyhedral as regards the initial complex $\bar{E}_r$, then so subdivide K and the C's that the sums of the cells of the latter not on N^i (their cuts by Φ^i in the sense of § 2), are subchains of $K - N^i$. This will then also hold when N^i is replaced by any N^j, $j < i$. We designate the cuts by an upper index j as C_p^{ij}, etc. We also have

$$C_p^{ij} \to \Gamma_{p-1}^{ij}, \tag{37}$$

a finite subcycle of Φ^j. From (36), and since $D_p^i \subset N^{i+1}$, we find by passing to the cuts by Φ^i,

$$C_{p+1}^{ii} \to C_p^{i+1, i} - C_p^{ii} + C_p'^{i},$$

where $C_p'^{i}$ is a subchain of Φ^i. The right hand side being a cycle its boundary is zero. Hence, owing to (37),

$$C_p^{\prime i} \to \Gamma_{p-1}^{ii} - \Gamma_{p-1}^{i+1,i},$$
$$C_p^{i+1,i} - C_p^{i+1,i+1} \to \Gamma_{p-1}^{i+1,i} - \Gamma_{p-1}^{i+1,i+1}.$$

Adding up, we find

(38) $$C_p^{\prime\prime i} \to \Gamma_{p-1}^{i+1} - \Gamma_{p-1}^{i}, \qquad \Gamma^j = \Gamma^{jj};$$

where $C_p^{\prime\prime i}$ is a finite subchain of $\overline{N^i - N^{i+1}}$. In the sequence $\{C_p^{\prime\prime i}\}$ only consecutive chains have cells in common, therefore

$$\Gamma_p = C_p^{11} + \sum C_p^{\prime\prime i}$$

is a proper infinite subchain of K. By (37) and (38) we find $\Gamma_p \to 0$, hence it is an infinite cycle. The corresponding cycle on L is $\gamma_{p-1} = \{\Gamma_{p-1}^i\}$ and Γ_p, γ_{p-1} satisfy (34). But

$$\bar{N}^i \supset C_p^i - C_p^{i,i} \to \Delta_{p-1}^i - \Gamma_{p-1}^i.$$

The chain $C_p^i - C_p^{i,i}$ tends to L as i increases indefinitely. Since it can be replaced by an ε_i-chain where ε_i tends to zero under the same circumstances, the boundary relation just written proves that $\gamma_{p-1} \approx \delta_{p-1}$ on L. Therefore Γ_p and δ_{p-1} are related by (35), where as a matter of fact mod Φ may be omitted.

When $p = 1$ everything that we have said holds provided that the boundary relations are taken mod Φ. In that case we have again (35) but mod Φ cannot be suppressed. Its role is thus seen to take care of the case $p = 1$.

42. Let us understand for the present by Γ_p a cycle mod Φ, and by γ_{p-1} the corresponding cycle on L satisfying (34).

THEOREM. *When $p < r$ the two homologies*

(39) $$\Gamma_p \approx 0 \; mod\, \Phi, \qquad \gamma_{p-1} \approx 0 \; on \; L$$

are equivalent.

We take for Γ_p a subcycle, and call Γ_{p-1}^i the part on Φ^i of the boundary of its cut by Φ^i, so that $\{\Gamma_{p-1}^i\}$ is a Vietoris sequence for γ_{p-1}. Now the first homology (39) implies

$$\Gamma_p^i \approx 0 \quad \text{mod}\,(\Phi + \Phi^i) \text{ on } K - N^i.$$

From this relation, plus the fact that Φ and Φ^i do not meet, follows that there is a finite subchain C_p^i of Φ^i, $\to \Gamma_{p-1}^i$ for every i, which in the Vietoris sense implies $\gamma_{p-1} = \{\Gamma_{p-1}^i\} \approx 0$ on L.

Conversely suppose that the second homology (39) holds. Then there exists an $N^{\alpha_i} \supset C_p^i \to \Gamma_{p-1}^i$, where α_i increases indefinitely with i. Therefore $\Gamma_p^i - C_p^i$ is a cycle mod Φ on $\bar{E}_r$, and since $p \neq r$, it is $\approx 0 \bmod \Phi$, or

$$\bar{E}_r \supset C_{p+1} \to \Gamma_p^i - C_p^i \quad \bmod \Phi.$$

Taking the cuts by Φ^{α_i}, which we designate by an upper index α_i, we have then

$$C_{p+1}^{\alpha_i} \to \Gamma_p^{\alpha_i} \quad \bmod \bar{N}^{\alpha_i},$$

which means that

$$\Gamma_p \approx 0 \quad \bmod \bar{N}^{\alpha_i}.$$

Since α_i increases indefinitely with i this implies the first homology (39).

For $p = r$ the situation is the same as regards the homologies

$$\Gamma_r \approx t\bar{E}_r \quad \bmod \Phi, \qquad \gamma_{p-1} \approx 0 \quad \text{on } L,$$

and it need not be taken up in detail.

43. Application to the sphere. Alexander's duality relation. The Jordan-Brouwer theorem. The preceding results find ready application to a closed (true) subset L of a combinatorial H_n which is also an M_n. All that is necessary is to take for E_n the combinatorial cell obtained by removing a point of $H - L$ from the sphere. The result of No. 42 combined with the duality relation (19), yields then the extension of the separation properties of Ch. III, No. 34, and in particular formula (17) of that Chapter, or Alexander's duality relation [8], for an arbitrary L,

$$(40) \qquad R_p(H_n - L) = R_{n-p-1}(L) + \delta_{p0} - \delta_{p,\,n-1}.$$

For $p = 0$ it gives the number of regions into which L disconnects H_n:

(41) $$R_0(H_n - L) = R_{n-1}(L) + 1.$$

When $n = 2$ this reduces to the following relation, first given by Brouwer [13],

$$R_0(H_2 - L) = R_1(L) + 1,$$

which describes the separation of the two-sphere or Euclidean plane by their closed sets. The fundamental relation (40) was proved by Alexander mod 2 and L a complex (not a subcomplex), then extended to any L by Alexandroff [8, 10]. Arguments of a related nature are also found in Alexandroff's investigations [9, 10] on the so-called Phragmén-Brouwer theorem, describing the separation of H_n by the sum of two closed sets in terms of the properties of their intersection.

44. THE JORDAN-BROUWER THEOREM. *Let L be an absolute $(n-1)$-circuit on H_n. Then:* (a) *L is orientable.* (b) $H - L = \mathfrak{R}^1 + \mathfrak{R}^2$, *the sum of two regions each totally bounded by L.* (c) *L is everywhere accessible from $\mathfrak{R}^i$.*

By *accessibility everywhere from* $\mathfrak{R}^i$ is meant that the points of L, which are end points of arcs on $\mathfrak{R}^i$, are everywhere dense on L.

We reproduce the very elegant proof which Alexander [8] derived from his duality relation.

Since H, L are (finite) complexes of different dimensions they cannot coincide, $H - L \neq 0$, and therefore it consists of a certain number of regions. But L being connected, $R_{n-1}(L; 2) = 1$, hence, by (41) mod 2,

$$R_0(H - L; 2) = R_{n-1}(L; 2) + 1 = 2;$$

so that there are just two regions, $\mathfrak{R}^1$, $\mathfrak{R}^2$. But

$$R_0(H - L) = R_0(H - L; 2) = 2.$$

Hence by (41), taken in the non-modular sense,

$$R_{n-1}(L) = R_0(H - L) - 1 = 1 = R_{n-1}(L; 2),$$

which proves the orientability of L.

For the rest we shall cover L with a complex K whose mesh $< \varepsilon$. Let E_{n-1} be any cell of K, and set $K' = K - E_{n-1}$. Since K is an absolute orientable circuit, every $(n-1)$-cycle made up of its $(n-1)$-cells must include them all. As a consequence, K' has no $(n-1)$-cycles, or $R_{n-1}(K') = 0$. From (41) follows then $R_0(H - K') = 1$, or $H - K'$ consists of a single region $\mathfrak{R}$. Let P be a point of $\mathfrak{R}^i$ and hence of $\mathfrak{R}$, Q a point of E_{n-1}. Since both $\subset \mathfrak{R}$, they can be joined by a polygonal line $\subset \mathfrak{R}$ and hence not intersecting K'. Its first intersection with L, when we follow it from P to Q, is a point Q' on E_{n-1} and except for its end point Q', the polygonal line $PQ' \subset \mathfrak{R}^i$. Since Q' is a point of any $(n-1)$-cell of a K_{n-1} whose mesh is ε, covering L, where ε is arbitrarily small, the set of all points Q' is everywhere dense on L, which proves (c).—On QQ' there are points of $\mathfrak{R}^i$ not farther than 2ε from any point of E_{n-1}. Hence $\bar{E}_{n-1}$ belongs to the boundary of both $\mathfrak{R}^1$ and $\mathfrak{R}^2$. Since it is an arbitrary closed $(n-1)$-cell of L, and all points of L are on such closed cells, this completes the proof of (b), and hence of the theorem.

45. **The pseudocycles of a closed set.** We return to the closed subset L of S_r, whose topology is henceforth our sole concern. Since L is a subset of the r-complex $\bar{E}_r$, its cycles above a certain dimension n are ≈ 0. We choose some fixed n having that property, which may or may not be the least integer for which it holds. As a matter of fact we may take for n the dimension of L, but its particular choice is unimportant.

Since S_r may be considered as on an $S_{r'}$ whatever $r' > r$, L has a homeomorph on any S_r whose r is sufficiently large. Since L can be replaced by its homeomorph, we may assume that it is a true subset of an S_r whatever r above a certain value. We shall choose $r > 2n+1$.

Departing at first slightly from Ch. VI, § 7, we define a *pseudocycle* g_{r-p} of L, as a chain of S_r whose boundary does not meet L, or what is the same, as a cycle $\operatorname{mod} S_r - L$. By $g \approx 0$, we understand $\approx 0 \operatorname{mod} S - L$, or that g is homologous to a chain which does not meet L. The corresponding Betti number will be called $\varrho_{r-p}(L)$. If we reduce the chain to a subchain of K, it becomes an infinite chain whose boundary is a finite cycle, and for the same reasons as usual, we can confine the discussion to that type throughout. Regarding the dimensions we shall be chiefly interested in comparing the p-cycles on L with the $(r-p)$-pseudocycles; therefore we take $0 \leqq p \leqq n$. Since $r > 2n+1$, we shall have $p < r-1$, $1 < r-p \leqq r$ and this will enable us to avoid more or less the awkward extreme dimensions.

Since we are only interested in the part of g_{r-p} near L, we may assume that $F(g_{r-p}) = G_{r-p-1} \subset E_r$. It is then an absolute (finite) subcycle of $K-L$.

Let now γ_p be a cycle on L, Γ_{p+1} its representative subcycle of $K \bmod \Phi$, C^i_{p+1} the cut of Γ_{p+1} by Φ^i, C^i_p the part of its boundary on Φ^i. We will assume for the moment that g_{r-p} has been reduced by the deformation theorem (No. 3) to a subchain of the dual K^* of K. By taking an initial K of suitably small mesh, the resulting displacement may be made as small as we please. Since G_{r-p-1} is finite it is on $K-N^i$ when i is sufficiently large. The intersection of C^i_{p+1} and g_{r-p} is then on the finite manifold $K-N^{i+1}$, and (19), (20) of Ch. IV may be applied to it. Since G does not meet $\Gamma_{p+1}-C^i_{p+1}$, these relations yield

$$(C^i_p \cdot g_{r-p}) = (-1)^{p+1}(C^i_{p+1} \cdot G_{r-p-1}) = (-1)^{p+1}(\Gamma_{p+1} \cdot G_{r-p-1}).$$

Therefore $(C^i_p \cdot g_{r-p})$ remains fixed when i exceeds a certain value. This value is by definition the index $(\gamma_p \cdot g_{r-p})$, or

$$(42) \qquad (\gamma_p \cdot g_{r-p}) = (-1)^{p+1}(\Gamma_{p+1} \cdot G_{r-p-1}).$$

The index $(g \cdot \gamma)$ is then defined by the permutation laws for chains on S_r.

46. LEMMA I. *Every* G_{r-p-1} *is the boundary of a* g_{r-p}.

This follows from the fact that E_r is an r-cell, and that G_{r-p-1}, a cycle of dimension >0 on E_r, bounds a chain g_{p-1} on the cell.

LEMMA II. *The homologies* $g_{r-p} \approx 0$, $G_{r-p-1} \approx 0$ *on* $K - \Phi$ *are equivalent.*

The first implies $S_r - L \supset g'_{r-p} \approx g_{r-p}$ on S_r. As $g' - g$ is an absolute cycle its boundary $\equiv 0$, or $F(g') = F(g) = G$. Now g' may have points not on E_r. In that case we slide the part of g' not on the cell along the rays issued from a point A on E_r, and thus reduce g' to a chain on E_r with the same boundary G. Therefore the second homology is a consequence of the first.

Let now the second homology hold. There is a finite subchain of $K - \Phi$, $g' \to G$, hence $g' - g$ is an absolute cycle. As above, it is ≈ 0 on E_r, so that $g \approx 0 \bmod K - L$, or the second homology implies the first.

THEOREM. *The cycles* γ_p, g_{r-p} *satisfy the duality theorems* $I, \cdots, IV$, *of No. 19.*

From the two preceding lemmas, the result of No. 42, and formula (42), we infer that the homologies and intersection properties for γ, g, are as for the dual types Γ, G. Hence our theorem is a consequence of the fact that it holds for these two types.

We point out explicitly: (a) The base $\{g^{\alpha}_{r-p}\}$ is semi-finite, since it is associated with the base $\{G^{\alpha}_{r-p-1}\}$ which has that property. (b) We have the duality relation

$$\varrho_{r-p}(L) = R_p(L).$$

The third proposition of No. 19 enables us to extend the definition of the pseudocycles. Henceforth we consider as a single pseudocycle, the total class of those for which the indices $(\gamma \cdot g)$ have the same value. If we wish to make the permutation laws for indices and the like independent of r, we may restrict r to odd or even integers. For the present, however, it is not necessary to be particularly careful about it.

§ 6. TRANSFORMATIONS OF COMPACT METRIC SPACES

47. The results of the preceding section will now be applied to the extension of the formulas for coincidences and fixed

points to compact metric spaces. The case of finite Betti numbers and subsets of S_r was treated in our recent paper [18]. The principal difficulties are caused by infinite Betti numbers and bases. We first consider subsets of Euclidean spaces, and then remove this restriction.

Let then L, L' be two closed subsets of Euclidean spaces undergoing certain continuous transformations. We may assume that they are immersed in spaces S_r, S'_r of the same dimension r. The set L may be identified in every respect with the L of § 5, whose general notations will be used here. In addition, the analogues of $K, n, \cdots$ for L' will be designated by $K', n', \cdots$. The study of the transformations leads us to introduce the product space $L \times L'$, immersed in the convex polyhedral cell $E_r \times E'_r$ of $S_r \times S'_r$. We shall take as the neighborhoods of $L \times L'$ on $S \times S'$, a subsequence of $\{N^i \times N'^j\}$. Any subsequence for which no two consecutive terms $N^i \times N'^j$ have the same i or j is satisfactory.

Just as for complexes, we can expect to solve the coincidence problem only for restricted transformations. Let T be a continuous transformation of L into L'. We understand by *extension* $\mathfrak{T}$ of T a sequence $\{\mathfrak{T}_h\}$ of continuous transformations of S_r into S'_r such that the transformation T_h of L into L' induced by $\mathfrak{T}_h$, tends uniformly to T with indefinitely increasing h.

ASSUMPTION. *Every transformation T to be considered has an extension $\mathfrak{T}$ such that $\mathfrak{T}_h$ has for image on $S \times S'$ a pseudocycle, Γ_r^h, of $L \times L'$, and that all the Γ's are homologous pseudocycles.*

We will now single out two dual types of transformations satisfying our assumption, and for which the coincidence problem can be solved.

The following new terms will be found convenient at this juncture. $\mathfrak{T}$ will be called a *finite* extension whenever $d(F(\Gamma_r^h); L \times L')$ has a positive lower bound when h tends to infinity.—Two pseudocycles Γ_r, Γ'_r of $L \times L'$ will be called *homologous on the set* A, whenever $\Gamma - \Gamma' \approx$ on A to a chain that does not meet $L \times L'$.

Our first type of transformation, T_1, is characterized by an extension $\mathfrak{T}_1 = \{\mathfrak{T}_{1h}\}$, called an *expansion*, and having the following property: If Γ_r^{1h} is the image of $\mathfrak{T}_{1h}$, then, for any i, all pseudocycles Γ_r^{1h} whose h is sufficiently high, are homologous on $N^i \times S'$, while the boundary of each is exterior to some $S \times N'$. Our second type, T_2, has an extension $\mathfrak{T}_2$ called a *contraction*, whose pseudocycles Γ_r^{2h} are all homologous on $S \times N'^i$ for h sufficiently high, while the boundary of each is exterior to some $N \times S'$.

The justification for these terms is immediate: A contraction is such that the totality of points of S transformed by $\mathfrak{T}_{2h}$ tends to L'. If the contraction is finite, there exists a neighborhood N of L, such that $\mathfrak{T}_{2h} \cdot N$ tends to L' when h increases indefinitely. The expansion acts in inverse manner, and in particular, if it is finite, there exists a neighborhood N' of L' such that $\mathfrak{T}_{1h}\, \Phi^i \subset S' - N'$, whatever i above a certain value. The similarity with the types T_{1b}, T_{1c} and T_{2b}, T_{2c} of Ch. VI, is clear.

It may be observed that the property of T of having an extension of specified type, is to some extent independent of the spaces in which L, L' are immersed. For let $S_r \subset S_{r+1}$, $S'_r \subset S'_{r+1}$. Let P be a point of S_r, QR a segment in S_{r+1} bisected by P and orthogonal to S_r. We define the extension $\mathfrak{T}' = \{\mathfrak{T}'_h\}$ of T associated with S_{r+1}, S'_{r+1}, by the condition that $\mathfrak{T}'_h \cdot \overline{QR} = \mathfrak{T}_h \cdot P$. It is then a simple matter to show that, by choosing for the length of QR a suitable function of P, $\mathfrak{T}'$ is of the same type as $\mathfrak{T}$. For example, if $\mathfrak{T}$ is a contraction, $\mathfrak{T}'$ will also be a contraction, if we take QR of constant length.

48. Suppose that we have T_1, T_2, such that their extensions $\mathfrak{T}_1$, $\mathfrak{T}_2$ are respectively an expansion and a contraction of which one, say $\mathfrak{T}_1$, is finite. We will then begin by taking $\{N'^i\}$ such that every $F(\Gamma_r^{1h})$ is exterior to $S \times N'^1$, and hence to every $S \times N'^i$. For every N'^i we select a $\mathfrak{T}_{2h_i}$ such that $\Gamma_r^{2h_i} \subset S \times N'^i$. Preserving in $\{\Gamma_r^{2h}\}$ only the elements whose h is an h_i, and replacing h_i by i, we shall then have a $\mathfrak{T}_{2i}$ such that, for every i, $\Gamma_r^{2i} \subset S \times N'^i$. We then choose

an N which, in similar manner we may consider as N^i, such that $F(\Gamma_r^{2i})$ is exterior to $N^i \times S'$. Finally, we select $\mathfrak{T}_{1i}$ such that $\Gamma_r^{1i} \subset N^i \times S'$. It follows that $\mathfrak{T}_{1i}$, $\mathfrak{T}_{2i}$ are a pair of transformations of type T_{1b}, T_{2b} for N^i, N'^i. Therefore we can determine the signed coincidences $(\Gamma_r^{1i} \cdot \Gamma_r^{2i})$ by the formulas of Ch. VI.

As i increases indefinitely, the coincidences of $\mathfrak{T}_{1i}$, $\mathfrak{T}_{2i}$ tend manifestly to those of T_1, T_2 on L, L'. If $\lim(\Gamma_r^{1i} \cdot \Gamma_r^{2i})$ exists, we may then consider it as the index of signed coincidences for T_1, T_2. In particular, if there are no coincidences, there will be none for $\mathfrak{T}_{1i}$, $\mathfrak{T}_{2i}$ when i exceeds a certain value and the limit in question will be zero. Hence when it is $\neq 0$ some coincidences must exist. A similar treatment evidently holds when $\mathfrak{T}_2$ is finite. *Therefore, when* T_1, T_2 *have extensions which are respectively an expansion and a contraction, one of them finite, the coincidence problem is reduced to finding lim* $(\Gamma_r^{1i} \cdot \Gamma_r^{2i})$ *as* i *increases indefinitely.*

49. Frequently there will exist an extension $\mathfrak{T} = \{\mathfrak{T}_h\}$ whose components $\mathfrak{T}_h$ are all one and the same transformation U, which coincides then with T on L, L'. The following two cases are particularly important:

I. *T is a single-valued transformation.* The existence of U is then an immediate consequence of a noted theorem on the extension of continuous functions, applied to the different coördinates, (Carathéodory [2], p. 617).

II. *L' is a copy of L, S' a copy of S, and the transformation is the image of the identity for L.* Then U is the image of the identity for S.

In the important problem of the fixed points of single-valued transformations, we have the two cases I, II together. The solution of the fixed point problem can be given only when the two transformations have extensions of opposite types, and one of them is finite. The extension in case II can be considered as an expansion or a contraction, but it is not finite. Hence in the fixed point problem it becomes important to obtain an extension of the other transformation which is a finite expansion or contraction.

A sufficient condition in order that every continuous single-valued transformation T of L into L' possess an extension which is a finite contraction, is that the identity for L' possess an extension $\{\mathfrak{F}_h\}$ over S' which is a finite contraction, where the $\mathfrak{F}$'s are single-valued over some fixed N'. For, U being as before, $\{\mathfrak{F}_h U\}$ is the required extension.

When L' is locally connected, the condition just mentioned is always fulfilled. For let N' be a polyhedral neighborhood of L', and cover $\bar{N}$ with a complex of mesh ε. Referring to Ch. II No. 24, we can choose $\bar{N}$ such that it will be deformable onto L, the points on L undergoing a displacement that tends to zero with ε. Take a sequence $\{\varepsilon_h\}$ monotonely tending to zero. If $\mathfrak{F}_h$ is the deformation corresponding to $\varepsilon = \varepsilon_h$, $\{\mathfrak{F}_h\}$ is an extension of the identity which is a finite contraction.

As regards the fixed point problem, it follows from what we have just said that *in order that our method be applicable to the fixed points of every continuous single-valued transformation of L into itself, it is sufficient that L be locally connected.*

50. In the paper [18] in which we treated these questions, we stated without proof (p. 278) that all sets possess the property just brought out for those that are locally connected. As a consequence, the fixed point formula loc. cit. was asserted to hold for all closed sets there considered. That some restriction is necessary, was pointed out to us by Hopf through the following example: Let γ, γ' be two concentric circles, with γ as the interior circle, O their common center, ζ a spiral in the region between γ and γ', asymptotic to both, and meeting every circle concentric to them in a single point, or not at all. We will assume that ζ winds around O in the positive direction as it tends to γ'. Take $L = \gamma + \gamma' + \zeta$, and consider the continuous single-valued transformations T, T', T'' of L into itself, defined as follows: Through T every point P of ζ is translated along the spiral in such manner that the angular rotation of the radius vector OP is a fixed positive angle $\omega < 2\pi$, while the circles undergo a rotation

of ω. Through T', on the other hand, γ' rotates by ω, γ by $-\omega$. Choose some fixed point A on ζ. Any point P of ζ will rotate by an amount that increases monotonely from 0 to ω as P describes the arc issued from A and asymptotic to γ', while it decreases monotonely from 0 to $-\omega$ as P describes the other arc. For T'' the rotations are reversed throughout. The fixed point index to be obtained later, (formula (29) of Ch. VI extended), will yield for all three transformations the index -1, since bases for the cycles are $\Gamma_0 = A$; $\Gamma_1^1 = \gamma$, $\Gamma_1^2 = \gamma'$. However, T has no fixed point, hence its index should be zero; T', T'' have A as a fixed point. For T' it has an index readily found to be $+1$, for T'' the index -1. We have thus a contradiction, showing that our method can at best be applied to T''. And in fact there is a corresponding finite contraction $\{\mathfrak{T}_h\}$. Let indeed δ be a circle with center O, between γ and γ', intersecting ζ at P, and take A equidistant from γ, γ'. Let δ' be the circle of center O through $P' = T'' \cdot P$. If QP is the diameter of δ through P, we choose $\mathfrak{T}_h$ such that $\mathfrak{T}_h \cdot (\delta - Q)$ is an arc of δ' bisected by P', whose angular measure is $2k\pi$, where if ϱ_0 is the half difference of the radii of γ and γ' and ϱ the absolute value of the excess of the radius of δ over their half sum,

$$k = \frac{h-1}{h}\,\frac{\varrho}{\varrho_0}, \qquad \varrho \leqq \frac{h}{h-1}\varrho_0; \qquad k = 1 \text{ otherwise.}$$

If OR is a variable radius vector whose length is fixed and exceeds the radius of γ', and V, V' are the points where it meets γ, γ', we take

$$\mathfrak{T}_h \cdot \overline{OV} = T'' \cdot V; \qquad \mathfrak{T}_h \cdot \overline{V'R} = T'' \cdot V'.$$

The extension of T'' thus defined is readily seen to be a finite contraction.

For the L of this example, then, the existence of a finite contraction is the property of some, but not all, single-valued transformations. In particular there is none for the identity of the type of $\{\mathfrak{I}_h\}$ of No. 49.

51. We must now digress for a moment in order to construct suitable bases for the cycles of N^i, N'^i.

Let N^i, N^j, N^k, $i<j<k$, be three neighborhoods of L. Since they coincide with open finite complexes, the maximum number s_j of p-cycles of N^j independent on N^i is finite. A maximal set on N^k as regards independence on N^i is also on N^j, hence $s_k \leqq s_j$. Therefore the numbers s_j for j variable and i fixed, do not increase with j, and their minimum for all values of p is reached, say, for $j = \alpha_i$. Then if $k > \alpha^i$, every p-cycle of N^{α_i} is homologous on N^i to a cycle of N^k. We can thus obtain a monotone increasing sequence i, α_i, $\alpha_{i+1}, \cdots$, such that every Γ_p of N^{α_j} is homologous on $N^{\alpha_{j-1}}$ to a cycle of $N^{\alpha_{j+1}}$. From this follows evidently that every Γ_p of N^{α_i} is the first term of a Vietoris sequence $\{\Gamma_p^j\}$ where

$$\Gamma_p^j \subset N^{\alpha_j}; \qquad \Gamma_p^{j+1} \approx \Gamma_p^j, \text{ on } N^{\alpha_{j-1}}$$

and all the Γ's are homologous to one another on N^i.

52. If we take into account the theorem of No. 42, we find that a certain base $\{\Gamma_{p+1}^\alpha\}$ for the infinite $(p+1)$-cycles mod Φ determines a base $\{\gamma_p^\alpha\}$ for the associated p-cycles of L. We will call $C_{p+1}^{\alpha j}$, $\Gamma_p^{\alpha j}$ the cut of Γ_{p+1}^α by Φ^j and the part of its boundary which is on Φ^j. It is a property of $\{\Gamma_{p+1}^\alpha\}$ that for j fixed and α above a certain value, $\Gamma_{p+1}^\alpha \subset N^j$. Since Φ is finite, $\Phi \subset K - N^j$ when j exceeds a certain value. For such a j (fixed) and α high enough, Γ_{p+1}^α does not meet Φ and $C_{p+1}^{\alpha h} \rightarrow \Gamma_p^{\alpha h}$. Take then $i>j$, and let t_i be the highest α such that $\Gamma_{p+1}^\alpha \not\subset N^i$. According to what precedes:

$$\text{(a)} \qquad \Gamma_p^{\alpha, i+h} \approx \Gamma_p^{\alpha i} \quad \textit{on } \bar{N}^i; \qquad \Gamma_p^{\alpha h} \approx 0 \quad \textit{on } N^i \textit{ for } \alpha > t_i.$$

Furthermore:

(b) *The cycles* $\Gamma_p^{\alpha h}$, $\alpha \leqq t_i$, *are independent on* N^i *for every* $h>i$.

For otherwise there exists a linear combination Γ_{p+1} of the cycles Γ_{p+1}^α, $\alpha \leqq t_i$, with cuts C_{p+1}^j and boundaries Γ_p^j of the latter on Φ^j, such that

$$N^i \supset D_{p+1} \rightarrow \Gamma_p^{i+1}.$$

Hence

$$D_{p+1} - C_{p+1}^{i+1} \to 0 \mod \Phi,$$

and therefore $\approx 0 \mod \Phi$ on E_r. As a consequence $C_{p+1} \approx$ a cycle on N^i, mod Φ. As this contradicts one of the basic constructive assumptions regarding $\{\Gamma_{p+1}^{\alpha}\}$ (Nos. 8, 9), (b) is proved.

Let Γ_p^{α} be some $\Gamma_p^{\alpha h}$, $h > i$, for every $\alpha \leqq t_i$. Owing to (a) and (b) the cycles Γ_p^{α} are independent on N^i. To have a base for the p-cycles of N^i we may have to add to them a certain number of other cycles Δ_p^{λ}, $\lambda = 1, 2, \cdots, s_i$, where s_i is finite, since N^i coincides with a finite complex. When i exceeds a certain value, N^i is related to some preceding N^h like N^{α_i} to N^i in No. 51. Hence there exists a cycle $\delta_p^{\lambda} = \{\Delta_p^{\lambda j}\}$ with Δ^{λ} as its first term, and the cycles $\Delta_p^{\lambda j}$ with fixed λ are all homologous on N^h. Furthermore, those whose j exceeds a certain value (the α_i of No. 51) are homologous on N^i itself. Now we have homologies

$$\delta_p'^{\lambda} = \delta_p^{\lambda} - \sum x_{\lambda\alpha} \gamma_p^{\alpha} \approx 0 \quad \text{on } L.$$

Taking into account (a) and (b), we find that $\delta_p'^{\lambda}$ can be represented by a sequence $\{\Delta_p'^{\lambda j}\}$ whose first term

$$\Delta_p'^{\lambda} = \Delta_p^{\lambda} - \sum_{\alpha=1}^{t_i} x_{\lambda\alpha} \Gamma_p^{\alpha}.$$

This shows that the s_i cycles $\Delta_p'^{\lambda}$ are likewise independent of the cycles Γ_p^{α} on N^i, and hence together with them they also constitute a base. It means that we may choose the Δ's such that every $\delta_p^{\lambda} \approx 0$ on L.

53. The treatment of the same question for the cycles mod Φ^i is now a simple matter. According to the duality theorems for finite manifolds we can find a base mod Φ^i, $\{g_{r-p}^{\alpha}\}$, $\alpha = 1, 2, \cdots, t_i$, $\{d_{r-p}^{\lambda}\}$, $\lambda = 1, 2, \cdots, s_i$, such that

$$\begin{aligned} (\Gamma_p^{\alpha} \cdot g_{r-p}^{\beta}) &= \delta_{\alpha\beta}; & (\Gamma_p^{\alpha} \cdot d_{r-p}^{\lambda}) &= 0; \\ (\Delta^{\lambda} \cdot g_{r-p}^{\beta}) &= 0; & (\Delta_p^{\lambda} \cdot d_{r-p}^{\mu}) &= \delta_{\lambda\mu}. \end{aligned}$$

In view of (a) we also have

$$(\Gamma_p^{\alpha j} \cdot g_{r-p}^{\beta}) = (\Gamma_p^{\alpha j} \cdot d_{r-p}^{\lambda}) = 0; \quad \alpha > t_i,\ j > i.$$

Hence the g's and d's are pseudocycles such that, for every α,

$$(\gamma_p^{\alpha} \cdot g_{r-p}^{\beta}) = \delta_{\alpha\beta}; \qquad (\gamma_p^{\alpha} \cdot d_{r-p}^{\lambda}) = 0.$$

Therefore: (a) the pseudocycles g_{r-p}^{α}, $\alpha \leqq t_i$, may be taken as the first t_i terms of the base $\{g_{r-p}^{\alpha}\}$ of No. 46; (b) $d_{r-p}^{\lambda} \approx 0$ as a pseudocycle. As usual, we take for d_{r-p}^{λ} a subchain of the finite complex $\bar{N}^i$. There exists, then, a polyhedral chain $d_{r-p}^{0\lambda}$ of $S_r - L$, such that $d_{r-p}^{\lambda} - d_{r-p}^{0\lambda} \approx 0$ on S_r. The part of $d_{r-p}^{0\lambda}$ on $\bar{N}^i$ is then a chain $\approx d_{r-p}^{\lambda} \bmod \Phi^i$ on $\bar{N}^i$, and which may therefore replace it throughout. This means in essence that we may assume that d_{r-p}^{λ} does not meet L. We have then proved:

THEOREM. *The cycles Γ_p^{α}, $\alpha \leqq t_i$, together with certain first cycles of Vietoris chains ≈ 0, constitute a base for the p-cycles of N^i. The cycles mod Φ^i, g_{r-p}^{α}, $\alpha \leqq t_i$, together with certain cycles not meeting L, constitute a base for the cycles mod Φ^i on $\bar{N}^i$. The two bases may be chosen associated canonical.*

54. We now return to the determination of the signed coincidences of the transformations T_1, T_2 of No. 48. Since $\mathfrak{T}_{1i}$, $\mathfrak{T}_{2i}$ are transformations of N^i into N'^i of the respective types T_{1b}, T_{2b}, and since the bases are associated canonical, the number of signed coincidences $(\Gamma_r^{1i} \cdot \Gamma_r^{2i})$ is given by (23) of Ch. VI. To determine it we need to know the transformations induced by $\mathfrak{T}_{1i}$ on the cycles of $\bar{N}^i$ mod Φ^i, and by $\mathfrak{T}_{2i}$ on the absolute cycles of N^i.

Since the d's, whose number is finite, do not meet L, we can select $\mathfrak{T}_{1i}$ such that Γ_r^{1i} does not meet the sets $d \times S'$. This is always possible, for whether $\mathfrak{T}_1$ is finite or not, $\mathfrak{T}_{1i}$ is selected after N^i, hence when the d's are already known. Under the circumstances every $\mathfrak{T}_{1i} \cdot d = 0$. Hence the equations of transformations for the relative cycles under $\mathfrak{T}_{1i}$ are

$$\left.\begin{aligned}&\mathfrak{T}_{1i}\, g^{\alpha}_{r-p} \approx \sum_{\beta=1}^{t'_i} \varphi^{pi}_{\alpha\beta}\, g'^{\beta}_{r-p} + \sum \omega_{\alpha\mu}\, d'^{\mu}_{r-p}\\ &\qquad\qquad (\alpha = 1, 2, \cdots, t_i)\\ &\mathfrak{T}_{1i}\, d^{\lambda}_{r-p} \approx 0\end{aligned}\right\} \bmod \Phi'^{i} \text{ on } \bar{N}'^{i}.$$

Since the pseudocycles d' are ≈ 0 on L', the first relations written as homologies between pseudocycles on L', are

$$\mathfrak{T}_{1i}\, g^{\alpha}_{r-p} \approx \sum \varphi^{pi}_{\alpha\beta}\, g'^{\beta}_{r-p}.$$

Referring to Ch. VI, § 4, we find immediately that, since all cycles Γ^{1i}_r of the type chosen are homologous mod $N^i \times \Phi'^i$ on $N^i \times \bar{N}'^i$, the coefficients $\varphi^{pi}_{\alpha\beta}$ for a given α, are independent of i. In an evident sense then, and as loc. cit. No. 21, these homologies are to be considered as the first t_i homologies of the transformations of the pseudocycles by T_1:

$$T_1 \cdot g^{\alpha}_{r-p} \approx \sum \varphi^{p}_{\alpha\beta}\, g'^{\beta}_{r-p}, \qquad (\alpha = 1, 2, \cdots, +\infty). \tag{43}$$

Thus φ^{pi} is the matrix of the first t'_i columns and t_i rows of φ^p, the terms of the same rows not in φ^{pi} being all zero.

There remains then the determination of the transformations induced by $\mathfrak{T}_{2i}$ on the absolute cycles. If we take care to choose for Γ^{α}_p, Γ'^{β}_p, cycles $\Gamma^{\alpha j}_p$, $\Gamma'^{\beta j}_p$ whose index j is sufficiently high, the cycles $\mathfrak{T}_{2i} \cdot \Gamma^{\alpha j}_p$ will be as near as we please to L' and hence will not meet the chains d'_{r-p} of the canonical base. As a consequence

$$\mathfrak{T}_{2i} \cdot \Gamma^{\alpha}_p \approx \sum_{\beta=1}^{t'_i} \psi^{pi}_{\alpha\beta}\, \Gamma'^{\beta}_p \text{ on } N'^i,$$
$$(\alpha = 1, 2, \cdots, t_i).$$

For the same reason as before the ψ's are independent of i. The preceding homologies are then to be considered, in an evident sense, as the first t_i homologies of the transformation which T_2 induces on the p-cycles:

$$T_2 \cdot \gamma^{\alpha}_p \approx \sum \psi^{p}_{\alpha\beta}\, \gamma'^{\beta}_p \text{ on } L', \tag{44}$$

the summation being infinite.

We have, in, addition for the cycles Δ:

$$\mathfrak{T}_{2\imath} \cdot \Delta_p^\lambda \approx \sum_{\beta=1}^{t_i'} \varrho_{\lambda\beta}^p \Gamma_p'^\beta + \sum \tau_{\lambda\mu}^p \Delta_p'^\mu \text{ on } N'^\imath,$$
$$\lambda = 1, 2, \cdots, s_i.$$

55. We are now ready to apply the coincidence formulas of Ch. VI. The transformation matrices for the pseudocycles and cycles of dual types are respectively

$$\left\| \begin{matrix} \varphi^{p\imath} & \omega^\beta \\ \mathbf{0} & \mathbf{0} \end{matrix} \right\|, \qquad \left\| \begin{matrix} \psi^{p\imath} & \mathbf{0} \\ \varrho^p & \tau^p \end{matrix} \right\|,$$

where the terms similarly placed stand for matrices with the same numbers of rows and columns. According to No. 53 the bases are canonical (i. e. they satisfy Th. II of No. 19). Hence the formula to be applied is (23), Ch. VI. To embody in it the effect of the parity of r, we shall designate the number of signed coincidences by $(-1)^r \Theta_i$ and we find, after replacing in that formula n by r and p by $r-p$,

$$\Theta_i = \sum_{p=0}^{n} (-1)^p \; trace \; \psi^{p\imath} (\varphi^{p\imath})'. \tag{45}$$

The summation for p is from 0 to n since the cycles on L whose dimension $> n$ are all ≈ 0 and hence $\psi^{p\imath} = 0$ for $p > n$.

Regarding this expression, it will be recalled that the number $t_i'^p$ of columns in $\varphi^{p\imath}$, $\psi^{p\imath}$ may be taken arbitrarily high. For $t_i'^p$ is merely the analogue of t_i^p attached to some $N'^\imath$ associated with N^i and $N'^\imath$ may be replaced by any interior neighborhood. Now φ^p has only a finite number of terms $\neq 0$ in each row, since $\{g_{r-p}'^\alpha\}$ is semi-finite. Therefore we may choose $N'^\imath$ such that $\varphi^{p\imath}$ will include all non-zero terms of the first t_i^p rows of φ^p and this for every p. Furthermore, for the same reasons the formal product $\psi^p (\varphi^p)'$ is an infinite matrix whose terms are all finite. The coefficient of $(-1)^p$ in Θ_i is then merely the sum of the first t_i^p terms in the principal diagonal of that product.

It is natural to call *trace* of an infinite matrix the series obtained by adding the terms of the principal diagonal in their right order. It is important to preserve that order, since we have no reason to expect the series to be absolutely convergent. If the series converges, its numerical value itself is called *trace* of the matrix. We have then

$$\Theta_i = \sum (-1)^p \,[\mathit{trace}\ \psi^p (\varphi^p)']_{t_i^p},$$

where the index t_i^p indicates the number of terms to be taken in the corresponding infinite series.

The limit of Θ_i appears then as the limit of a sum of $(n+1)$ series under the restriction that the number of terms taken in the pth series must be t_i^p, as $i = 1, 2, 3, \cdots$. While the sequences $\{t_i^p\}$ are characteristic for L, they present a certain indetermination as we shall now show.

56. Consider the two sets of integers t_{i-1}^p for N^{i-1} and t_i^p for N^i. We have $t_{i-1}^p \leqq t_i^p$. Suppose that the inequality holds for one or more values of p. If N^{0i} is any neighborhood between N^{i-1} and N^i, and t^p the corresponding integers t_i^p, we shall have $t_{i-1}^p \leqq t^p \leqq t_i^p$. We propose to show that if the inequality holds, N^{0i} may be so chosen that the set t^p coincides with certain intermediary sets between t_{i-1}^p and t_i^p. Such a set exists only if $\sum (t_i^p - t_{i-1}^p) > 1$, and this we shall assume for the present.

Now referring to No. 51, we find that when $t_{i-1}^p < t_i^p$ there exists a cycle $\gamma_p \approx 0$ on N^{i-1} but not on N^i. Let N^{*i} designate the closed complex left on removing from some complex covering $\bar{N}^i$, also called $\bar{N}^i$, the $\bar{N}^i$-neighborhood of Φ^i. As usual we may assume that neighborhood normal, and then, owing to its sliding properties, we find, as in Ch. III No. 31, that we can replace in what we have just said regarding the numbers t, the neighborhoods N^i by the complexes N^{*i}. The insertion of N^{0i} above is then equivalent to the insertion of some closed complex between the N^*'s. This we shall now do by adding the zero-cells, the one-cells, etc. of $N^{*i-1} - N^{*i}$, one at a time to N^{*i} until we have N^{*i-1}.

The question is now reduced to this: Let the simplicial cell E_q have for boundary Γ_{q-1} a subcycle (sphere) of a complex $\mathfrak{K}$ whose dimension $\leqq q$. Assuming that E_q has no points on $\mathfrak{K}$, what are the relations between the Betti numbers of $\mathfrak{K}$ and $\mathfrak{K}' = \mathfrak{K} + E_q$?

Every q-cycle of $\mathfrak{K}'$ is of the form $\Gamma_q = mE_q + C_q$, where C_q is a subchain of $\mathfrak{K}$. If there exists a $C_q \to \Gamma_{q-1}$, there exists a Γ_q, and since the dimension of $\mathfrak{K}' \leqq q$, $\Gamma_q \not\approx 0$. On the other hand if $\Gamma'_q = m'E_q + C'_q$ is another such cycle, $m'\Gamma_q - m\Gamma'_q$ is a cycle of $\mathfrak{K}$. Hence, in the case considered $R_q(\mathfrak{K}') = R_q(\mathfrak{K}) + 1$. The only boundary relation satisfied for $\mathfrak{K}'$ but not for $\mathfrak{K}$ is $E_q \to \Gamma_{q-1}$. It yields only $\Gamma_{q-1} \approx 0$, on $\mathfrak{K}'$ and since $C_q \to \Gamma_{q-1}$, the homology holds also on $\mathfrak{K}$. Therefore here $R_p(\mathfrak{K}') = R_p(\mathfrak{K})$ for $p < q$.

If no chain $C_q \to \Gamma_{q-1}$ on $\mathfrak{K}$, $\Gamma_{q-1} \not\approx 0$ on $\mathfrak{K}$. Hence on passing from $\mathfrak{K}$ to $\mathfrak{K}'$ we have no new q-cycle but a new homology $\Gamma_{q-1} \approx 0$. Therefore R_{q-1} is decreased by one and the other Betti numbers remain the same.

Now through the addition of E_q the numbers t^p can only decrease. Hence it leaves them unchanged in the first case, and may decrease t^q by one unit in the second. But our procedure requires that we add first all the zero-cells, then all the one-cells, etc., and when we are through we shall have returned to N^{i-1}. Therefore it will result in first lowering t_i^0 to t_{i-1}^0, then t_i^1 to t_{i-1}^1, etc. Thus if $\{t_{i-1}^p\}$, $\{t_i^p\}$ are two consecutive sets there can be found an N^{0i} whose set $\{t^p\}$ is as follows:

$$t^p = t_{i-1}^p, \quad i < q; \qquad t^p = t_i^p, \quad i > q;$$

$t^q =$ any integer between t_{i-1}^q and t_i^q including the extremes.

As for q it is any number between 0 and n.—This shows clearly the degree of indetermination in the sets $\{t^p\}$.

REMARK. The argument based on the successive addition of cells to a complex, is frequently useful. It is due to Alexander who applied it on various occasions, notably in the proof of his duality theorem.

57. We have seen (No. 53) that the geometric coincidences of $\mathfrak{T}_{1i}$ and $\mathfrak{T}_{2i}$ tend towards those of T_1 and T_2 on L, L'. If T_1, T_2 have no coincidences, $\mathfrak{T}_{1i}$, $\mathfrak{T}_{2i}$ will have none when i is sufficiently large and hence $\lim \Theta_i = 0$. *Therefore coincidences occur certainly for T_1, T_2 on L, L' when Θ_i does not converge to zero.*

When the convergency to zero takes place, the partial sum Θ_i is strictly zero if a sufficiently large number of terms is taken in the series *trace* $\psi^p (\varphi^p)'$. When the Betti numbers of L are finite, ψ^p is finite, Θ_i has a fixed value Θ for sufficiently large values of i and we have

$$(46) \qquad \Theta = \sum (-1)^p \, trace\, \psi^p (\varphi^p)'.$$

This expression also holds when the trace series are convergent, even with infinite Betti numbers.

It is a simple matter to show that *trace* $\psi^p (\varphi^p)'$ is invariant when we pass from $\{\gamma_p^\alpha\}$, $\{g_{r-p}^\alpha\}$ to any other pair of canonical bases such that the first is totally infinite. Therefore Θ is a topological invariant.

58. Referring to No. 49 we find that the results just obtained, can be applied to the problem of the fixed points of a continuous transformation T of L into itself possessing an extension $\mathfrak{T}$ which is a finite expansion or contraction. The identity I is then extended into the identity $\mathfrak{I}$ for S, and $\mathfrak{I}$ is considered as an expansion or a contraction according as $\mathfrak{T}$ is a contraction or an expansion. We have in the first case every $\varphi^p = 1$, in the second every $\psi^p = 1$.

59. **Extension to arbitrary compact metric spaces.**

Let L be a compact set on $\mathfrak{H}$. If we apply the affine transformation

$$x_i' = \frac{1}{4}\left(\frac{1}{i} + x_i\right)$$

we reduce L to an "interior" subset of $\mathfrak{H}$, i. e. to a subset such that $0 < x_i < \frac{1}{i}$ for every i. We will designate by $\mathfrak{H}_i$, L^i the projections of $\mathfrak{H}$, L on $x_{i+1} = x_{i+2} = \cdots = 0$.

Then a neighborhood N^i of L^i on $\mathfrak{H}_i$ is the projection of a subset $\mathfrak{N}^i$ of $\mathfrak{H}$, which carries L. We can and shall choose the N's such that $\bar{N}^{i+1} \subset \mathfrak{N}^i$ and that $\mathfrak{N}^i$ converges to L on $\mathfrak{H}$ with increasing i. If we observe that due to the convexity of $\mathfrak{H}$, every cycle of dimension >0 bounds on it, we find that an arbitrary γ_p on L can be represented by a sequence $\{\Gamma_p^i\}$, $\Gamma_p^i \subset \mathfrak{H}_i$, $\Gamma_p^i \approx \Gamma_p^h$ on $\mathfrak{H}_i$, $h > i$. A totally infinite base $\{\gamma_p^\alpha\}$ is then constructed as follows: For every i there is at most a finite number of cycles Γ_p^i independent on N^i. We take, then, a first group of cycles γ_p^α, $\alpha = 1, 2, \cdots, t_1^p$, consisting of cycles whose representatives $\Gamma_p^{\alpha 1}$ are independent on N^1, a second group $\gamma^{t_1^p+\beta}$, $\beta = 1, 2, \cdots, t_2^p$, whose representatives $\Gamma_p^{t_1^p+\beta, 2}$ are independent of the cycles $\Gamma_p^{\alpha 1}$ on N^2, etc.

60. We will now introduce pseudocycles for L. Let us first agree to orient $\mathfrak{H}_i$ by the coördinates $x_1, \cdots, x_i$ as named. If E_q is any convex cell on N^i the locus of all points of $\mathfrak{H}_{i+r}$ projected on E_q is a convex E_{q+r} of $\mathfrak{H}_{i+r}$. Let $y_1, \cdots, y_q$ be a coördinate system orienting E_q in a certain way. Then we can choose $y_1, \cdots, y_q, x_{i+1}, \cdots, x_r$ as coördinates on E_{q+r}, and agree to orient the cell by these $q+r$ coördinates in the order just named. We will call E_{q+r} the *antiprojection* of E_q on N^{i+r}. If E_{q+r}^j is the antiprojection of E_q^j and

$$C_q = \sum u_j E_q^j, \qquad C_{q+r} = \sum u_j E_{q+r}^j,$$

C_{q+r} is called the *antiprojection* of C_q.

Returning now to the base $\{\gamma_p^\alpha\}$, suppose that representatives $\Gamma_p^{\alpha i} \not\approx 0$ on N^i appear first for $i = \nu$. On N^ν there can be found t_ν^p convex cycles $G_{\nu-p}^\beta$ mod Φ^ν (= the boundary of N^ν), such that

$$(\Gamma_p^{\alpha\nu} \cdot G_{\nu-p}^\beta) = \delta_{\alpha\beta}; \quad \alpha, \beta = 1, 2, \cdots, t_\nu^p.$$

Taking account of the fact that $\Gamma_p^{\alpha\nu} \approx \Gamma_p^{\alpha, \nu+r}$, on $\mathfrak{H}_\nu$, it is a simple matter to show that if $G_{\nu-p+r}^\beta$ is the antiprojection of $G_{\nu-p}^\beta$,

$$(\Gamma_p^{\alpha, \nu+r} \cdot G_{\nu-p+r}^\beta) = \delta_{\alpha\beta}, \quad \beta = 1, 2, \cdots, t_\nu^p,$$

for every α and r. Similarly on $\mathfrak{H}_{\nu+1}$ we have $t^p_{\nu+1} - t^p_\nu$ new cycles $\Gamma \not\approx 0$, and hence new cycles mod $\Phi_{\nu+1}$, G^β_{r+1-p} giving rise to a sequence of antiprojections $G^\beta_{\nu-p+r}$. Thus corresponding to every γ^α_p we shall have an obvious sequence $\{G^\alpha_{i-p}\} = g^p_\alpha$ which we call pseudocycle in an evident sense. On each $\mathfrak{H}_i$ we have between the representatives $\Gamma^{\alpha i}_p$, G^α_{i-p} of cycles and pseudocycles the relations

$$(\Gamma^{\alpha i}_p \cdot G^\beta_{i-p}) = \delta_{\alpha\beta}$$

which we express by

$$(47) \qquad (\gamma^\alpha_p \cdot g^p_\beta) = \delta_{\alpha\beta}.$$

Homologies between pseudocycles are then readily introduced and $g \approx 0$ means that some representative of g on an $\mathfrak{H}_i$ does not meet the corresponding projection L^i of L.

61. On going over the treatment of transformations of subsets of Euclidean spaces it will be seen that every step may now be extended to coincidences of transformations T_1, T_2 of two compact metric spaces L, L', subsets of two Hilbert space parallelotopes $\mathfrak{H}$, $\mathfrak{H}'$. We assume that they possess extensions $\mathfrak{T}_1$, $\mathfrak{T}_2$ to $\mathfrak{H}$, $\mathfrak{H}'$ defined as previously by certain sequences $\{\mathfrak{T}_{1h}\}$, $\{\mathfrak{T}_{2h}\}$, whose images on $\mathfrak{H} \times \mathfrak{H}'$ are pseudocycles $\{G^{1h}_i\}$, $\{G^{2h}_i\}$ of $L \times L'$. Assuming here also that $\mathfrak{T}_1$ is an expansion, $\mathfrak{T}_2$ a contraction, one of them being finite, we can so dispose of the situation that: (a) The chains G^{1i}_i, G^{2i}_i represent a pair of transformations of $\bar{N}^i$ into $\bar{N}'^i$ of the respective types T_{1b}, T_{2b} for these two i-manifolds. (b) Their number of signed coincidences is again given by (45). (c) When i increases indefinitely, their actual coincidences tend to those of T_1, T_2 on L, L'. The situation is then exactly as before and leads to the same consequences.

62. The results just established are embodied in the

FUNDAMENTAL COINCIDENCE AND FIXED POINT THEOREM. *Let L, L' be two arbitrary compact metric spaces, T_1, T_2 two continuous transformations possessing extensions to some Euclidean or Hilbert parallelotopes in which homeomorphic copies of L, L' may be immersed. Let one of the extensions, say $\mathfrak{T}_1$ be an* *expansion, the other* $\mathfrak{T}_2$ *a contraction, one of them being finite. Let* ψ^p, φ^p *be the infinite matrices of the transformations of the cycles by* T_2 *and pseudocycles by* T_1 *respectively, of two associated canonical bases, and* Θ_i *a partial sum of the terms of the series*

$$\Theta = \sum (-1)^p \text{ trace } \psi^p (\varphi^p)' \tag{48}$$

taken in certain rectangular arrays. Then if Θ_i *does not converge to zero the two transformations possess at least one coincidence on* L, L'. *A continuous transformation* T *of* L *into itself having a finite extension which is an expansion or a contraction, and inducing upon the pseudocycles or cycles transformations with matrices* φ^p *or* ψ^p, *respectively, possesses at least one fixed point provided that, under the same circumstances, respectively one or the other of the series*

$$\Theta = \sum (-1)^p \text{ trace } \varphi^p, \quad \Theta = \sum (-1)^p \text{ trace } \psi^p \tag{49}$$

does not converge to zero. This is likewise true for single-valued transformations and the second series whenever L *is locally connected. The series are in any case formal topological invariants of* L, L' *and the transformations. When the Betti numbers of sufficiently high dimension are all zero, the numbers* Θ *are finite and may be considered as the numbers of signed coincidences and fixed points.*

COROLLARY. *Let* L *be a locally connected compact metric space whose Betti numbers* R_p, $p>0$ *are all zero, and whose* $R_0 = 1$ *(continuum). Every continuous single-valued transformation of* L *into itself has at least one fixed point.*

For every $\psi^p = 0$ except ψ^0, and ψ^0 reduces to 1; for if γ_0 is the fundamental zero-cycle of L (a single point), $T \cdot \gamma_0 \approx \gamma_0$, on L. It follows that $\Theta = 1 \neq 0$.

The corollary just given is the extension to locally connected sets with zero Betti numbers of Brouwer's first fixed point theorem for cells. It constitutes a partial affirmative answer to a question proposed to us two years ago by Alexandroff.

63. **Important special cases.** I. The *closed cell* of Hilbert space, i. e. the homeomorph of the parallelotope $\mathfrak{H}$. Since

$\mathfrak{H}$ is convex, it is locally connected, and its Betti numbers are evidently as demanded by the Corollary which holds therefore for the cell.

This case is the topological translation of fixed point theorems due to Birkhoff-Kellogg [1] (1922), by means of which they were able to prove with little effort a mass of existence theorems of analysis. See also on these questions Schauder [1]. In view of the much wider scope of our fundamental theorem, further progress can probably be made in the same direction today.

II. Let G be an abstract continuous topological group whose image is a connected *topological* M_n i. e. a compact metric space whose determining neighborhoods are n-cells. From a well known group property it follows that M_n can be transformed into itself by a homeomorphism T without fixed points. Since M_n is locally connected, the Corollary informs us here that M_n has at least one non-zero Betti number $R_p(M_n)$ whose index $p > 0$.

As a concluding remark, we may state that the extension of the fundamental theorem to open subsets of compact metric spaces, in a manner similar to that of Ch. VI, presents no real difficulty. In carrying out this extension the relative cycles and Betti numbers as defined in No. 38, Remark III, will come into play, but otherwise there is no new argument.— Together with the results of No. 27 we have, then, completely extended the theory developed in Ch. VI.

CHAPTER VIII

APPLICATIONS TO ANALYTICAL AND ALGEBRAIC VARIETIES

In the applications of Analysis Situs, one rarely meets configurations conveniently cut into cells, such as we have been studying. It is, then, a practical question, to find when an assigned configuration may be broken up into the cells of a complex or a manifold. The most important class of point sets that possess that property consists of the *analytical varieties*, characterized by local analytical parametric representations. As a matter of fact, the functions of the representations may be taken merely single-valued, with a sufficient number of derivatives, but the relative simplicity of analytical functions justifies giving them the preference. Furthermore the passage from analytical functions to the more general class is by a fairly standard procedure of real function theory.

The first investigations on Analysis Situs (Riemann, Betti, Poincaré) were confined to analytical varieties, or rather to the more general type just mentioned. In his first paper on Analysis Situs (1895) [2] Poincaré dealt exclusively with them. He attempted to develop the subject on strictly analytical lines, without the process of cutting into cells. Up to a certain point he met with success, in particular as regards orientation and the Kronecker index, and the definitions of homology and of the Betti numbers. He did not succeed, however, in proving the duality theorem in this fashion and missed torsion (Heegaard [1]). It was precisely in order to overcome these difficulties that, in his subsequent papers, Poincaré introduced the complex and the manifold.

These historical remarks indicate that the direct treatment of analytical varieties, while perhaps the more natural and intuitive procedure, presents serious practical difficulties.

We will suppose, for the moment, that it has been shown, as recently done by Van der Waerden [1], that an analytical variety is a manifold. The question arises to investigate the topology of the variety directly and effectively from the analytical representation. This is the question which will occupy us in the present chapter. We had planned it substantially in the present form, intending to integrate and complete our earlier writings on the same topic (see notably our monographs [13, 16]), when Van der Waerden's article appeared (January 1930), more or less on the same topic. It is not surprising, therefore, that there will be found numerous points of contact with him.

The varieties on which information is most extensive are the algebraic varieties. Some of the most difficult and hidden properties of algebraic surfaces are inextricably interwoven with their topology, as may be seen from a perusal of [13, 16], also from Severi's lithographed Lectures [5] which appeared as we were going to press. These properties are cursorily examined in the last section of the chapter.

§ 1. Analytical Varieties.

1. Let S_n be the space of the real coördinates $x_1, \cdots, x_n$, $\mathfrak{R}$ a region of S_n, $F_i(x_1, \cdots, x_n) = F_i(x)$, $i = 1, 2, \cdots, r$, a set of real functions holomorphic on $\mathfrak{R}$. It may be shown exactly as in the case of complex functions of complex variables (Weierstrass; compare Osgood [1] pp. 91, 132), that the system

$$F_i(x) = 0 \tag{1}$$

represents a subset of $\mathfrak{R}$, which consists of an enumerable aggregate of connected subsets, any one of which say V_p, is characterized by the property that the Jacobian matrix

$$\left\| \frac{\partial F_i}{\partial x_j} \right\| \tag{2}$$

is of rank at most $n-p \leqq n$ on V_p; the subset of V_p for which the rank does not assume its maximum value is a true subset W of V_p, which is the sum of an enumerable set of varieties analogous to V but with a smaller p. A V_0 consists of isolated points. On V_p itself, the r functions F_i are functionally dependent upon $n-p$ of the set. Therefore, if we are only concerned with representing V_p, we may assume that the system has the form

$$F_i(x) = 0, \quad i = 1, 2, \cdots, n-p. \tag{1'}$$

However, we are not assured that (1′) will represent V_p alone over $\mathfrak{R}$. To have a representation of V_p alone we must confine ourselves to a small neighborhood N of a point of V_p, that is reduce $\mathfrak{R}$ to N, and then a suitable system (1′) will represent V_p alone on N.

If $Q:(a_1, \cdots, a_n)$ or (a), is any point of $V_p - W$ on $\mathfrak{R}$, there is a neighborhood of Q on V_p which is an *analytical* p-cell, that is, a cell parametrically represented by equations

$$x_i - a_i = \varphi_i(u_1, \cdots, u_p), \quad i = 1, 2, \cdots, n; \quad |u_i| < \varepsilon, \tag{3}$$

where the φ's are holomorphic in the region $|u_i| < \varepsilon$ and have a Jacobian matrix of rank p at $(u) = 0$. We say that V_p is a p *dimensional analytical variety,* or more simply *p-variety* and that W is its *singular locus.*

The definition can be extended to a (finite or enumerable) sum of analytical varieties V_{p_i}, $p_i \leqq p$. We likewise designate this more general variety by V_p and consider as its singular locus the set of all its points on $\mathfrak{R}$ that do not have an E_p for neighborhood. The type of variety first defined, which constitutes a connected point-set, is said to be *irreducible*, the more general type *reducible.*

A V_n of S_n is a sum of regions of S_n together with some varieties of dimensions $< n$.

2. We may similarly introduce the p-varieties on a real projective n-space which we shall designate by Σ_n. From

our point of view, Σ_n is best defined as the totality of sets of real numbers $x_0, \cdots, x_n$ not all zero, two proportional sets being considered as defining the same point of Σ_n. If we remove any Σ_{n-1} from Σ_n, the residual space is Euclidean. A V_p of Σ_n is then a set which becomes a "Euclidean" p-variety in the previous sense on some $\Sigma_n - \Sigma_{n-1}$ of our projective space. Its singular locus W is the subset of V_p which becomes the singular locus on the Euclidean variety. The analytical p-cell is defined as previously, except that the differences $x_i - a_i$ are replaced by kx_i, where k is a factor of proportionality, and the Jacobian matrix of the φ's by its projective equivalent

$$\left\| \begin{matrix} \varphi_0, \varphi_1, & \cdots, & \varphi_n \\ \dfrac{\partial \varphi_0}{\partial u_j}, & \cdots, & \dfrac{\partial \varphi_n}{\partial u_j} \end{matrix} \right\|,$$

whose rank must now be $p+1$.

3. The connection between analytical varieties and combinatorial topology rests upon a simple proposition often taken for granted. We believe that its first proof is the one recently published by Van der Waerden [1] and it will be substantially followed here. Similar results with weaker hypotheses have just been obtained by S. S. Cairns [1].

We shall first introduce analytical elements and complexes. By an *analytical p-element* ε_p is meant an open subset of a V_p, such that $\bar{\varepsilon}_p$ is also on V_p, and that the boundary $\bar{\varepsilon}_p - \varepsilon_p$ consists of a finite number of non-intersecting elements of dimension $< p$. An ε_0 is a point. An *analytical p-complex* κ_p is made up of ε's like an ordinary complex of cells. From the elementary properties of varieties we find readily that (a) the intersection of several κ's is a κ; (b) the parallel projection in a suitable direction of a κ on an S_{n-1} of S_n is likewise a κ.[a]

We shall define κ as finite or infinite accordingly as the number of its ε's is finite or not. Finally, we shall say that κ *covers* a given point set, if it coincides with the set.

THEOREM. *Let the analytical complex κ_p, $p < n$, be on the bounded region $\Re_n$ of S_n, and let $\varepsilon^i_{q_i}$ be its elements. There* *exists a simplicial* K_n *covering* $\mathfrak{R}_n$ *with* p- *and* q_i-*subcomplexes covering* $\varkappa_p$ *and* $\varepsilon^i_{q_i}$. *When* $\varkappa_p$ *is finite and the boundary of* $\mathfrak{R}_n$ *is a finite* $\varkappa_{n-1}$, *the covering complexes are all finite and we may replace the region by its closure.*

The required theorem being trivial for $n = 0$, we use induction on n. We assume first that the boundary of $\mathfrak{R}_n$ is a finite $\varkappa_{n-1}$ and also that $\varkappa_p$ is finite. Then the point set $\varkappa_{n-1} + \varkappa_p$ can be covered with a unique complex $\varkappa'_{n-1}$ having a subcomplex covering $\varkappa_p$. We therefore do not introduce any restrictions if we assume that $\varkappa_p$ is a $\varkappa_{n-1}$ having a subcomplex covering the boundary of $\mathfrak{R}_n$.

The projection of $\mathfrak{R}_n$ on an S_{n-1} of S_n is an $\mathfrak{R}_{n-1}$ of S_{n-1} whose boundary is a finite $\varkappa'_{n-2}$ and the projection of $\varkappa_{n-1}$ will be a $\varkappa'_{n-1}$ covering $\overline{\mathfrak{R}}_{n-1}$. The conditions for the coincidence of two points on the same or on two distinct ε's of $\varkappa_{n-1}$ are expressed by the vanishing of functions holomorphic over certain regions of S_{n-1}, so that the coincident points will be distributed into sets of analytical elements. Owing to this and to obvious reasons of continuity we can cover $\overline{\mathfrak{R}}_{n-1}$ with a finite $\varkappa''_{n-1}$ having the following property. If Q is a point on ε'' of $\varkappa''_{n-1}$ and $P^1, \cdots, P^t$ the points of $\varkappa_{n-1}$ projected on Q, the number t is independent of the position of Q on ε'', and when Q describes ε'', P^i remains on a unique element ε^i of $\varkappa_{n-1}$.

By the hypothesis of the induction the theorem holds on S_{n-1}. Therefore if $\varkappa''_{n-2}$ is the complex obtained on removing the $(n-1)$-elements from $\varkappa''_{n-1}$, we can cover $\overline{\mathfrak{R}}_{n-1}$ with a simplicial K'_{n-1} having an $(n-2)$-subcomplex covering $\varkappa''_{n-2}$ and its elements, in the manner prescribed by the theorem. K'_{n-1} and its cells will be related to $\varkappa_{n-1}$ like $\varkappa''_{n-1}$ and its elements ε''. Therefore the property just obtained holds when Q describes any cell E'_q of K'_{n-1}. But when Q describes E'_q, the points P^i and the open segments $P^i P^{i+1}$ describe q- and $(q+1)$-cells whose boundaries are ordinary $(q-1)$- and q-spheres. There results a K_n covering $\mathfrak{R}_n$ and behaving as required, except that its cells are not simplexes. However, since their boundaries are ordinary spheres, the first derived of K_n will fulfill all the conditions of the theorem.

When the boundary of $\mathfrak{R}_n$ is not a $\varkappa_{n-1}$ we approximate it by a sequence of regions $\{\mathfrak{R}_n^i\}$, each having a $\varkappa_{n-1}$ for boundary, and such that $\bar{\mathfrak{R}}_n^i \subset \mathfrak{R}_n^{i+1}$. We construct as above a K_n^1 covering $\bar{\mathfrak{R}}_n^1$ and a K_n^{i+1} covering $\bar{\mathfrak{R}}_n^{i+1} - \mathfrak{R}^i$ such that $\sum_{i=1}^{h} K_n^i$ behaves as required relatively to $\bar{\mathfrak{R}}^h$. The infinite sum of all the K's will have again a first derived such as demanded by the theorem. A similar treatment may be applied to the infinite $\varkappa_p$.

4. Corollaries. I. *If W is the singular locus of V_p, $V_p - W$ is a p-manifold.*

For every point of $V_p - W$ has a neighborhood which is a p-cell whose diameter is arbitrarily small, hence the covering open complex is a manifold (Ch. III, § 4).

II. *The Theorem and Corollary hold for analytical varieties on projective spaces.*

The same proofs apply throughout, with parallel projection replaced by central projection.

5. A particularly interesting type of variety is the *algebraic* variety, or variety defined by equations (1) with the F's all polynomials. In projective n-space, Σ_n, they are obtained by equating to zero a certain number of homogeneous polynomials. Real algebraic varieties represent the real sheets of complex algebraic varieties, or simply "algebraic" varieties, the word complex being usually omitted: an algebraic variety is complex unless otherwise stated. The preceding results imply the

Theorem. *The real sheets of algebraic varieties are manifolds. When they are bounded and without singularities they are absolute manifolds and can be covered with finite complexes.*

6. **Orientation of analytical varieties.** It follows from Corollary I of No. 4 that a connected subset of $V_p - W$ is a simple p-circuit. We will assume at first that $V_p - W$ is connected, so that it is itself a simple circuit.

From the construction of No. 3 plus an elementary induction, we find that the K_n of that number can be so chosen that each p-cell is on a single analytical p-cell. Then, by subdividing further, if necessary, the K_p covering V_p, and still calling the new subdivision K_p, we can so choose it that any two closed p-cells of K_p, with a common $(p-1)$-face and exterior to any assigned neighborhood of W, are on a single analytical p-cell.

We will first orient the individual cells, and then the circuit itself, when it is orientable.

7. (a) *Orientation of* E_p. Let H_{p-1} be its boundary sphere, and $u_1, \cdots, u_p$ the variables of the parametric representation for an analytical cell carrying E_p. We can consider E_p as a cell of the S_p of the variables $u_1, \cdots, u_p$, and adopt for the present the language corresponding to straightness for that space. Thus a simplex is now the image of a simplex of S_p, etc. Since E_p and H_{p-1} are complexes, we can take them as the K, L of Ch. II, § 5. We wish to construct a cycle Δ_p, such as the Δ's of Ch. II, No. 26, approximating $\bar{E}_p$ oriented in a definite way as a cycle mod H_{p-1}. In the present instance, the construction loc. cit. merely amounts to this: Take a sufficiently narrow neighborhood N of H on $\bar{E}$; subdivide S_p into cells whose diameter is less than the width of N. The sum of all closed p-cells on E (hence not meeting H) can be oriented so as to become a p-cycle mod N. That p-cycle, or its opposite, is Δ_p. An assigned orientation will determine which of the two must be chosen, and conversely, the choice of Δ_p determines the orientation of the cell.

Since the situation is referred to a Euclidean space, it is essentially as in Ch. I, § 6, and the orientation of E_p is determined by the order in which the local coördinates are named. We can, however, enlarge the range of the method loc. cit., and instead of comparing merely affine transformations we can compare the parametric representations of all the elements E_p. Instead of the signs of the determinants of the transformations from one set of variables to another we now take the signs of the Jacobians. That is, if $u_1, \cdots, u_p$ and $v_1, \cdots, v_p$ are two sets of variables associated with two analytical cells both $\supset E_p$, and

$$J = \frac{D(v_1, \cdots, v_p)}{D(u_1, \cdots, u_p)},$$

then the orientations are concordant when $J > 0$, non concordant otherwise. The Jacobian, however, while more general than the determinant, is strictly local in its effect.

8. (b) *Orientation of the circuit.* The whole problem is to correlate the orientations of two cells E_p, E'_p with a common $(p-1)$-face E_{p-1}. The sum $E_p + E'_p + E_{p-1}$ is a p-cell whose boundary H'_{p-1} is a sphere. We must now orient E'_p in such manner that the new p-cell is a circuit mod H'. By assumption there is an analytical p-cell of $V - W$ which carries both E_p and E'_p. Therefore we can proceed exactly as above and use as indicatrix of the new Δ any cell of the former Δ corresponding to E_p. There results an orientation for the new circuit, and hence for E'_p.

The preceding scheme evidently amounts to orienting E'_p like E_p by the order in which we name the variables u_i of the representation of the analytical p-cell carrying both. By means of the Jacobians, we can then correlate this mode of orienting E'_p with any other associated with any analytical cell that carries it, whether it carries E_p or not. If we proceed thus step by step, we find that we compare Jacobians of the variables for pairs of overlapping analytical cells. Starting from any element and returning to it, if we come back to the initial variables always in the same order, that is with a $J > 0$, $V - W$ is orientable; if that is not the case it is non-orientable. In the orientable case an orientation of $V - W$ is assigned by specifying the order of naming the variables for a single analytical element.

When $V - W$ consists of several disconnected parts, each of them is a circuit and we treat it as above. If all these circuits are orientable, so is the manifold. In the orientable case a choice of orientations for the component circuits defines a unique fundamental p-cycle mod W for each. Hereafter V_p will designate indifferently the variety itself or the chain sum of these p-cycles.

The ultimate orientation scheme that we have obtained is essentially the same as given by Poincaré in his first paper on Analysis Situs [2].

§ 2. Intersections of Analytical Varieties

9. We will first treat a preliminary question. Let E_r be an analytical cell on another E_p, $u_1, \cdots, u_r$ and $v_1, \cdots, v_p$ their parameters, and let their equations be

$$E_r\colon\ x_i = f_i(u); \qquad E_p\colon\ x_i = \varphi_i(v).$$

We will suppose that the point P, $(u) = 0$, of E_r is also the point $(v) = 0$, its coördinates being x_i^0. Since the Jacobian matrix of the f's is of rank r and that of the φ's of rank p, we can express the v's as power series of p of the differences $x_i - x_i^0$, about their values zero, the Jacobian matrix being of rank p. Substituting from the equations of E_r, we obtain for the latter a representation in terms of the v's:

$$v_i = \psi_i(u), \quad i = 1, 2, \cdots, p,$$

where the ψ's are holomorphic about $(u) = 0$ and have a Jacobian matrix of rank r at that point.

Let for example the Jacobian determinant of $\psi_1, \psi_2, \cdots, \psi_r$ be $\neq 0$. Then we can solve for the u's in terms of $v_1, \cdots, v_r$, and substitute in the last $p - r$ functions ψ, so that they will assume the form ω_i, where

$$w_i = v_{r+i} - \omega_i(v_1, \cdots, v_r) = 0, \quad i = 1, 2, \cdots, p - r.$$

The Jacobian of the variables $v_1, \cdots, v_r$, $w_1, \cdots, w_{p-r}$ relatively to the set $v_1, \cdots, v_p$, is $+1$. Therefore we can choose as parameters on E_p near P the variables v, w and referred to these, the equations $w_i = 0$ will be satisfied on E_r about P. Therefore, with a slight change in notation:

Theorem. *Let E_r be an analytical cell on another, E_p, and P a point of E_r. There exist analytical subcells E_r', E_p'*

of the two given cells, and parametric representations, such that: (a) $E'_p \supset E'_r \supset P$. (b) *Referred to the parameters* $v_1, \cdots, v_p$ *of* E'_p *the equations of* E'_r *are* $v_{r+1} = \cdots = v_p = 0$, *so that* $v_1, \cdots, v_r$ *are the parameters for* E'_r.

10. Suppose now that two analytical cells E_r, E_s are on the analytical E_p and have in common an analytical E_t, $t = r + s - p$. Their tangent spaces along E_t are then in position of maximum generality with respect to one another. Replacing if need be the cells by subcells, if P is a point of E_t, we can choose, according to the above, variables $u_1, \cdots, u_{r-t}$, $v_1, \cdots, v_{s-t}$, $w_1, \cdots, w_t$ for E_p, such that: (a) The w's are the parameters for E_t, the u's and w's those for E_r, the v's and w's those for E_s, and all three sets the parameters for E_p. (b) The equations of the cells are

$$E_r\colon\ v_i = 0; \qquad E_s\colon\ u_i = 0; \qquad E_t\colon\ u_i = 0,\ v_j = 0.$$

Let the parameters u, w; w, v; u, w, v; w constitute respective indicatrices of $\beta_r E_r$, $\beta_s E_s$, $\beta_p E_p$, $\beta_t E_t$. The cell E_t will be considered as the oriented intersection of E_r and E_s, if we have as in Ch. IV, § 1,

$$\beta_r \beta_s \beta_p \beta_t = +1.$$

When $t = 0$, $s = p - r$, we have a Kronecker index

$$(E_r \cdot E_{p-r}) = \beta_t,$$

as in Ch. IV. This is the same index as derived from Ch. IV, for on E_p, about P, everything is as if we were dealing with flat intersecting cells on S_p, and as shown in Ch. IV, § 6, the rule for convex cells, here used, yields the same result as that of Ch. IV, § 1.

The extension to intersections of more than two cells is obvious.

11. We now return to our analytical V_p, with the singular locus $W_{p'}$. Since $V - W$ is an M_p, we can apply to chains on it the intersection theory of Ch. IV. We take it to be

orientable and oriented, with the usual observation regarding the non-orientable case and mod 2. The oriented manifold is a certain p-chain, which we also designate by V_p.

Let V_r, V_s be analytical orientable varieties on V_p with $W_{r'}$, $W_{s'}$ as their respective singular loci. We will suppose that they intersect in an orientable V_t on $V_p - W_{p'}$, with a singular locus $W_{t'}$ such that, if P is any point of $V_t - W_{t'}$ there pass through it four cells E_r, E_s, E_t, E_p, respectively on $V_r - W_{r'}, \cdots$, related to one another as the cells of same designations in No. 10. The two intersecting varieties are then said to be *in general position relatively to one another.*

Under the circumstances, we define as *intersection chain of* V_r *and* V_s the chain V_t, oriented by the indicatrices such as E_t just considered. We say "indicatrices" and not "indicatrix", because $V_t - W_{t'}$, may consist of several disconnected parts each requiring its indicatrix.

Let us suppose that L is a closed locally connected subset of $\bar{V}_p$ which includes the boundary points of the manifold, and let the manifolds $V_r - W_{r'}$, $V_s - W_{s'}$, fail to meet each other's boundaries on $\bar{V}_p - L$. Then it is a simple matter to show that V_t itself is a member of the homologous family about V_t which defines the intersection cycle $V_r \cdot V_s$ in the sense of Ch. IV, § 3. All that is necessary is to reason as in Ch. IV, No. 46, with an analytical cell E_{p-t} intersecting V_t at a point, in place of the cell E_{n-s}^{**k} there considered.

12. All this can be extended to the case where V_r, V_s intersect in several orientable varieties V_t^i, concerning which nothing further is specified. We may consider the closure of each V^i as a variety and count it separately, which is equivalent to assuming that the manifold part of every V^i is a simple orientable, and oriented, circuit. Then there exists an analytical E_{p-t}^i on $V_p - W_{p'}$, intersecting V_t^i at a single point but no other V_t. We may assume that E^i does not meet the boundary of V^i. Then $(V_t^i \cdot E_{p-t}^i) = \lambda_i$ is a well defined integer, and we set

$$V_r \cdot V_s = \sum \lambda_i V_t^i.$$

One surmises that the λ's are independent of the particular cells E^i_{p-t} by means of which they are determined. This is readily shown as follows. Suppose that

$$V_r \cdot V_s \sim \sum \mu_i V^i_t.$$

Then we would have

$$\sum (\lambda_i - \mu_i) V^i_t \sim 0,$$

mod as narrow a neighborhood of the V's as we please. But the chain whose boundary is at the left is deformable onto $\sum V^i_t$, since the latter is a complex and hence locally connected. Therefore

$$\sum (\lambda_i - \mu_i) V^i_t \sim 0 \text{ on } \sum V^i_t.$$

This homology between the t-cells of a K_t is impossible, hence $\lambda_i = \mu_i$ for every i.

The analytical intersections just determined can take everywhere the place of polyhedral intersections; for example in approximating chains, etc.

13. It is worth while to examine more closely, at least for the Kronecker index, the analytical implications of our discussion.

Let V_r, V_{p-r} be in general position on V_p. Then their intersection on $V_p - W_{p'}$ consists of a finite number of isolated points, and if P is one of them, the two varieties have indicatrices E_r, E_{p-r}, through P, referred to parameters $u_1, \cdots, u_r$ for the first, $u_{r+1}, \cdots, u_p$ for the second, such that:

(a) The equations of the cells are

$$E_r\colon\ u_{r+i} = 0, \quad i \leqq p-r; \qquad E_{p-r}\colon\ u_i = 0, \quad i \leqq r.$$

(b) The u's are parameters for an analytical p-cell on V_p through P. We will suppose that the orders of the two sets are such as to orient the manifolds in preassigned manner. Let $u_1, \cdots, u_p$ be the order of these parameters by which E_p is oriented. The index is ± 1 according as V_p has for indicatrix $\pm E_p$. If $v_1, \cdots, v_p$ are preassigned parameters for some analytical p-cell which is an indicatrix of V_p through P, we have

$$(E_r \cdot E_{p-r}) = \frac{J}{|J|}, \qquad J = \frac{D(u_1, \cdots, u_p)}{D(v_1, \cdots, v_p)}.$$

In other words, the index has the sign of J.

Let V_r and V_{p-r} intersect at the points P^i. Then

$$(V_r \cdot V_{p-r}) = \sum \frac{J(P^i)}{|J(P^i)|}.$$

The index is then *the difference between the number of points where* $J > 0$ *and where* $J < 0$.

It is as convenient as it is instructive to have a method for obtaining the index directly in terms of representations such as (2), by equations and inequalities. Since we consider only points of the sets $V - W$, the inequalities such as $\varphi > 0$ are satisfied *a priori* under our assumptions. It will be sufficient to take up the case where $p = n$ and V_p is a region of S_n. Let the equations be

$$V_r: \ F_i = 0, \ i \leqq n - r; \qquad V_{n-r}: \ F_{n-r+i} = 0.$$

Assuming again that the varieties are in general position, the intersections are isolated, and if P, (x^0), is one of them,

$$J^* = \frac{D(F_1, \cdots, F_n)}{D(x_1, \cdots, x_n)} \neq 0$$

at P. It follows that any one of the three systems

$$\begin{aligned} &F_i = 0, \quad i \leqq n - r, \qquad F_{n-r+i} = u_{n-r+i}; \\ &F_i = u_i, \quad i \leqq n - r, \qquad F_{n-r+i} = 0; \\ &F_i = u_i, \quad i = 0, 1, 2, \cdots, n; \end{aligned}$$

can be solved for the (x)'s in the form of holomorphic functions about $(u) = 0$, taking the values x_i^0 at $(u) = 0$. This means that the F's themselves can be taken as the u's of No. 10, however with reversal of the order of the two intersecting cells. Thus the order of the equations for each of the two cells is associated with an orientation for the other. Granting that the orientation is that of an indicatrix, the contribution of P to the index $(E_{n-r} \cdot E_r)$ has the sign of $J^*(x^0)$.

REMARK. Suppose that V_r alone is given. We can then find equations of a V_{n-r} through P such that $(E_{n-r} \cdot E_r) = 1$. The orientation of E_{n-r}, hence that of E_r itself, is determined by the order in which the equations of E_r are written. This mode of associating an orientation for V_p with the order of writing its equations was also given by Poincaré [2].

In view of the theory developed in Ch. IV, the extension to several intersecting varieties is obvious.

14. **Histortcal note.** The Kronecker index was considered for the first time in its non-elementary form in a most interesting paper published by Kronecker in 1869, [1]. This fully justifies attaching his name to that index. Kronecker was interested in the following problem: *Let* $F_i(x_1, \cdots, x_n)$, $i = 0, 1, \cdots, n$, *be functions such that every equation* $F_i = 0$ *separates* S_n *into two regions where* F_i *takes opposite signs, one of the two being bounded, say* $\mathfrak{R}^i$. *To find the number of solutions of the system*

$$F_0 = \cdots = F_{i-1} = F_{i+1} = \cdots = F_n = 0$$

on $\mathfrak{R}^i$.

To bring the problem within the scope of our theory, we shall assume that the F's are holomorphic wherever considered, and never vanish simultaneously. To normalize the problem, it is also convenient to choose the signs of the F's such that $F_i < 0$ on $\mathfrak{R}^i$. Finally, in order to make our intersection theory applicable we must assume that

$$J_i = \frac{D(F_0, \cdots, F_{i-1}, F_{i+1}, \cdots, F_n)}{D(x_1, \cdots, x_n)} \neq 0$$

for any solution of the required nature. Kronecker imposes all these conditions except analyticity and merely assumes that the F's are continuous, single-valued and possess all their first partial derivatives wherever they are needed.

Let then P^h, $h = 1, 2, \cdots, s$, be the points determined by the roots of our system. The numbers $J_i(P^h)$ are all $\neq 0$, so that, say π are positive, $\varkappa$ negative. Let $\varrho = \pi - \varkappa$ Kronecker proves:

THEOREM. *The number ϱ is independent of i.*

Thus ϱ depends solely upon the set of $n+1$ functions and their order. Kronecker calls it the *characteristic* of the system $F_0, \cdots, F_n$.

The region $\mathfrak{R}^i$ is a sum of n-circuits. Let us orient each concordantly with S_n, and let V^i_{n-1} be the boundary variety $F_i = 0$, positively related to $\mathfrak{R}^i$. This will normalize the geometric situation. Having operated likewise on all the V's let V^{hi}_{n-1} be the subchain of V^h_{n-1} which is on $\bar{\mathfrak{R}}^i$, that is $V^{hi}_{n-1} = V^h \cdot \bar{\mathfrak{R}}^i$. Then by reference to No. 13, we find that

$$\varrho = \pm (V^{0i} \ldots V^{i-1,i} \cdot V^{i+1,i} \ldots V^n)$$

where the sign depends upon n alone.

To show that ϱ is independent of i, Kronecker had recourse to the curve of intersection of all but two of the V's, and used a transformation of indices analogous to that represented by formula (20) of Ch. IV. Incidentally, he gave the first instance of an oriented intersection of dimensions >0,—naturally in analytical form only.

The independence of ϱ from i becomes obvious when we introduce the looping coefficients. We have, indeed, $\mathfrak{R}^i \to V^i_{n-1}$. Therefore, by an evident extension of the definitions of Ch. IV, § 5,

$$\varrho = \pm (V^0 \cdots V^{i-1} \cdot \bar{\mathfrak{R}}^i \cdot V^{i+1} \cdots V^n) = \pm Lc(V^0 V^1 \cdots V^n),$$

the sign being again dependent solely upon n and not upon the V's or the functions F. Without mentioning the looping coefficients we can also obtain the same result by considering S_n as immersed in some S_{n+1} and paraphrasing the treatment of Ch. IV, No. 51. The number ϱ becomes then the index $(\Delta^0 \cdots \Delta^n)$ of certain conical n-varieties of S_{n+1} whose bases are the V's.

15. The question considered by Kronecker is in intimate relation with the theory of vector distributions on S_n. We have already considered vector distributions on an M_n in connection with fixed points of deformations (Ch. VI, No. 31). Certain notions there considered are readily extended. A vector field or vector distribution on S_n is simply a set of n point-functions or components $F_i(x)$, $i = 1, 2, \cdots, n$, whose range is a certain set on S_n.

Let the range be a bounded region $\mathfrak{R}$ with a V_{n-1}, $F_0(x) = 0$, for its boundary, and let the F's be holomorphic on $\mathfrak{R}$, and never vanish simultaneously on V_{n-1}. Then the points of indetermination of the field are merely the solutions of the system $F_i(x) = 0$, $i = 1, 2, \cdots, n$ on $\mathfrak{R}$. The number of "signed" indeterminations is precisely the index ϱ of No. 14.

It is possible to interpret ϱ in terms of the degree of a certain transformation after the manner of Ch. VI, No. 32, as done by Hadamard [1]. Let P be any point of V_{n-1}, O a fixed point, OQ a vector equal to the vector of our distribution at P, H_{n-1} a Euclidean sphere of center O and radius r. Since the distribution has no point of indetermination on V_{n-1}, OQ has a positive lower bound and we take r less than that lower bound. Then OQ meets H_{n-1} at a unique point Q', and the correspondence between Q, Q', defines a continuous single-valued transformation T of V_{n-1} into H_{n-1}, whose degree is ϱ, provided that certain orientation conventions are observed. The scheme is of course the same as in Ch. VI, No. 32, but applied to the whole region $\mathfrak{R}$ and not merely locally on the given manifold. This is due to the fact that on S_n we are able to transport vectors at a distance. There is an evident extension to the general manifolds possessing the so-called *Levi-Civita parallelism* of modern differential geometry.

When the vectors on V_{n-1} all point inward or outward, the number $\pm\varrho$ is equal to the Euler characteristic of the region (Ch. VI, No. 31). When the behavior on the boundary is arbitrary it is not possible to give a strictly topological evaluation of the index, for the interpretation of ϱ at the end of No. 14 shows clearly that it is in large measure arbitrary. There exist, however some interesting results regarding it.

I. *Theorem of Poincaré-Bohl.* (See Hadamard loc. cit.) This proposition asserts that if two distributions whose general vectors are A, B, do not have the same index relatively to V_{n-1}, then there exists a positive k such that $A + kB = 0$ for some point of V_{n-1}; when their indices are not in the ratio $(-1)^n$, there exists a negative k having the same property.

This proposition is of importance in the applications to differential equations and dynamics.

II. Morse [5] has recently studied at length vector distributions on a general Riemannian manifold (manifold with a metric defined by a quadratic differential form). He has given definite relations between the index ϱ and certain indices attached to the vector distribution on V_{n-1} which is generated by the orthogonal projections on V_{n-1} of the vectors of the assigned distribution, then indices of similar projections on certain varieties V_{n-2}, V_{n-3}, etc.

III. Morse [3] has also investigated at length the singularities of a *gradient vector*, that is, of a vector whose components are the partial derivatives $\frac{\partial F}{\partial x_i}$ of a point function $F(x)$. He obtained a set of characteristic integers for the various types of singular points in the general case, and some very interesting inequalities between the type numbers. He has since applied these results with great success in the calculus of variations [6, 7, 8]. These investigations extend earlier work by Poincaré [1] and Birkhoff [2]. Further investigations on the same subject have been made recently by Brown [1,2].

§ 3. Complex Varieties

16. When the ranges of functions and variables are all complex numbers we have the complex cartesian or projective spaces and their varieties. We shall designate them by

German letters as $\mathfrak{A}_p$, $\mathfrak{B}_p$ etc., the dimension being now the complex dimension, defined as in No. 1.

The space of particular interest, and which shall alone occupy us for the present, is the complex projective n-space. Let $x_0, \cdots, x_n$ be homogeneous coördinates of reference so chosen that $x_0 = 0$ is an assigned $\mathfrak{S}_{n-1}$ of $\mathfrak{S}_n$. We can introduce cartesian coördinates

$$y_j = y'_j + i y''_j = \frac{x_j}{x_0}$$

for $\mathfrak{S}_n - \mathfrak{S}_{n-1}$, then define the analytical elements as before, and analytical manifolds by reference to the real coördinates y', y'', or to any similar system.

Among the varieties are found the *complex* p-varieties. They are defined by means of equations (1) with the F's holomorphic complex functions of the (complex) coördinates y_j. A variety of that nature is designated by $\mathfrak{B}_p$ and it is a simple matter to show that it is a V_{2p}.

The most interesting type of $\mathfrak{B}_p$ is the algebraic p-variety. It is defined as in No. 5, except that the coefficients of the polynomials and the variables now range over the whole complex domain.

As regards complex varieties, the analytical p-cell is to be replaced throughout by the complex p-cell $\mathfrak{E}_p$, which is a real $2p$-cell. It is defined as in No. 1, except that the ranges of variables and functions are complex.

17. We shall now extend the covering theorem of No. 3. We first cover $\mathfrak{S}_n$ with a K_{2n}.

Given any $\mathfrak{S}_{n-1}$ of $\mathfrak{S}_n$, the space $\mathfrak{S}_n - \mathfrak{S}_{n-1}$ may be considered in a well known sense as an S_{2n}. The real simplexes of any such S_{2n} shall be called the simplexes of $\mathfrak{S}_n$. Every point of $\mathfrak{S}_n$ is on some $\bar{\sigma}_{2n}$, and since the space is compact and separable, it may be covered with a finite number of simplexes $\bar{\sigma}^i_{2n}$ that may, however, overlap. Now σ^1 can be taken as the $\mathfrak{R}$ of No. 3, and its intersections with the other simplexes and their faces as parts of the analytical varieties there considered. Therefore $\bar{\sigma}^1$ can be covered with a K_{2n} such that the parts of the faces of various dimensions of the other σ's that are on $\overline{\sigma}^1$ are covered by subcomplexes of K_{2n}^1 whose dimension is the same. We can assume that $\overline{\sigma}^2 \not\subset \overline{\sigma}^1$, else we would discard it. Let then ζ_{2n}^2 represent the region left when the part of $\overline{\sigma}^2$ on $\overline{\sigma}^1$ is removed from σ^2, i. e. let $\zeta^2 = \sigma^2 - \overline{\sigma}^1 \cdot \sigma^2$. We reason with $\overline{\zeta}^2$ exactly as above, and cover it with a K_{2n}^2 which has subcomplexes covering the cells of K_{2n}^1 on $F(\zeta^2)$. As a consequence, $K_{2n}^1 + K_{2n}^2$ will cover $\overline{\sigma}^1 + \overline{\sigma}^2$. Its $2n$-cells are obtained by taking $2n$-simplexes and covering some of their faces with simplicial subdivisions so that $K^1 + K^2$ is merely convex. We proceed in the same manner until all σ's have been brought into play. At the end we shall have a convex complex $\sum K_{2n}^i$ covering $\mathfrak{S}_n$. A first derived of that complex will be simplicial.

If we have a set of analytical varieties on $\mathfrak{S}_n$, we reason in the same way, merely taking care to construct the complexes $K_{2n}^1, K_{2n}^2, \cdots$ such as to have subcomplexes covering the portions of the varieties on $\overline{\sigma}^1$, $\overline{\zeta}^2$, etc. The modifications when $\mathfrak{S}_n$ is replaced by some region of the space are insignificant. Therefore:

THEOREM. *The covering theorem of No. 3 holds for $\mathfrak{S}_n$, even when $\mathfrak{R}$ is $\mathfrak{S}_n$ itself. In particular, a variety defined by functions holomorphic over the whole space can be covered with a finite subcomplex of a finite complex covering $\mathfrak{S}_n$ itself.*

COROLLARIES. I. *$\mathfrak{S}_n$ is an absolute M_{2n}.*

II. *Let an algebraic $\mathfrak{B}_n$ have $\mathfrak{W}$ for singular locus. Then $\overline{\mathfrak{B}_n - \mathfrak{W}}$ is an M_{2n} and $\overline{M} = \mathfrak{B}$. Furthermore $\overline{M}$ is a sum of a finite number of absolute circuits.*

In both cases there are finite covering complexes. All this follows at once from what precedes plus the fact that every point of $\mathfrak{S}_n$ or of $\overline{\mathfrak{B}_n - \mathfrak{W}}$ has a complex E_n ($=$ an E_{2n}) for neighborhood, and that $\mathfrak{W}$ is of dimension $\leqq 2n-2$.

18. We shall now examine the orientation of complex varieties. Consider first the complex cell $\mathfrak{E}_p$:

$$x_i = \varphi_i(u_1, \cdots, u_p),$$

and let us suppose that it possesses a point P whose $x_0 \neq 0$. If we set

$$y_j = \frac{x_j}{x_0}; \quad \psi_j = \frac{\varphi_j}{\varphi_0},$$

we have for $\mathfrak{E}_p$ a representation as a cell of complex Euclidean n-space

$$y_j = \psi_j(u_1, \cdots, u_p).$$

Separating real and complex parts, let us write $y_j = y_j' + iy_j'', \cdots$. Then $\mathfrak{E}_p$ is homeomorphic to the analytical $2p$-cell of real S_{2n}:

$$y_j' = \psi_j'(u_1', u_1'', \cdots, u_p', u_p'')$$
$$y_j'' = \psi_j''(u_1', u_1'', \cdots, u_p', u_p'').$$

We shall agree once and for all to orient the analytical cell and concordantly with it $\mathfrak{E}_p$, by naming the real parameters u', u'' in the following order: $u_1', u_1'', u_2', u_2'', \cdots, u_p', u_p''$.

THEOREM. *The orientation of $\mathfrak{E}_p$ is independent of the order of the complex parameters $u_1, u_2, \cdots, u_p$.*

For when two of them are permuted there results an even permutation in the assigned order of the real parameters u', u''.

Suppose now that we make a complex change of variables

$$v_h = \omega_h(u_1, \cdots, u_p), \qquad h = 1, 2, \cdots, p,$$

where at $(u) = 0$ the ω's are holomorphic, and the Jacobian

$$J = \left\| \frac{\partial \omega_h}{\partial u_j} \right\| \neq 0.$$

Our convention demands that the cell be oriented by the order $v_1', v_1'', \cdots, v_p''$ of the new real parameters. *The new orientation is the same as the old.* For we have the well known relation, proved by elementary operations on the Jacobian,

$$\frac{D(v_1', v_1'', \cdots, v_p'')}{D(u_1', u_1'', \cdots, u_p'')} = J\bar{J};$$

and since $J \neq 0$, the Jacobian at the left is > 0. Referring to No. 7, our assertion follows.

Henceforth it will be understood that all complex analytical cells are to be oriented in the invariant manner just introduced.

Suppose now that we have a complex $\mathfrak{V}_p$, with its singular locus $\mathfrak{W}$. If we take any connected subset of the manifold $\mathfrak{V}-\mathfrak{W}$, the result just proved shows that the indicatrices are concordant in the parts common to any pair of p-cells, since the Jacobians corresponding to the passage from one set of variables to the other are always >0. Referring to No. 8, it follows that the connected subset of $\mathfrak{V}-\mathfrak{W}$ considered is an orientable circuit. Therefore (Lefschetz [5]):

THEOREM. *Every complex variety is an orientable manifold. In particular algebraic varieties and complex projective spaces are all orientable manifolds.*

Chosing an $\mathfrak{E}_p$ on any connected part of $\mathfrak{V}-\mathfrak{W}$, oriented in the manner just prescribed, as indicatrix for that part, we have a unique fundamental $2p$-cycle mod $\mathfrak{W}$ for each part. Henceforth we designate indifferently by $\mathfrak{V}$ the complex variety itself or the chain sum of these $2p$-cycles. This enables us to introduce, wherever the requisite conditions of Ch. IV, § 4, are fulfilled, the chain intersections of two or more complex varieties and their Kronecker indices. We shall now examine some special but highly important intersection properties of complex varieties.

The association of a chain with $\mathfrak{V}$ is as in No. 9, except that, due to the unique choice of indicatrices, the chain $\mathfrak{V}$ introduced here is unique.

19. **Intersections.** Suppose that we have the complex cells $\mathfrak{E}_r$, $\mathfrak{E}_{p-r}$ on $\mathfrak{V}_p-\mathfrak{W}$, intersecting at a point Q and in general position. Then, as in No. 9, we can replace the two cells by subcells, to be designated by the same letters, and such that if $u_1, \cdots, u_r$ are the parameters for the first, $u_{r+1}, \cdots, u_p$ those for the second, there exists an $\mathfrak{E}_p$ on $\mathfrak{V}_p-\mathfrak{W}$ through Q, with $u_1, \cdots, u_p$ for parameters. Since the orientations of the three cells are given by the orders $u_1', u_1'', \cdots, u_r''$; $u_{r+1}', \cdots, u_{2p}''$ and $u_1', \cdots, u_{2p}''$, we always have

$$(\mathfrak{E}_r \cdot \mathfrak{E}_{p-r}) = +1. \tag{4}$$

20. Let now $\mathfrak{B}_r$ and $\mathfrak{B}_{p-r}$ have an isolated intersection Q on $\mathfrak{B}_p - \mathfrak{W}$. The natural definition for the geometric multiplicity with which Q is to be counted among the intersections, is to single out a sufficiently small neighborhood of Q say on $\mathfrak{B}_r$, then to displace it slightly so as to have intersections at non-singular points and without contact. Their number μ is then by definition the multiplicity. Here we have a basic theorem (Lefschetz [2, 5]):

THEOREM. *The geometric multiplicity μ of an isolated intersection Q of $\mathfrak{B}_r$ and $\mathfrak{B}_{p-r}$ on $\mathfrak{B}_p - \mathfrak{W}$ is positive, and equal to the contribution of Q to the Kronecker index $(\mathfrak{B}_r \cdot \mathfrak{B}_{p-r})$.*

Let $\mathfrak{B}_p - \mathfrak{W} \supset \mathfrak{E}_p \supset Q$, the cell carrying no other intersection of $\mathfrak{B}_r$ and $\mathfrak{B}_{p-r}$. As far as our proof is concerned, we may confine the intersecting varieties to $\mathfrak{E}_p$ and assume them irreducible on it, and represented about Q in the form

$$\mathfrak{B}_r: \quad \varphi_i(u) = 0, \qquad i = 1, 2, \cdots, p-r;$$
$$\mathfrak{B}_{p-r}: \quad \psi_i(u) = 0, \qquad i = 1, 2, \cdots, r.$$

We may cover $\mathfrak{B}_r$ and $\mathfrak{B}_{p-r}$ with complexes having Q for vertex and such that all closed cells of which it is a vertex are on $\mathfrak{E}_p$, then restrict the varieties to the sums of those cells alone, which we denote by $\mathfrak{B}_r^*$, $\mathfrak{B}_{p-r}^*$. Ultimately, then, we have to prove

$$(\mathfrak{B}_r^* \cdot \mathfrak{B}_{p-r}^*) = \mu > 0. \tag{5}$$

Since Q is not on the boundaries of the varieties considered as oriented chains, the new operations leading to the starred chains introduce no boundary cells incident with Q, hence $Q \not\subset F(\mathfrak{B}_r^*)$ or $F(\mathfrak{B}_{p-r}^*)$. It follows by reasoning as in Ch. IV, No. 42, that if $\mathfrak{B}_{p-r}^{*\prime}$ is obtained from $\mathfrak{B}_{p-r}^*$ by a sufficiently small deformation

$$(\mathfrak{B}_r^* \cdot \mathfrak{B}_{p-r}^{*\prime}) = (\mathfrak{B}_r^* \cdot \mathfrak{B}_{p-r}^*). \tag{6}$$

Hence in proving (5) we can replace $\mathfrak{B}_{p-r}^*$ by a $\mathfrak{B}_{p-r}^{*\prime}$ obtained from it by a small translation in the space of the variables u_i.

21. The proof rests upon the following proposition: *Let $\mathfrak{V}_r$, $\mathfrak{V}_s$ of $\mathfrak{S}_p$ intersect at the isolated point Q. If $\mathfrak{V}_s$ undergoes a small arbitrary translation it will continue to intersect $\mathfrak{V}_r$ in the vicinity of Q when $s = p - r$, and will not when $s < p - r$.*

(An arbitrary translation is one whose components w_i are not constrained to satisfy a certain number of analytical relations.)

The result concerning $s = p - r$ is a special case of a theorem given in Osgood [1], p. 135. Let then $r + s < p$. When the sum is zero the proposition is obvious. We assume it proved for values up to $r + s - 1$ of the sum, and show that it holds for $r + s$. By a first small arbitrary translation impressed upon $\mathfrak{V}_s$ we replace it by $\mathfrak{V}'_s$ such that $\mathfrak{V}_r$ and $\mathfrak{V}'_s$ do not intersect each other's singular loci about Q. We are then back to the initial case with Q a non-singular point of the varieties. We then have about Q parametric representations:

$$u_i = F_i(x_1, \cdots, x_r); \qquad u_i = G_i(y_1, \cdots, y_s).$$

If the required result were false we would have a compatible system in the variables u, x, y about the values zero,

$$u_i = F_i, \qquad u_i - w_i = G_i,$$

whatever the w's. Hence, about the values zero, we could represent the w's in the form: $w_i = F_i - G_i$, that is in terms of less than p variables, which is absurd since they are independent.

22. Going back to our problem, according to what has just been proved we can assume that $\mathfrak{V}'_{p-r}$ actually intersects $\mathfrak{V}_r$ in a certain number (>0) of isolated points, but that the varieties do not intersect each other's singular loci in the region considered. The problem is thus reduced to the case of a single intersection Q which is non-singular for $\mathfrak{V}_r$ or $\mathfrak{V}_{p-r}$.

Under the circumstances, about Q we may assume for $\mathfrak{V}_r$ the representation $u_{r+i} = 0$, i. e. $\varphi_i(u) \equiv u_{r+i}$, and we impress upon $\mathfrak{V}_{p-r}$ a small translation

$$u'_i = u_i - w_i, \qquad i \leqq r, \qquad u'_{r+i} = u_{r+i}.$$

The varieties have then a variable intersection Q' about which I say that they are in general position. In the contrary case we find that

$$\frac{D(\psi_1, \cdots, \psi_r)}{D(u_1, \cdots, u_r)} = 0,$$

when every $u_{r+i} = 0$. Therefore (Osgood [1], p. 165), the ψ's vanish on a $\mathfrak{V}_t, t > 0$, of the space $u_{r+i} = 0$, which passes through Q. This means that $\mathfrak{V}_r$ and $\mathfrak{V}_{p-r}$ intersect in $\mathfrak{V}_t$ through Q, which is not the case. Therefore the varieties $\mathfrak{V}_r$, $\mathfrak{V}'_{p-r}$ are in general position about Q'.

It follows from No. 19 that each of the points Q' contributes $+1$ to the index (6). Their number $\mu > 0$ as we have seen, and by what precedes,

$$(7) \qquad (\mathfrak{V}^*_r \cdot \mathfrak{V}^{*\prime}_{p-r}) = \mu > 0.$$

From (7) and (6) follows (5), and since μ is the number of isolated ordinary intersections of one of the varieties with the other slightly translated, our theorem is proved.

23. If we have two varieties $\mathfrak{V}_r$, $\mathfrak{V}_{p-r}$ intersecting in isolated points of $\mathfrak{V}_p - \mathfrak{W}$ with geometric multiplicities μ_i, we will have, summing for all the intersections,

$$(\mathfrak{V}_r \cdot \mathfrak{W}_{p-r}) = \sum \mu_i.$$

This proves the

THEOREM. *If two complex varieties intersect in isolated points the total number of intersections, each counted with its geometric multiplicity, is equal to their Kronecker index.*

It is always understood that the only intersections considered are those on the manifold $\mathfrak{V}_p - \mathfrak{W}$.

As an application let us suppose that we have the situation of Ch. III, No. 29, with $K - L$ a submanifold of the manifold $\mathfrak{V}_p - \mathfrak{W}$. We will briefly call a chain coincident with a $\mathfrak{V}_r$, a complex r-chain. Then if $\mathfrak{V}_r$ and $\mathfrak{V}_{p-r}$ are of the dual types $K - L^i$ mod L^j, and intersect in isolated points, it follows from (12) that $(\mathfrak{V}_r \cdot \mathfrak{V}_{p-r}) \neq 0$, hence (Ch. IV, § 2) $\mathfrak{V}_r \not\sim 0$, in its appropriate sense, and likewise for $\mathfrak{V}_{p-r}$. Therefore:

THEOREM. *If the complex cycle $\mathfrak{V}_r$ of type $K-L^i$ mod L^j is intersected in isolated points by a complex $(p-r)$-cycle of dual type, it is $\not\approx 0$ on $K-L^i$ mod L^j.*

This theorem has important applications in algebraic geometry (§ 4).

24. We have considered so far only questions related to the Kronecker index, but the more general intersections can be treated now with ease. For let $\mathfrak{V}_q$, $\mathfrak{V}_r$ have for intersections varieties $\mathfrak{V}_s^i$ on $\mathfrak{V}_p-\mathfrak{W}$, $s = q+r-p$. Then we define the λ's as in No. 12, except that the cells are now complex. As a consequence, from (5) follows that every $\lambda > 0$. Then, here as in No. 12:

$$\mathfrak{V}_q \cdot \mathfrak{V}_r = \sum \lambda_i \mathfrak{V}_s^i,$$

and λ_i is the geometric multiplicity with which $\mathfrak{V}_s^i$ is counted in the intersection of $\mathfrak{V}_q$ and $\mathfrak{V}_r$. Here also the coefficients λ_i depend solely upon the intersecting varieties.

§ 4. APPLICATIONS TO ALGEBRAIC GEOMETRY

25. In the applications of Analysis Situs to algebraic geometry, it is highly important to bear in mind that the basic group in dealing with any configuration is not the homeomorphic group, but the birational group. To take a very simple example, consider the complex cubic curve

$$x^3 + y^3 - xy = 0.$$

It is a $\mathfrak{V}_1$ with a double point at the origin O. The point O has a neighborhood on $\mathfrak{V}_1$ consisting of two 2-cells. The singular variety $\mathfrak{W} = O$, and $\mathfrak{V}_1 - O$ is an M_2 but $\mathfrak{V}_1$ is not. However, from the point of view of the intrinsic algebraic properties of $\mathfrak{V}_1$, it may be replaced by a birational transform. Now

$$x = \frac{x'}{1+x'^3}, \qquad y = \frac{x'^2}{1+x'^3}$$

is a birational transformation of $\mathfrak{V}_1$ into the x-axis, that is, into an H_2 which is an absolute M_2. Thus the modification incurred is topologically improper, but algebraically satisfactory. Regarding algebraic curves we have the following basic.

THEOREM. *Every irreducible algebraic curve is birationally transformable into a non-singular irreducible algebraic curve of some* $\mathfrak{S}_n$. *The transform curve is an orientable absolute* M_2, *whose topological type is unique.*

The Betti number $R_1(M_2)$ is even (Ch. IV, No. 60), and if $2p$ is its value, p is the *genus* of the curve. That genus is a birational invariant of the initial curve, and it determines in unique manner the type of M_2. It is then also in a sense a topological invariant of the curve itself. All these facts are classical in algebraic geometry, and are taken up at length in any major treatise on the subject (Picard, Cours d'Analyse, vol. II; Severi-Loeffler, Vorlesungen über algebraische Geometrie; etc.).

26. The situation is much less precise as regards algebraic surfaces and varieties. There are invariants of several types, some topologically, but not birationally, invariant, and *vice versa.* Altogether it cannot be said that perfection has as yet been achieved in clearing up ideas along that line. The present Chapter, however, taken together with our monographs of the Borel series [5], and of the Memorial series [16], will put the subject on a fairly solid basis. We will merely describe some of the main results, and refer the reader to our monographs for the rest.

To avoid complications we shall consider only non-singular varieties on $\mathfrak{S}_n$. This is, however, more a matter of convenience than of necessity. For in the first place it is a simple matter to dispose of the singular $(p-1)$-varieties, somewhat after the manner indicated for the multiple points of an algebraic curve. In the second place it is often possible to find suitable parametric representations for the isolated singularities, and by means of them to introduce ideal points which turn $\mathfrak{V}_p$ into an absolute manifold: the cone is an example.

27. Henceforth it will be understood that $\mathfrak{V}_q$ stands for an algebraic variety. That variety is in any case a $2q$-cycle

and cycles depending upon such cycles will be called *algebraic.* We have first the important

THEOREM. *Every* $\mathfrak{V}_q$ *on* $\mathfrak{V}_p$ *is* $\not\sim 0$ *on* $\mathfrak{V}_p$.

For there exists always an $\mathfrak{S}_{n-q}$ intersecting $\mathfrak{V}_q$ in isolated points, hence $\mathfrak{V}_q \not\sim 0$ on $\mathfrak{S}_n$ (No. 23) and *a fortiori* on $\mathfrak{V}_p$.

The set of all algebraic $2q$-cycles is a submodulus of the modulus of all $2q$-cycles. We designate it by $\mathfrak{A}_q$. Since $\mathfrak{V}_p$ can be covered by a finite complex its Betti numbers are all finite. Therefore any $1 + R_{2q}(\mathfrak{V}_p)$ elements of $\mathfrak{A}_q$ satisfy a homology whose coefficients are integers. Taking as the zero-modulus the modulus of all bounding $2q$-cycles, it is not difficult to show that since the modulus of all $2q$-cycles has a base, this is also the case for $\mathfrak{A}_q$. Therefore $\mathfrak{A}_q$ possesses a Betti number $R_{2q}(\mathfrak{A}_q)$ and likewise torsion coefficients. We can similarly consider its Betti numbers mod 2 or mod m. Regarding the absolute homology characters of $\mathfrak{A}_{p-1}$ much interesting information is at hand, and we shall examine that modulus presently. Concerning the other moduli, or even the characters of $\mathfrak{A}_{p-1}$ mod 2, mod m, practically nothing is known.

28. The homology theory of the modulus $\mathfrak{A}_{p-1}$ is in the closest possible relationship with a fundamental algebro-geometric notion due to Severi [2], the notion of *algebraic dependence* between the curves on a surface, or the hypersurfaces (= the $\mathfrak{V}_{p-1}$'s) on a $\mathfrak{V}_p$. Let $\mathfrak{V}_{p-1}^i$, $i = 1, 2, \cdots, r$, be assigned hypersurfaces of $\mathfrak{V}_p$, and let us suppose that there can be found a new hypersurface $\mathfrak{V}_{p-1}^0$, and a continuous algebraic system of hypersurfaces $\{\mathfrak{V}_{p-1}\}$, such that for some values of its parameters it breaks up into a variety $\lambda_0 \mathfrak{V}^0 + \sum \lambda_i \mathfrak{V}^i$, for some others into another variety $\lambda_0 \mathfrak{V}^0 + \sum \mu_i \mathfrak{V}^i$. Then Severi states that the varieties $\mathfrak{V}^i$ are algebraically dependent, and writes for them the symbolic relation of algebraic dependence,

$$\sum (\lambda_i - \mu_i)\, \mathfrak{V}_{p-1}^i = 0.$$

The convention is made that such relations are to be treated like homologies: transposition of terms is allowed, etc.

EXAMPLES. I. Let $\mathfrak{V}_p = \mathfrak{S}_n$ and let $\mathfrak{V}_{n-1}^{\mu}$ be a hypersurface of order μ, $\mathfrak{S}_{n-1}$ a "hyperplane" of the space. The system of all hypersurfaces of order μ is a $\{\mathfrak{V}_{n-1}\}$ with $\mu\mathfrak{S}_{n-1}$ and $\mathfrak{V}_{n-1}^{\mu}$ among its elements. Therefore

$$\mathfrak{V}_{n-1}^{\mu} = \mu\mathfrak{S}_{n-1}.$$

II. Let $\mathfrak{V}_2$ be an ordinary quadric, G^1, G^2 two generators of opposite systems. Then it is a well known theorem that every algebraic curve D on $\mathfrak{V}_2$ can be cut out by a surface $\mathfrak{V}_2^{\mu}$ passing, say ν times, through one of the generators. But $\mathfrak{V}_2^{\mu}$ cuts out on $\mathfrak{V}^2$ the system cut out by $\mu\mathfrak{S}_2$. Since a suitable tangent $\mathfrak{S}_2$ cuts out $G^1 + G^2$, we have

$$D = \mu(G^1 + G^2) - \nu G^1 = \mu G^1 + \lambda G^2.$$

29. The preceding examples exhibit the clearest possible resemblance between algebraic dependence and homology. Algebraic dependence gives rise to an algebraic modulus, provided that so-called *virtual* hypersurfaces are counted in. We then have:

THEOREM I. *The algebraic modulus coincides with the homology modulus* (*Lefschetz* [2, 5]).

In his investigations on integrals of total differentials attached to an algebraic surface, Picard ([1], vol. II) had occasion to introduce what might be called "transcendental" equivalence between its algebraic curves with an associated Betti number ϱ, since then designated as the *Picard number*. It is one unit less than the least number of algebraic curves that are the logarithmic singularities of an integral of total differentials with an integrand rational on the surface. A similar number can be defined for any $\mathfrak{V}_p$, and we have:

THEOREM II. *The Betti number* R_{2p-2} $(\mathfrak{A}_p)$ *is equal to the Picard number* ϱ *of* $\mathfrak{V}_p$, (*Severi* [2]).

Further important theorems are also:

THEOREM III. *Every zero-divisor of* $\mathfrak{V}_p$ *for the dimension* $2p-2$ *is* $\sim$ *to a cycle of* $\mathfrak{A}_{p-1}$. *Therefore the torsion coefficients of* $\mathfrak{V}_p$ *and of* $\mathfrak{A}_p$ *for the dimension* $2p-2$ *are the same* (*Lefschetz*).

THEOREM IV. *Let* $\mathfrak{G}$ *designate an arbitrary hyperplane section of* $\mathfrak{V}_p$. *When* $p > 2$, *the relations*

$$\mathfrak{V}_{p-1} = \mathfrak{V}'_{p-1}, \qquad \mathfrak{V}_{p-1}\cdot\mathfrak{G} = \mathfrak{V}'_{p-1}\cdot\mathfrak{G} \text{ on } \mathfrak{G},$$
$$\mathfrak{V}_{p-1} \sim \mathfrak{V}'_{p-1}, \qquad \mathfrak{V}_{p-1}\cdot\mathfrak{G} \sim \mathfrak{V}'_{p-1}\cdot\mathfrak{G} \text{ on } \mathfrak{G},$$

are all equivalent (*Severi, Lefschetz*).

THEOREM V. *Let the variety be an algebraic surface with* G *as an arbitrary hyperplane section. A necessary and sufficient condition in order that the algebraic curves* $D^1, \cdots, D^r$ *be dependent is that the matrix*

$$\left\| \begin{matrix} (D^i \cdot D^j) \\ (D^i \cdot G) \end{matrix} \right\|$$

be of rank $< r$ (*Severi* [2]).

The algebraic part of these theorems is due to Severi and we first gave the topological part in [2]. A summary and detailed references will be found in our monograph [16], complete proofs in [5].

In combination with the intersection theory of Ch. IV, the algebraic moduli may be used to clear up the numerous questions of enumeration that arise in algebraic geometry. For many of them this has never been possible before. The most important instances are:

I. *Questions of enumeration for an* $\mathfrak{S}_n$. We refer to theorems similar to those of Bezout. As an example, it is not an easy matter to prove algebraically that when a $\mathfrak{V}_k$ and $\mathfrak{V}_{n-k}$ on $\mathfrak{S}_n$ whose orders are respectively m, m' intersect in isolated points, the total number of points, each counted with its multiplicity, is mm'. Topologically it is a matter of a few lines, as we have shown in [16]:

(a) $\mathfrak{S}_n$ is an orientable $2n$-circuit, hence $R_{2n}(\mathfrak{S}_n) = 1$, and it has no torsion coefficients for the dimension $2n-1$.

(b) $\mathfrak{S}_n - \mathfrak{S}_{n-1}$ is a Euclidean S_{2n} and therefore an ordinary E_{2n}. Hence every cycle mod $\mathfrak{S}_{n-1}$ is deformable onto $\mathfrak{S}_{n-1}$. It follows that every $\Gamma_{2n-1} \sim 0$ on $\mathfrak{S}_n$, and that if $\mathfrak{S}_{n-1} \supset \Gamma_k \sim 0$ on $\mathfrak{S}_n$, likewise $\Gamma_k \sim 0$ on $\mathfrak{S}_{n-1}$. Hence, by (a):

$$R_p(\mathfrak{S}_n) = \begin{cases} 1 \text{ for } p \text{ even and } \leqq 2n, \\ 0 \text{ for } p \text{ odd,} \end{cases}$$

and $\mathfrak{S}_n$ has no torsion coefficients. Moreover, an $\mathfrak{S}_k$ is a base for all $2k$-cycles.

(c) From (b) follows $\mathfrak{V}_k \sim \mu \mathfrak{S}_k$. But since $\mathfrak{V}_k$ is of order m there is an $\mathfrak{S}_{n-k}$ intersecting it in m isolated points. Therefore

$$(\mathfrak{V}_k \cdot \mathfrak{S}_{n-k}) = m = \mu(\mathfrak{S}_k \cdot \mathfrak{S}_{n-k}) = \mu$$

and hence $\mathfrak{V}_k \sim m\mathfrak{S}_k$. Similarly $\mathfrak{V}_{n-k} \sim m'\mathfrak{S}_{n-k}$, hence finally

$$(\mathfrak{V}_k \cdot \mathfrak{V}_{n-k}) = mm'(\mathfrak{S}_k \cdot \mathfrak{S}_{n-k}) = mm'.$$

II. *Poncelet's continuity principle; Schubert's enumeration principle.* This principle has had a great vogue since Poncelet applied it with much success in projective geometry. It will be remembered that Poncelet, in a manner that seemed rather reckless in his day, did not hesitate to introduce imaginary elements in questions of enumeration as if they were real. Later Schubert was guilty of a similar recklessness in far more sophisticated form; and many authors have used that mode of reasoning wherein abstraction is made of all difficulties that may occur when coefficients of algebraic equations are allowed to vary with abandon. Poncelet's initial procedure has been justified long since, but it is only recently that Schubert's so-called *Kalkül der abzählenden Geometrie* has been put on a firm basis. It has been shown in fact by Van der Waerden [1] (1929) that the enumeration formulas of Schubert are essentially homologies between certain algebraic cycles, and relations between Kronecker indices resulting from our intersection theory. Somewhat earlier Severi ([3], 1912; [4], 1916), had shown that Schubert's results could be reduced in most cases to the algebraic intersection theorems which follow from his theory of equivalence. It is, however, in the nature of Severi's theory that only intersections between the hypersurfaces of a $\mathfrak{V}_p$ can enter into it, whereas, in general, intersections between all sorts of algebraic cycles might occur, and in that case our intersection theory alone can clear up the matter.

The doubtful and cumbersome cases which have been brought forth against the principle of Schubert are to be explained off by amplifying the basic $\mathfrak{V}_p$ as indicated at the end of No. 27, so as to turn it into an absolute $\mathfrak{M}_{2p}$. This is really implicit in the explanation given by Severi [3] of the examples proposed by Kohn and Study as contradicting Schubert's principle.

III. *Enumeration in the theory of multiply periodic functions.* In that theory there occur important formulas, due largely to Poincaré and Wirtinger, giving the number of common zeros of sets of multiply periodic functions and the like. We have shown [2, 5] that all these formulas are reducible to questions of intersection on a $2p$-ring, where p is the genus of the functions, or the rank of their period matrix. Indeed, by means of our intersection theory, the derivation of the formulas in question becomes wholly elementary. Many other important results can be derived in that field by means of relatively simple topological considerations.

IV. *Algebraic correspondences.* Let D, D', be two irreducible algebraic curves, and T_1 an algebraic correspondence or transformation of one into the other. Then $D \times D'$ is an algebraic surface which is an M_4, and T_1 is represented on it by an algebraic curve Γ^1. If T_2 is another such correspondence, and Γ^2 its image on $D \times D'$, then $(\Gamma^1 \cdot \Gamma^2)$ is the number of the coincidences unsigned, that is, each counted with its multiplicity. The value of the index is given by formula (22) of Ch. VI. The sum in (22) consists of three terms, the extreme terms being very simple. Let μ_1, μ_1' be the degrees of T_1 and T_1^{-1}, μ_2, μ_2' the same numbers for T_2. They are called the *indices* of the transformations. Using the double interpretation of the index we recall that for example μ_1 represents the number of times that $T_1 \cdot D$ covers D', while μ_1' is the number of points of D' derived by T_1 from a given point of D. Therefore, in the notation of Ch. VI, § 5,

$$\mu_1 = f^2, \quad \mu_1' = f^0; \quad \mu_2 = g^2, \quad \mu_2' = g^0.$$

Since any point is a zero-cycle we have

$$\alpha^0 = \alpha^2 = \beta^0 = \beta^2 = 1.$$

Hence the extreme terms in formula (22) are $\mu_1\mu_2' + \mu_2\mu_1'$. If $\{\gamma^i\}$ and $\{\delta^i\}$ are the bases for the one-cycles on D and D',

$$\alpha = \| (\gamma^i \cdot \gamma^j) \|, \quad \beta = \| (\delta^i \cdot \delta^j) \|,$$

and $\boldsymbol{f}$, $\boldsymbol{g}$ are the transformation matrices for the same cycles. The coincidence formula yields here,

$$(\Gamma^1 \cdot \Gamma^2) = \mu_1\mu_2' + \mu_2\mu_1' - \mathit{trace}\ \alpha^{-1}\boldsymbol{g}\beta\boldsymbol{f}'.$$

This is simply, under another form, the noted coincidence formula due to A. Hurwitz and obtained by him by rather difficult manipulations of θ functions. If T is an algebraic correspondence between the points of D itself, with μ, μ' for indices and $\boldsymbol{f}$ as transformation matrix of the cycles, we have, by (28) of Ch. VI, for its fixed points the expression

$$(\Gamma \cdot \Gamma^0) = \mu + \mu' - \mathit{trace}\,\boldsymbol{f},$$

where Γ^0 represents the identity on the product surface. The simplest possible case and also the most frequent in practise, is when $\boldsymbol{f}$ is a multiple of the identity. The number v such that $\boldsymbol{f} = -v\boldsymbol{1}$, is then called the *valence* of T, and T is said to be a correspondence with valence. The fixed point formula reduces then to the classical Cayley-Brill formula:

$$(\Gamma \cdot \Gamma^0) = \mu + \mu' + 2vp,$$

where p is the genus of $D(2p = R_1(D))$.

Regarding this important question the reader will consult with profit Severi [1], where the preceding notions are used with great advantage in their geometric form. Further details concerning the topological treatment will be found in our paper [10], a summary of the work along that line and full references in [13], Ch. 16.

BIBLIOGRAPHY

Alexander, J. W.

1. Sur les cycles des surfaces algébriques et sur une définition topologique de l'invariant de Zeuthen-Segre. Rendic. dei Lincei (2) 23 (1914), 55–62.
2. A proof of the invariance of certain constants of analysis situs. Trans. Am. M. S. 16 (1915), 148–154.
3. Normal forms for one- and two-sided surfaces. Annals of Math. (2) 16 (1915), 158–161.
4. Note on two three-dimensional manifolds with the same group. Trans. Am. M. S. 20 (1919), 339–342.
5. Note on Riemann spaces. Bull. Am. M. S. 26 (1920), 370–372.
6. On transformations with invariant points. Trans. Am. M. S. 23 (1922), 89–95.
7. A proof of Jordan's theorem about a simple closed curve. Annals of Math. (2) 21 (1919), 180–184.
8. A proof and extension of the Jordan-Brouwer separation theorem. Trans. Am. M. S. 23 (1922), 333–349.
9. Invariant points of a surface transformation of a given class. Trans. Am. M. S. 25 (1923), 173–184.
10. A lemma on a system of knotted curves. Proc. Nat. Acad. 9 (1923), 93–95.
11. On the deformation of an n-cell. Proc. Nat. Acad. 9 (1923), 406–407.
12. On the subdivision of three-space by a polyhedron. Proc. Nat. Acad. 10 (1924), 6–8.
13. An example of a simply connected surface bounding a region which is not simply connected. Proc. Nat. Acad. 10 (1924), 8–10.
14. Remarks on a point set constructed by Antoine. Proc. Nat. Acad. 10 (1924), 10–12.
15. New topological invariants expressible as tensors. Proc. Nat. Acad. 10 (1924), 99–101.
16. On certain new topological invariants of a manifold. Proc. Nat. Acad. 10 (1924), 101–103.
17. Topological invariants of a manifold. Proc. Nat. Acad. 10 (1924), 493–494.

18. On the intersection invariants of a manifold. Proc. Nat. Acad. 11 (1925), 143–146.

19. Combinatorial analysis situs, I. Trans. Am. M. S. 28 (1926), 301–329.

20. Topological invariants of knots and links. Trans. Am. M. S. 30 (1928), 275–306.

21. The combinatorial theory of complexes. Annals of Math. (2) 31 (1930), 294–322.

22. (With Veblen) Manifolds of n dimensions. Annals of Math. (2) 14 (1913), 163–178.

23. (With Briggs, G. B.) On types of knotted curves. Annals of Math. (2) 28 (1927), 562–586.

Alexandroff, P.

1. Zur Begründung der n-dimensionalen mengentheoretischen Topologie. Math. Ann. 94 (1925), 296–308.
2. Simpliziale Approximationen in der allgemeinen Topologie. Math. Ann. 96 (1925), 488–511.
3. Über kombinatorische Eigenschaften allgemeiner Kurven. Math. Ann. 96 (1925), 512–554.
4. Darstellung der Grundzüge der Urysohnschen Dimensionstheorie. Math. Ann. 98 (1926), 31–63.
5. Über den allgemeinen Dimensionsbegriff und seine Beziehung zur elementaren geometrischen Anschauung. Math. Ann. 98 (1926), 617–636.
6. Über die Dualität zwischen den Zusammenhangszahlen einer abgeschlossenen Menge und des zu ihr komplementären Raumes. Göttinger Nachr. (1927).
7. Une définition des nombres de Betti pour un ensemble fermé quelconque. Comptes Rendus 184 (1927), 317–319.
8. Sur la décomposition de l'espace par des ensembles fermés. Comptes Rendus 184 (1927), 425–428.
9. Une généralisation nouvelle du théorème de Phragmén-Brouwer. Comptes Rendus 184 (1927), 575–577.
10. Untersuchungen über Gestalt und Lage abgeschlossener Mengen beliebiger Dimension. Annals of Math. (2) 30 (1928), 101–187.
11. Sur la théorie de la dimension. Comptes Rendus 190 (1930), 1102–1105.
12. (With Urysohn, P.) Mémoire sur les espaces topologiques compacts. Verhand. der Akad. van Wetensch. Amsterdam, Deel XIV (1929), 1–96.

Antoine, L.

1. Sur l'homéomorphie de deux figures et de leurs voisinages. Journal de Math. (8) 4 (1921), 221–325.
2. Sur les ensembles parfaits partout discontinus. Comptes Rendus 173 (1921), 284–285.

Artin, E.

1. Theorie der Zöpfe. Hamburger Abhand. 4 (1925), 47–72.

Baer, R.

1. Isotopie von Kurven auf orientierbaren, geschlossenen Flächen und ihr Zusammenhang mit der topologischen Deformation der Flächen. Journal für Math. 159 (1928), 101-116.
2. Zur Einführung des Schaarbegriffs. Journal für Math. 160 (1929), 199-207.
3. Zur Topologie der Gruppen. Journal für Math. 160 (1929), 208-226.

Bankwitz, C.

1. Über die Torsionszahlen der alternierenden Knoten. Math. Ann. 103 (1930), 145-162.

Bilz, E.

1. Beitrag zu den Grundlagen der kombinatorischen Analysis Situs. Math. Zeitsch. 18 (1923), 1-41.

Birkhoff, G. D.

1. Proof of Poincaré's geometric theorem. Trans. Am. M. S. 14 (1913), 14-22.
2. Dynamical systems with two degrees of freedom. Trans. Am. M. S. 18 (1917), 199-300.
3. An extension of Poincaré's last geometric theorem. Acta Mat. 47 (1926), 279-311.
4. Dynamical Systems. Amer. Math. Soc. Colloquium Publications IX. New York (1927).
5. (With Kellogg, O. D.) Invariant points in function space. Trans. Am. M. S. 23 (1922), 96-115.
6. (With Smith, P. A.) Structure analysis of surface transformations. Journal de Math. (9) 7 (1928), 345-379.

Bochner, S.

1. Der Dimensionsbegriff für Punktmengen in Kartesischen Räumen. Math. Ann. 102 (1930), 768-784.

Boy, W.

1. Über die Curvatura integra und die Topologie geschlossener Flächen. Math. Ann. 57 (1903), 151-184.

Brahana, H. R.

1. Systems of circuits on two-dimensional manifolds. Annals of Math. (2) 23 (1921), 144-168.
2. Regular maps on an anchor ring. American Journal 48 (1926), 225-240.
3. Regular maps and their groups. American Journal 49 (1927), 268-284.

Briggs, G. B.

1. (See Alexander 23.)

Brouwer, L. E. J.

1. On continuous vector distributions on surfaces. Amsterdam Proc. 11 (1909), 850-858; 12 (1910), 716-734; 13 (1910), 171-186.
2. On continuous one-to-one transformations of surfaces into themselves. Amsterdam Proc. 11 (1909), 788-798; 12 (1909), 286-297; 13 (1911),

767–777; 14 (1911), 300–310; 15 (1912), 352–360; 22 (1920), 811–814; 23 (1920), 232–234.
3. Die Theorie der endlichen kontinuierlichen Gruppen unabhängig von den Axiomen von Lie. Math. Ann. 67 (1909), 246–267.
4. Zur Analysis Situs. Math. Ann. 68 (1910), 422–434.
5. Beweis des Jordanschen Kurvensatzes. Math. Ann. 69 (1910), 169–175.
6. Beweis der Invarianz der Dimensionszahl. Math. Ann. 70 (1911), 161–165.
7. Über Abbildungen von Mannigfaltigkeiten. Math. Ann. 71 (1912), 97–115.
8. Beweis der Invarianz des n-dimensionalen Gebiets. Math. Ann. 71 (1912), 305–313.
9. Beweis des Jordanschen Satzes für n-dimensionalen Raum. Math. Ann. 71 (1912), 314–319.
10. Über Jordansche Mannigfaltigkeiten. Math. Ann. 71 (1912), 320–327, 598.
11. Zur Invarianz des n-dimensionalen Gebiets. Math. Ann. 72 (1912), 55–56.
12. Beweis des ebenen Translationssatzes. Math. Ann. 72 (1912), 37–54.
13. Beweis der Invarianz der geschlossenen Kurve. Math. Ann. 72 (1912), 422–425.
14. Sur la notion de "classe" de transformations d'une multiplicité. Proc. 5th Intern. Math. Congr. 2 (1912), 9–10.
15. On looping coefficients. Amsterdam Proc. 15 (1912), 113–122.
16. Über den natürlichen Dimensionsbegriff. Jahrb. für Math. 142 (1913), 146–152.
17. Über topologische Involutionen. Amsterdam Proc. 21 (1919), 1143–1145.
18. Aufzählung der periodischen Transformationen des Torus. Amsterdam Proc. 21 (1919), 1352–1356.
19. Enumération des surfaces de Riemann régulières de genre un. Comptes Rendus 168 (1919), 677–678.
20. Enumération des groupes finis de transformations topologiques du tore. Comptes Rendus 168 (1919), 845–848.
21. Sur les points invariants des transformations topologiques des surfaces. Comptes Rendus 168 (1919), 1042–1044.
22. Enumération des classes de représentations du plan projectif. Comptes Rendus 170 (1920), 834–835.
23. Über die periodischen Transformationen der Kugel. Math. Ann. 80 (1921), 39–41.
24. Über die Minimalzahl der Fixpunkte bei den Klassen von eindeutigen stetigen Transformationen der Ringflächen. Math. Ann. 82 (1921), 94–96.
25. Aufzählung der Abbildungsklassen endlichfach zusammenhängender Flächen. Math. Ann. 82 (1921), 280–286.
26. Über den natürlichen Dimensionsbegriff. Amsterdam Proc. 26 (1923). 795–800.

27. Zum natürlichen Dimensionsbegriff. Math. Zeitsch. 21 (1924). 312–314.
28. On the n-dimensional simplex-star in R_n. Amsterdam Proc. 27 (1924), 778–780.
29. Intuitionistischer Beweis des Jordanschen Kurvensatzes. Amsterdam Proc. 28 (1925), 503–508.
30. Intuitionistische Einführung des Dimensionsbegriff. Amsterdam Proc. 29 (1926), 855–863.
31. On transformations of projective spaces. Amsterdam Proc. 29 (1926), 864–865.

Brown, A. B.
1. Relations between the critical points and curves of a real analytic function of two independent variables. Annals of Math. (2) 31 (1930), 251–270.
2. Relations between the critical points of a real analytic function of n independent variables. American Journal 52 (1930), 449–456.
3. Coalescence of parts of a complex. Proc. Nat. Acad. 16 (1930), 401–406.
4. An extension of the Alexander duality theorem. Proc. Nat. Acad. 16 (1930), 407-408.

Cairns, S. S.
1. The cellular division and approximation of regular spreads. Proc. Nat. Acad. 16 (1930), 488–490.

Carathéodory, C.
1. Über die Begrenzung einfach zusammenhängender Gebiete. Math. Ann. 73 (1913), 323–370.
2. Vorlesungen über reelle Funktionen, 2nd edit. Leipzig (1927).

Chittenden, E. W.
1. On the metrization problem and related problems in the theory of abstract sets. Bull. Am. M. S. 33 (1927), 13–34.
2. On general topology and the relation of the properties of the class of all continuous functions to the properties of space. Trans. Am. M. S. 31 (1929), 290-322.

Chuard, J.
1. Questions d'analysis situs. Palermo Rendic. 46 (1921–22), 185–224.

van Danzig, D.
1. Über topologisch homogene Kontinua. Fundamenta Mat. 14 (1930), 102-125.

Dehn, M.
1. Über die Topologie des drei-dimensionalen Raumes. Math. Ann. 69 (1910), 137-168.
2. Über unendliche diskontinuierliche Gruppen. Math. Ann. 71 (1911), 116–144.
3. Transformationen der Kurven auf zweiseitigen Flächen. Math. Ann. 71 (1912), 413–421.
4. Die beiden Kleeblattschlingen. Math. Ann. 75 (1914), 402–413.

5. (With Heegaard) Analysis Situs. Encykl. der M. W. III 1^1 AB 3, 153–220.

v. Dyck, W.

1. Beiträge zur Analysis Situs I. Ein- und zwei-dimensionale Mannigfaltigkeiten. Math. Ann. 32 (1888), 457–512.
2. Beiträge zur Analysis Situs II. Mannigfaltigkeiten von n Dimensionen. Math. Ann. 37 (1890), 273-316.

Errera, A.

1. L'origine et les problèmes de l'analysis situs. Revue de l'Univ. de Bruxelles, 7–8 (1922), 1–15.
2. Sur les polyèdres réguliers de l'analysis situs. Bruxelles (1922).
3. Le théorème de Jordan pour les polygones. Mathesis 36 (1922), 13–16.
4. Une démonstration du théorème de Jordan. Mathesis 36 (1922), 374–386.

Feigl, G.

1. Über einige Eigenschaften der einfachen stetigen Kurven, I. Math. Zeitsch. 27 (1927), 161–186.
2. Fixpunktsätze für spezielle n-dimensionale Mannigfaltigkeiten. Math. Ann. 98 (1928), 355–390.
3. Geschichtliche Entwicklung der Topologie. Jahresb. der D. M. V. 37 (1928), 273–286.

Flexner, W. W.

1. (With Lefschetz). On the duality theorems for the Betti numbers of topological manifolds. Proc. Nat. Acad. 16 (1930), 530–533.

Frankl, F.

1. Topologische Beziehungen in sich kompakter Teilmengen euklidischer Räume zu ihren Komplementen sowie Anwendung auf die Primendentheorie. Wiener Berichte 136 (1927), 689-699.
2. (With Pontrjagin, L.) Ein Knotensatz mit Anwendung auf die Dimensionstheorie. Math. Ann. 102 (1930), 785–789.

Franklin, P.

1. (With Wiener, N.) Analytic approximations to topological transformations. Trans. Am. M. S. 28 (1926), 762–785.

Fréchet, M.

1. Les Espaces Abstraits. Paris, Gauthiers-Villars (1926).

Furch, R.

1. Orientierung von Hyperflächen im projektiven Raum. Hamburger Abhand. 1 (1922), 210–212.

Hadamard, J.

1. Sur quelques applications de l'indice de Kronecker. Note at end of J. Tannery's Introduction à la Théorie des Fonctions d'une Variable. Vol. 2, 437–477. Paris (1910).

Hausdorff, F.

1. Grundzüge der Mengenlehre. First edition, Leipzig (1914), Veit und Co. Second edition, Leipzig (1927), Walter de Gruyter.

Heegaard, P.

1. Sur l'Analysis Situs. Bull. Soc. M. de F. 44 (1916), 161–242. Translated from Forstudier til en topologisk Teori for de algebraiske Fladers Sammenhang. Diss. Kopenhagen (1898).
2. (See Dehn 5.)

Hildebrandt, T. H.

1. The Borel theorem and its generalizations. Bull. Am. M. S. 32 (1926), 423–474.

Hopf, H.

1. Zum Clifford-Kleinschen Raumproblem. Math. Ann. 95 (1925), 313–339.
2. Die Curvatura integra Clifford-Kleinscher Raumformen. Göttinger Nachr. (1925), 131–141.
3. Über die Curvatura integra geschlossener Hyperflächen. Math. Ann. 95 (1925), 340–367.
4. Abbildungsklassen n-dimensionaler Mannigfaltigkeiten. Math. Ann. 96 (1926), 209–224.
5. Vectorfelder in n-dimensionalen Mannigfaltigkeiten. Math. Ann. 96 (1926), 225–250.
6. Über Mindestzahlen von Fixpunkten. Math. Zeitsch. 26 (1927), 762–774.
7. A new proof of the Lefschetz formula on invariant points. Proc. Nat. Acad. 14 (1928), 149–153.
8. On some properties of one-valued transformations of manifolds. Proc. Nat. Acad. 14 (1928), 206–214.
9. Eine Verallgemeinerung der Euler-Poincaréschen Formel. Göttinger Nachr. (1928), 127–136.
10. Zur Topologie der Abbildungen von Mannigfaltigkeiten. Math. Ann.; I. Neue Darstellung der Theorie des Abbildungsgrades für topologische Mannigfaltigkeiten, 100 (1928), 579–608; II. Klasseninvarianten von Abbildungen, 102 (1929), 562–623.
11. Über die algebraische Anzahl von Fixpunkten. Math. Zeitsch. 29 (1929), 493–525.

Hotelling, H.

1. Three dimensional manifolds of states of motion. Trans. Am. M. S. 27 (1925), 329–344; 28 (1926), 479–490.

Hurewicz, W.

1. Über ein topologisches Theorem. Math. Ann. 101 (1929), 210–218.
2. Ein Theorem der Dimensionstheorie. Annals of Math. 21 (2)(1930), 176–180.

van Kampen, E. R.

1. Eine Verallgemeinerung des Alexanderschen Dualitätssatzes. Amsterdam Proc. 31 (1928), 899–905.
2. Zur Isotopie zweidimensionaler Flächen im R_4. Hamburger Abhand. 6 (1928), 216.
3. Die kombinatorische Topologie und die Dualitätssätze. The Hague (1929). (Leyden thesis.)

v. Kerékjártó, B.

1. Beweis des Jordanschen Kurvensatzes. Math. Naturw. Ber. d. Ungarischen Akad. d. Wiss. 38 (1919), 194–198.
2. Über Transformationen des ebenen Kreisringes. Amsterd. Akad. Versl. 28 (1919).
3. Über die Brouwerschen Fixpunktsätze. Math. Ann. 80 (1919), 29–32.
4. Über Transformationen des ebenen Kreisringes. Math. Ann. 80 (1919), 33–35.
5. Über die periodischen Transformationen der Kreisscheibe und der Kugelfläche. Math. Ann. 80 (1919), 36–38.
6. Zur Theorie der mehrdeutigen stetigen Abbildungen. Math. Zeitsch. 8 (1920), 310–319.
7. Zur Gebietsinvarianz. Math. Nat. Ber. d. Ung. Akad. d. Wiss. 39 (1921), 220–221.
8. Kurvenscharen auf Flächen. Gött. Nachr. (1922), 71–79.
9. Zur Topologie von Kurven und Kurvenscharen. Math. Nat. Ber. d. Ung. Akad. d. Wiss. 39 (1922), 306–313.
10. Vorlesungen über Topologie. I. Flächentopologie. Berlin, Springer (1923).
11. Hauptsatz der Flächentopologie bei unendlich hohem Zusammenhang. Jahresb. der D. M. V. 31 (1923), 98–99.
12. The plane translation theorem of Poincaré. Acta Litt. ac. Scient. 4 (1928), 86–102.
13. Démonstration élémentaire du théorème de Jordan sur les courbes simples. Szeged Acta 5 (1930), 56–59.

Knaster, B.

1. (With Kuratowski, C., and Masurkiewicz, S.). Ein Beweis des Fixpunktsatzes für n-dimensionale Simplexe. Fundamenta Mat. 14 (1929), 132–137.

Kneser, H.

1. Kurvenscharen auf geschlossenen Flächen. Jahresb. der D. M. V. 30 (1921), 83–85.
2. Reguläre Kurvenscharen auf den Ringflächen. Mat. Ann. 91 (1924), 135–154.
3. Die Topologie der Mannigfaltigkeiten. Jahresb. der D. M. V. 34 (1925), 1–14.
4. Eine Bemerkung über dreidimensionale Mannigfaltigkeiten. Göttinger Nachr. (1925), 128–130.
5. Die Deformationssätze der einfach zusammenhängenden Flächen. Math. Zeitsch. 25 (1926), 362–372.
6. Glättung von Flächenabbildungen. Math. Ann. 100 (1928), 609–617.
7. Die kleinste Bedeckungszahl innerhalb einer Klasse von Flächenabbildungen. Math. Ann. 103 (1930), 347–358.

König, D.

1. Zur Analysis Situs der Doppelmannigfaltigkeiten und der projektiven Räume. Internat. Cong. of Math., Cambridge (1912), 1–5.

2. Über Ein- und Zweiseitigkeit mehrdimensionaler Mannigfaltigkeiten. Archiv der Math. u. Phys. (3) 19 (1912), 214–224.
3. Az Analysis Situs Elemei. Budapest (1918).
4. Sur les rapports topologiques d'un problème d'analyse combinatoire. Acta Litt. ac. Scient. 2 (1924), 32–38.

Kronecker, L.
1. Über Systeme von Funktionen mehrerer Variabeln. Werke, Vol. 1, 175–227. Leipzig (1869).

Künneth, H.
1. Zur topologischen Untersuchung geometrischer Gebilde. Sitzb. der Bayer. Akad. d. Wiss. (1922), 213–220.
2. Zur Bestimmung der Fundamentalgruppe einer Produktmannigfaltigkeit. Sitzb. der Phys.-Med. Soz. in Erlangen 54, 55 (1922–23), 190–196.
3. Über die Bettischen Zahlen einer Produktmannigfaltigkeit. Math. Ann. 90 (1923), 65–85.
4. Über die Torsionszahlen von Produktmannigfaltigkeiten. Math. Ann. 91 (1924), 125–134.

Kuratowski, C.
1. (See Knaster 1.)

Landsberg, G.
1. Beiträge zur Topologie geschlossener Kurven mit Knotenpunkten und zur Kronekerschen Charakteristikentheorie. Math. Ann. 70 (1911), 563–579.

Lebesgue, H.
1. Sur l'invariance du nombre de dimensions d'un espace et sur le théorème de M. Jordan relatif aux variétés fermées. Comptes Rendus 152 (1911), 841–844.
2. Sur les correspondances entre les points de deux espaces. Fundamenta Mat. 2 (1921), 256–285.

Lefschetz, S.
1. Algebraic surfaces, their cycles and integrals. Annals of Math. (2) 21 (1920), 225–258; 23 (1922), 33.
2. On certain numerical invariants of algebraic varieties with application to Abelian varieties. Trans. Am. M. S. 22 (1921), 327–482.
3. Report on curves traced on algebraic surfaces. Bull. Am. M. S. 29 (1923), 242–258.
4. Continuous transformations of manifolds. Proc. Nat. Acad. 9 (1923), 90–93.
5. L'analysis situs et la géométrie algébrique. Paris, Gauthiers-Villars (1924).
6. Intersections of complexes on manifolds. Proc. Nat. Acad. 11 (1925), 287–289.
7. Continuous transformations of manifolds. Proc. Nat. Acad. 11 (1925), 290–292.

8. Intersections and transformations of complexes and manifolds. Trans. Am. M. S. 28 (1926), 1–29.
9. Transformations of manifolds with a boundary. Proc. Nat. Acad. 12 (1926), 737–739.
10. Correspondences between algebraic curves. Annals of Math. (2) 28 (1927), 342–354.
11. Manifolds with a boundary and their transformations. Trans. Am. M. S. 29 (1927), 429–462.
12. The residual set of a complex on a manifold and related questions. Proc. Nat. Acad. 13 (1927), 614–622, 805–807.
13. (With National Research Committee on Rational Transformations) Selected topics in algebraic geometry, Chs. 15, 16, 17. Washington (1928).
14. Closed point-sets on a manifold. Annals of Math. (2) 29 (1928). 232–254.
15. Duality relations in topology. Proc. Nat. Acad. 15 (1929), 367–369.
16. Géométrie sur les surfaces et les variétés algébriques. Mémorial des Sc. Math. 40, Paris, Gauthiers-Villars (1929).
17. Les transformations continues des ensembles fermés et leurs points fixes. Comptes Rendus 190 (1930), 99–100.
18. On transformations of closed sets. Annals of Math. (2) 32 (1930), 271–280.
19. (See W. W. Flexner, 1).

Lenness, N. J.
1. Theorems on the Simple Finite Polygon and Polyhedron. American Journal 33 (1911), 37–62.
2. Curves in Non-Metrical Analysis Situs with Application in the Calculus of Variations. American Journal 33 (1911), 287–326.

Levi, F.
1. Streckenkomplexe auf Flächen. Math. Zeitsch. 16 (1923), 148–158.

Lindenbaum, A.
1. Contributions à l'étude de l'espace métrique I. Fundamenta Mat. 8 (1926), 209–222.

Löbell, F.
1. Ein Satz über die eindeutigen Bewegungen Clifford-Kleinscher Flächen in sich. Journal f. Math. 162 (1930), 114–125.
2. Zur Frage der Struktur der geschlossenen geodätischen Linien in den offenen Clifford-Kleinschen Flächen mit positiver Charakteristik. Journal f. Math. 162 (1930), 125–134.

Lüroth, J.
1. Über Abbildung von Mannigfaltigkeiten. Math. Ann. 63 (1907), 222–238.

Lusternik, L.
1. Topologische Grundlagen der allgemeinen Eigenwerttheorie. Monatsh. f. Math. u. Phys. 37 (1930) 125–130.

2. (With Schnirelmann, L.) Méthodes topologiques dans les problèmes variationnels. Publ. of the res. inst. for math. and mechanics of the first state Univ. of Moscow (1930), 68 pages. (In Russian.)

Lwowski, W.

1. Konstruktion geschlossener einseitiger Flächen mit geschlossenen Doppellinien. Matematičeski Sbornik (1925), 325–364.
2. Über geschlossene zweiseitige Flächen. Journal of the Leningrad Math. Soc. (Russian) (1927), 169–181.

Masurkiewicz, S.

1. (See Knaster 1.)

Mayer, W.

1. Über abstrakte Topologie. Mon. f. M. u. Ph. 36 (1929), 1–42, 219–258.

Menger, K.

1. Dimensionstheorie. Leipzig und Berlin (1928).

Moore, R. L.

1. Report on continuous curves from the viewpoint of analysis situs. Bull. Am. M. S. 29 (1923), 289–302.

Morse, M.

1. Recurrent geodesics on a surface of negative curvature. Trans. Am. M. S. 22 (1921), 84–100.
2. A fundamental class of geodesics on any closed surface of genus greater than one. Trans. Am. M. S. 26 (1924), 25–60.
3. Relations between the critical points of a real analytic function of n independent variables. Trans. Am. M. S. 27 (1925), 345–396.
4. The analysis and analysis situs of regular n-spreads in $(n+r)$-space. Proc. Nat. Acad. 13 (1927), 813–817.
5. Singular points of vector fields under general boundary conditions. American Journal 51 (1929), 165–178.
6. The foundations of a theory in the calculus of variations in the large. Trans. Am. M. S. 30 (1928), 213–274.
7. The critical points of functions and the calculus of variations in the large. Bull. Am. M. S. 35 (1929), 38–54.
8. The foundations of the calculus of variations in the large in m-space (first paper). Trans. Am. M. S. 31 (1929), 379–404.

Nagy, J. v. Sz.

1. Zur Topologie der geschlossenen Kurven auf der Sphäre. Szeged Acta 5 (1930), 1–12.

Newman, M. H. A.

1. On the foundations of combinatory analysis situs. Amsterdam Proc.; I. Definitions and elementary theorems, 29 (1926), 611–626; II. Theorems on sets of elements, 29 (1926), 627–641; Additions and corrections, 30 (1927), 670–673.
2. A property of 2-dimensional elements. Amsterdam Proc. 29 (1926), 1401–1405.

3. On the superposition of n-dimensional manifolds. Journal of London M. S. 2 (1927), 56–64.
4. Topological equivalence of complexes. Math. Ann. 98 (1927–28), 399–412.
5. Combinatory topology of convex regions. Proc. Nat. Acad. 16 (1930), 240–242.

Nielsen, J.
1. Die Isomorphismen der allgemeinen, unendlichen Gruppe mit zwei Erzeugenden. Math. Ann. 78 (1918), 385–397.
2. Über fixpunktfreie topologische Abbildungen geschlossener Flächen Math. Ann. 81 (1920), 94–96.
3. Über die Minimalzahl der Fixpunkte bei den Abbildungstypen der Ringflächen. Math. Ann. 82 (1920–21), 83–93.
4. Über topologische Abbildungen geschlossener Flächen. Hamburger Abhand. 3 (1924), 246–260.
5. Zur Topologie der geschlossenen Flächen. Vorträge 6. skand. Math. Kong., Kopenhagen (1925).
6. Untersuchungen zur Topologie der geschlossenen zweiseitigen Flächen. Acta Mat.; I. 50 (1927), 189–358; II. 53 (1929), 1–76.

Osgood, W. F.
1. Lehrbuch der Funktionentheorie, vol. II, second edition. Leipzig, Teubner (1929).

Picard, E.
1. (With Simart, G.) Théorie des fonctions algébriques de deux variables indépendentes. Paris, Gauthiers-Villars. Vol. I (1897), Vol. II (1906).

Poincaré, H.
1. Sur les courbes définies par une équation différentielle. Journal de Math. (3) 7 (1881), 375–424; (3) 8 (1882), 251–296; (4) 1 (1885), 167–244; (4) 2 (1886), 151–217.
2. Analysis situs. Journal de l'Ec. Polyt. (2) 1 (1895), 1–123.
3. Complément à l'analysis situs. Palermo Rendic. 13 (1899), 285–343.
4. Deuxième complément à l'analysis situs. Proc. London M. S. 32 (1900), 277–308.
5. Sur certaines surfaces algébriques. Troisième complément à l'analysis situs. Bull. Soc. M. de F. 30 (1902), 49–70.
6. Sur les cycles des surfaces algébriques. Quatrième complément à l'analysis situs. Journal de Math. (5) 8 (1902), 169–214.
7. Cinquième complément à l'analysis situs. Palermo Rendic. 18 (1904). 45–110.
8. Sur un théorème de géométrie. Palermo Rendic. 33 (1912), 375–407.

Pontrjagin, L.
1. Zum Alexanderschen Dualitätssatz. Göttinger Nachr. (1927), 315–322.
2. Sur une hypothèse fondamentale de la théorie de la dimension. Comptes Rendus 190 (1930), 1105–1107.
3. (See Frankl 2.)

Radó, T.
1. Über den Begriff der Riemann'schen Fläche. Acta Litt. ac. Scient. 2 (1925), 101-121.
Reidemeister, K.
1. Knoten und Gruppen. Hamburger Abhand. 5 (1926), 7-22.
2. Elementare Begründung der Knotentheorie. Hamburger Abhand. 5 (1926), 24-32.
3. Über unendliche diskrete Gruppen. Hamburger Abhand. 5 (1926), 33-39.
4. Fundamentalgruppe und Überlagerungsräume. Göttinger Nachr. (1928), 69-76.
5. Über Knotengruppen. Hamburger Abhand. 6 (1928), 56-64.
6. Knoten und Verkettungen. Math. Zeitsch. 27 (1928), 713-729.
Riemann, B.
1. Theorie der Abelschen Funktionen. Journal für Math. 54 (1857), 115-155.
2. Gesammelte mathematische Werke. Leipzig (1892).
Schauder, J.
1. Zur Theorie stetiger Abbildungen in Funktionalräumen. Math. Zeitsch. 26 (1927), 47-65; 417-431.
Schnirelmann, L.
1. Über eine neue kombinatorische Invariante. Monatsh. f. Math. u. Phys. 37 (1930) 131-134.
2. (See Lusternik 2.)
Schoenflies, A.
1. Die Entwicklung der Lehre von den Punktmannigfaltigkeiten II. Jahresb. der D. M. V. Ergänz. II. Leipzig (1908).
2. Bemerkungen zu dem vorstehenden Aufsatz des Herrn L. E. J. Brouwer. Math. Ann. 68 (1910), 435-444.
Severi, F.
1. Sulle corrispondenze fra i punti di una curva algebrica e sopra certe classi di superficie. Torino Mem. 54 (1903).
2. Sulla totalità delle curve algebriche tracciate sopra una superficie algebrica. Math. Ann. 62 (1906).
3. Sul principio della conservazione del numero. Palermo Rendic. 33 (1912), 313-327.
4. Sui fundamenti della geometria numerativa e sulla teoria delle caratteristiche. Veneto Atti 75 (1916), 1121-1162.
5. Conferenze di geometria algebrica, raccolte da B. Segre (Lithographed Lectures). Rome (1930).
Simart, G.
1. (See Picard 1.)
Smith, P. A.
1. (See Birkhoff 5.)

Sperner, E.

1. Neuer Beweis für die Invarianz der Dimensionszahl und des Gebietes. Hamburger Abhand. 6 (1928), 265–272.

Steinitz, E.

1. Konfigurationen der projektiven Geometrie. Encykl. der M. W. III 1^{1}, AB 5a, 481–516.

Tietze, H.

1. Über die topologischen Invarianten mehrdimensionaler Mannigfaltigkeiten. Monatsh. f. Math. u. Phys. 19 (1908), 1–118.
2. Einige Bemerkungen über das Problem des Kartenfärbens auf einseitigen Flächen. Jahresb. der D. M. V. 19 (1910), 155–159.
3. Sur les réprésentations continues des surfaces sur elles-mêmes. Comptes Rendus 157 (1913), 509–512.
4. Über stetige Abbildungen einer Quadratfläche auf sich selbst. Palermo Rendic. 38 (1914), 247–304.
5. Über den Richtungssinn und seine Verallgemeinerung. Jahresb. der D. M. V. 29 (1920), 95–123.
6. Über Analysis Situs. Hamburger Mat. Einzelschr. 32 (1923), 8.
7. Zur Topologie berandeter Mannigfaltigkeiten. Monatsh. f. Math. u. Phys. 33 (1923), 15–17.
8. Beiträge zur allgemeinen Topologie. I. Axiome für verschiedene Fassungen des Umgebungsbegriffs. Math. Ann. 88 (1923), 290–312. II. Über die Einführung uneigentlicher Elemente. Math. Ann. 91 (1924), 210–224. III. Über die Komponenten offener Mengen. Monatsh. f. Math. u. Phys. 33 (1923), 15–17.
9. Zur Topologie berandeter Mannigfaltigkeiten. Monatsh. f. Math. u. Phys. 35 (1928), 25–44.

Tychonoff, A.

1. Über einen Metrisationssatz von P. Urysohn. Math. Ann. 95 (1926), 139–142.

Ursell, H. D.

1. Intersection of complexes. Journal London Math. Soc. 3 (1928), 38–48.

Urysohn, P.

1. Über die Metrisation der kompakten topologischen Räume. Math. Ann. 92 (1924), 275–293.
2. Zum Metrisationsproblem. Math. Ann. 94 (1924), 309–315.
3. Mémoire sur les multiplicités cantoriennes. Fundamenta Mat. 7 (1925), 30–140; 8 (1926), 225–359.
4. (See Alexandroff 11.)

Veblen, O.

1. Theory of plane curves in non-metrical analysis situs. Trans. Am. M. S. 6 (1905), 83–98.
2. Decomposition of n-space by a polyhedron. Trans. Am. M. S. 14 (1913), 65–72; 15 (1914), 506.

3. An application of modular equations in analysis situs. Annals of Math. (2) 14 (1913), 86–94.
4. On the deformation of an n-cell. Proc. Nat. Acad. 3 (1917), 654–656.
5. Analysis Situs. The Cambridge Colloquium, Part II. New York (1922).
6. The intersection numbers. Trans. Am. M. S. 25 (1923), 540–550.
7. Invariance of the Poincaré numbers of a discrete group. Bull. Am. M. S. 30 (1924), 405–406.
8. (See Alexander 22.)
9. (With Young) Projective Geometry, Vol. II (by Oswald Veblen) Boston (1918).

Vietoris, L.
1. Über den höheren Zusammenhang kompakter Räume und eine Klasse von zusammenhangstreuen Abbildungen. Math. Ann. 97 (1927), 454–472.
2. Über die Symmetrie der Zusammenhangszahlen kombinatorischer Mannigfaltigkeiten. Monatsh. f. Math. u. Phys. 35 (1928), 165–174.
3. Zum höheren Zusammenhang der kompakten Räume. Math. Ann. 101 (1929), 219–225.
4. Erzeugung der regulären Unterteilung von simplizialen Komplexen durch wiederholte Zweiteilung. Monatsh. f. Math. u. Phys. 37 (1930), 97–102.
5. Über die Homologiegruppen der Vereinigung zweier Komplexe. Monatsh. f. Math. u. Phys. 37 (1930) 159–162.

van der Waerden, B. L.
1. Topologische Begründung des Kalküls der abzählenden Geometrie. Math. Ann. 102 (1929), 337–362.
2. Kombinatorische Topologie. Jahresb. der D. M. V. 39 (1930), 121–139.

Weyl, H.
1. Strenge Begründung der Charakteristikentheorie auf zweiseitige Flächen. Jahresb. der D. M. V. 25 (1916), 265–278.
2. Die Idee der Riemannschen Fläche. Leipzig (1923). Second Edition.
3. Analysis situs combinatorio. Revista Mat. 5 (1923), 209–218, 241–249, 273–279.

Wiener, N.
1. (See Franklin 1.)

Wilson, W.
1. The equivalence in R_n of the n-dimensional simplex star and the spherical neighborhood. Amsterdam Proc. 27 (1924), 773–777.
2. The normal, not absolutely closed, topological space. Amsterdam Proc. 29 (1926), 462-464.
3. Representation of a simplicial manifold in a locally simplicial manifold. Amsterdam Proc. 29 (1926), 1129–1133.
4. Afbeelding van Ruimten. Proefschrift (Thesis), Amsterdam (1928).
5. Representation of Manifolds. Math. Ann. 100 (1928), 552–578.

ADDENDA

Page 92

This example only shows that a set may be locally connected up to a certain order r and fail for orders above r. The full problem raised by the example is as yet (1965) unanswered.

Page 122

The outline of the proof of Theorem I holds only for simplicial or convex cells, which is sufficient for ulterior purposes. The theorem as stated is proved by A. W. Tucker, Ann. of Math. (2), Vol. 34 (1933), pp. 191-243.

Page 364

See Koopman and Brown, Trans. Amer. Math. Soc., Vol. 34 (1932), pp. 231-252, and an elaboration of the proof of the theorem on page 364 by Lefschetz and Whitehead, Trans. Amer. Math. Soc., Vol. 35 (1933), April issue.

INDEX